Green Energy and Technology

For further volumes:
http://www.springer.com/series/8059

Photo credits Debajit Palit

Subhes C. Bhattacharyya · Debajit Palit
Editors

Mini-Grids for Rural Electrification of Developing Countries

Analysis and Case Studies from South Asia

Springer

Editors
Subhes C. Bhattacharyya
De Montfort University
Leicester
UK

Debajit Palit
The Energy and Resources Institute
New Delhi
India

ISSN 1865-3529 ISSN 1865-3537 (electronic)
ISBN 978-3-319-04815-4 ISBN 978-3-319-04816-1 (eBook)
DOI 10.1007/978-3-319-04816-1
Springer Cham Heidelberg New York Dordrecht London

Library of Congress Control Number: 2014939387

Printed on acid-free paper

Springer is part of Springer Science+Business Media (www.springer.com)

Foreword

Globally, there are still 1.3 billion people who lack access to electricity. Over 85 % of the people without electricity access live in rural areas. More strikingly, almost 40 % of these people are from South Asia, especially Bangladesh, India and Pakistan. These numbers will not change substantially by the year 2030 unless new strategies and intensified efforts are made at the national and international levels. Taking up the importance of energy issues for sustainable development, the United Nations General Assembly (UNGA) has declared the decade 2014–2024 to be the Decade of Sustainable Energy for All, underscoring the importance of energy issues for sustainable development and for the elaboration of the post-2015 development agenda. The UNGA resolution stressed the need to improve access to reliable, affordable, economically viable, socially acceptable and environmentally sound energy services and resources for sustainable development. The World Energy Outlook 2011 stresses the fact that to achieve the universal electrification objective by the year 2030, in terms of technology choice, grid extension will cater to 30 % of the population to be covered, whereas the balance 70 % will be served by mini-grids or off-grid systems.

As individuals, communities, policy makers and investors, we can all do something about this dismal situation. In TERI, we took the lead and initiated a major global effort in 2008, called Lighting a Billion Lives, which had impacted more than one million lives of people till the end of 2013 through provisioning of various types of decentralised energy interventions. New ways are also being found by many stakeholders to make off-grid renewable energy technologies more affordable for the poor and to bring electricity to households, communities and enterprises in far-flung areas where the grid has either not reached, or power supplied is erratic and unreliable.

This edited volume, which has been developed as part of the project, Off-grid Access System in South Asia, is an effort by the authors to contribute to this discourse on sustainable energy for all and assist in scaling-up of the dissemination of decentralised energy technologies in a sustainable manner. This research project specifically focused on the South Asian region, which continues to be plagued with lack of access to clean and sustainable energy supply. The multi-disciplinary research team of this project from universities and institutions in India and the UK have been working extensively over the past four years—in both analytical work as well as documenting the lessons learnt and best practises established—and has compiled this volume. The aim was to highlight the socio-technical, economic and governance-related lessons from various mini-grid projects in South Asia and to share the state-of-art knowledge and research outcomes with the wider research and practitioner community.

I believe this volume will prove to be a valuable addition to the literature on rural energy initiatives and will assist in promoting mini-grid and off-grid electrification efforts and act as an important contribution to the United Nations' Decade of Sustainable Energy for All.

R. K. Pachauri
Director General, TERI

Preface

Access to sustainable energy for all remains an important topic going by the current discussions at the international level. Globally, there is no universal approach to the provision of access to electricity and a number of factors and actors determine what approach would be most suitable for a given community. This edited volume is an effort to contribute to this discourse on sustainable energy for all and forms part of the dissemination activity of a Research Councils UK funded project on off-grid electrification, called OASYS South Asia, which is a collaborative research work of five partner organisations, namely De Montfort University (DMU), Manchester University, Edinburgh Napier University from UK, and The Energy and Resources Institute (TERI) and TERI University from India. The consortium has since 2009 undertaken a significant amount of research on rural electrification, especially off-grid electrification, in South Asia. Based on this work, it has identified mini-grid-based rural electrification as one of the options for extending the electricity access in developing countries. The project has also undertaken a set of demonstration activities in India to showcase alternative ways of delivering electricity in rural areas through mini-grids and to develop a better first-hand understanding of the challenges in doing so. This edited volume brings together studies and research carried out by the consortium to share the knowledge with the wider research and practitioner community.

This volume contains 15 chapters divided into two parts: Part I provides the background understanding of mini-grids while Part II provides seven case studies from South Asia covering different countries, technologies, and business models. Following the multidisciplinary focus of the project, the book covers various dimensions relevant for any mini-grid-based electricity supply. It also provides practical guidelines for design and implementation of mini-grid projects alongside more academic research studies. We hope this mix of presentation will impact a wider section of the interested readers.

The work reported here has been discussed internally and in various workshops organized as part of the research activity. These were held in Delhi, Bhubaneswar, and Leicester between 2012 and 2014. The chapters have thus benefited from the inputs and comments of a large number of participants from the academia as well as those involved in practice with mini-grid-based electrification and their financing. A part of the content has appeared in various peer-reviewed journals as well. These have been appropriately acknowledged in the relevant chapter.

As the editors of the volume we would like to thank all the contributors to this volume for their continued support and hard work. We are particularly thankful to the stakeholders who participated in our case studies and provided valuable insights. We would like to thank the reviewers of the journals who provided valuable comments to improve the quality of the work. We also thank Elsevier for allowing us to reuse the materials published in their journals. We also thank TERI, New Delhi, India and Practical Action, UK for allowing us to use their diagrams here. We are also grateful to our respective institutes, DMU and TERI, for encouraging us to edit this volume. Despite of all the support from different quarters, errors, if any, are ours.

We are also indebted to the workshop participants for sharing various thoughts that have enriched our experience. The demonstration project in Odisha (India) has benefited from the support of a large number of individuals, too many to mention here but we are thankful to them for their continued support, without which the work could not be taken forward. We are grateful to the funding agencies Engineering and Physical Science Research Council and Department of International Development of United Kingdom and Rural Electrification Corporation, Government of India for their support to this initiative.

We also thank the publisher—Springer for agreeing to publish this volume despite the specialized nature of the work that still faces limited academic attention.

We hope this volume will prove to be a valuable addition to the literature on rural electrification and will help to promote mini-grid-based sustainable electrification solutions worldwide. We believe that it would benefit researchers and practioners, as well as donors and other stakeholders involved in policy-making in enhancing electricity access in rural areas of the developing world.

Last but not the least, we would like to thank our respective families: Subhes thanks his spouse (Debjani) and daughter (Saloni) while Debajit thanks his spouse (Dipanwita) and daughter (Roshni) for their support and cooperation in completing this work over the winter of 2013–2014.

March 2014

Subhes C. Bhattacharyya
Debajit Palit

Contents

Introduction

Subhes C. Bhattacharyya and Debajit Palit

Abstract This chapter introduces the mini-grid option as a decentralised approach to electrification of rural areas in developing countries. It also provides an overview of the content of the book.

1 Background

Since the Johannesburg Summit in 2002, the critical role played by energy in achieving sustainable development has been well recognised in the energy policy literature (see for example, [5, 6, 12, 14] and a consensus seems to exist that without affordable, reliable and clean energy services to the population, sustainable development cannot be achieved. Yet, the situation in terms of energy access has not changed much even after a decade, and billions of people are without access to such vital services. According to IEA [7], even by 2030, this problem will not diminish unless actions are taken urgently. Recognising the challenge a global initiative, Sustainable Energy for All, was launched by the UN Secretary General, Mr Ban Ki Moon, to provide universal access to energy by the year 2030. The United Nation's decision to declare 2012 as the 'International Year of Sustainable Energy for All' has once again caught the global attention to this problem.

According to IEA [8], 1.3 billion people in the world did not have access to electricity in 2011 while 2.7 billion people did not have access to clean cooking energies. The access problem has a distinct regional dimension—sub-Saharan Africa and Developing Asia are two distinct regions where the problem is acute.

S. C. Bhattacharyya (✉)
Institute of Energy and Sustainable Development, De Montfort University, Leicester, UK
e-mail: subhesb@dmu.ac.uk

D. Palit
Social Transformation Division, The Energy and Resources Institute, New Delhi, India

S. C. Bhattacharyya and D. Palit (eds.), *Mini-Grids for Rural Electrification of Developing Countries*, Green Energy and Technology,
DOI: 10.1007/978-3-319-04816-1_1, © Springer International Publishing Switzerland 2014

600 million people in sub-Saharan (SS) and North Africa lack access to electricity while 615 million in Developing Asia face the same problem [8]. Near 100 % electrification in China has significantly improved the electricity access figure in Developing Asia but the cooking energy challenge persists. The regional averages also mask the severity of the problem faced by many countries. For example, 97 % of the population of Burundi, Liberia and Chad, 95 % of Rwanda, Central African Republic and Sierra Leone lacked electricity access in 2008 [13].

Further, energy access is predominantly a rural problem. 1.1 billion (out of 1.3 billion or 85 %) lacking electricity access are found in rural areas. This disparity is acute in low-income countries in general and in sub-Saharan Africa in particular. UNDP-WHO [13] indicated that 87 and 89 % of the rural population of LDC (Least Developed Countries) and SS-Africa lack access to electricity respectively.

There is a long tradition of supporting rural electrification programmes both by international organisations and national governments but as Cook [4] indicates, the trend depended on the political thinking of a given time. The wave of state-led infrastructure development for rural areas in the 1960s was responsible for initial efforts in this area in most countries but the liberalisation policies of the 1980s and 1990s reduced state support in favour of private sector initiatives. Specific funds were often created to take care of non-market issues related to reforms, thereby creating a transparent regulatory mechanism for supporting social and public goods dimension. But, rural electrification received a setback as the private investment interests were not often compatible with the rural market conditions. The return of state intervention to address the market failure issue is visible globally once again as the focus on electricity access has gained momentum.

Although grid extension has emerged as the preferred mode of electrification in all successful cases, off-grid electrification has often been considered either as a temporary solution (a pre-electrification option) or as an inferior solution. While off-grid options have received favour and support of international organisations and donor agencies, and a few technologies such as solar home systems (SHS) have emerged as leaders in this segment, there has been a relatively limited penetration of this option globally. High cost, limited application and poor performance of some of the technologies as well as the image of 'inferior or temporary' nature of such options have hindered the development.

It is difficult to trace off-grid developments due to small-scale operation and absence of any systematic regulatory reporting requirement for such operations. However, off-grid solutions have been promoted where the grid has not reached or is unlikely to reach in the near future. Two modes of operation are prevalent: stand-alone systems and local-grid systems. The local-grid systems often rely on diesel generators or hydropower. According to the World Bank [15], portable 5–10 kW diesel generators are widely used as the conventional solutions. However, heavy reliance on diesel for small-scale power generation imposes cost burden on the utilities (more importantly on oil importing countries). The price fluctuations in the international market affect the overall cost of production and the viability of the business. This, in turn, imposes a heavy subsidy burden on the government.

Local-grid system was also developed in hydro-dominated areas. For example, China Statistical Yearbook [3] reported that more than 27,000 hydropower stations are operating in rural China with a total installed capacity of about 14 GW. Many hydropower plants were initially developed using a local grid system and then connected to the main grid. In the stand-alone category, the solar photovoltaic systems (in local grid or in battery charging systems) and the SHS have emerged as the preferred off-grid technology for rural areas [10]. IFC [10] estimates that SHS has provided electricity access to between 0.5 and 1 million households in developing countries and that through various projects supported by the World Bank and the IFC group, more than 1.3 million solar PV systems have been installed worldwide. The difference in the two figures may be due to abandonment, retirement at the end-of life and errors in estimation. In fact, the south Asian country of Bangladesh deploys over 70,000 SHS per month, with over 2.6 million SHS already installed in the country by November 2013, making it one of the most successful off-grid SHS programmes. Despite such an impressive growth, SHS is catering to only a miniscule share of the non-electrified population (just a few million households as against about 300 million households without electricity), which raises concerns about its ability to provide sustainable supply of electricity services.

Off-grid solutions appear to cater to limited needs of the consumers for lighting and some entertainment through radio/TV connections. Many international organisations (e.g. Lighting Africa) and governments are promoting lighting only using small solar lamps and are not really providing electricity services. This is also reflected in the multi-tier framework developed by the World Bank under SE4ALL. Limited efforts have been found where these solutions have promoted productive use of energy for income generation. Similarly, limited effort has so far gone into hybrid off-grid solutions to provide a reliable and affordable solution, because of system complexity, added cost and over-emphasis on lighting-only solutions.

2 Mini-Grid as an Option

Off-grid electrification, especially mini-grids, has received attention in recent times, to complement grid-based electrification, as a possible electrification option for increasing connection rates in rural areas of developing countries, including providing energy for productive uses. IEA [7] estimates that to achieve the universal electrification objective by 2030, in terms of technology choice, grid extension will cater for 30 % of the cases, whereas the remaining 70 % will come from mini-grids or off-grid systems in the proportion of 65:35. Szabo et al. [11] using a spatial least-cost analysis framework identified that in many parts of Africa cost of decentralised off-grid options can be cheaper than grid extension and that if the affordability of consumers can be increased or cost of supply is reduced, off-grid options can surely play an important role. In a similar study, Bazilian et al. [2] also suggest that to provide universal basic electricity access, most rural areas in Africa will need off-grid supply systems, either based on diesel generators or solar PV systems.

Although grid extension remains the preferred mode of electrification, it faces a number of challenges in many countries. First, the generating capacity is inadequate in many countries and even the urban areas remain poorly supplied. Even if the grid is extended, the rural consumers are unlikely to receive a reliable and high quality supply at the current prices for grid electricity. Second, grid extension is capital-intensive and depending on the geographical location and remoteness, the cost can vary between $6,700 and $19,000 per kilometre [1]. Given the poor financial health of utilities in many developing countries, it is not difficult to imagine that grid extension to rural areas will not be at the top of their priorities. Third, for countries with low level of electrification, the priority will be to connect the urban areas and peri-urban areas first, thereby neglecting the rural needs to the future. A mini-grid can offer a solution in such cases.

On the other hand mini-grids face competition from stand-alone systems such as SHS. The gap left by the grid can be satisfied by stand-alone systems, particularly in areas with scattered population. However, the main limitation comes from the restricted scope of application, the maintenance burden on the users, lack of income generating opportunities and relatively high cost of installations. Generally, a small set of activities can be undertaken with these systems and the direct use of electricity for productive purposes remains limited. Also, the consumer has the responsibility of system maintenance and replacement of components on expiry of life. It is too much to expect from users in remote areas to be well equipped with the required knowledge and expertise to maintain such advanced technologies themselves. The individual systems are generally costly and an equivalent village-scale system is likely to be cheaper than the sum of all individual systems in a village. The economies of scale and scope can work here to provide a comparable service, if not better.

The advantage of mini-grids comes from a number of factors: (1) A high quality and reliability of supply can be ensured through proper design, operation and maintenance of the system. (2) The system can cater to non-domestic demand and can provide income-generating opportunities for improving the quality of life. (3) Where renewable energy technologies are used for power generation, the mini-grid avoids greenhouse gas emissions and contributes to climate mitigation efforts. (4) Such green mini-grids are not dependent on fossil fuel price fluctuations and can thus enhance energy supply security. Renewable energy-based mini-grids can also use locally available resources, which may otherwise have limited economic value. But their use in electricity generation can create an opportunity for local income generation in the fuel supply chain, thereby contributing to the local economic development. Therefore, mini-grids can be considered as a potential solution to the energy access problem.

Although mini-grids are receiving considerable attention now, there is a general lack of information and understanding about mini-grids. IED [9] suggests that reliable information about mini-grids is difficult to find. The green mini-grids and hybrid options are relatively new and less understood. As most of the recent activities are at the pilot project level, there is limited operating experience of such systems and appropriate mechanisms for sharing such experiences are lacking.

3 Purpose and Coverage of this Book

This edited book aims to bridge the knowledge gap found in this emerging area. Based on the research carried out by a team of researchers from the United Kingdom and India under a UKRC-funded research project (Decentralised Off-grid Electricity Generation in Developing Countries: Business Models for Off-grid Electricity Supply, EP/G063826/2, OASYS South Asia for short), this book attempts to cover the multidimensional perspectives related to any mini-grid-based rural electrification with a special focus on South Asia. The research project specifically focused on this region, which continues to be plagued with lack of energy access. This book brings state-of-the art knowledge and presents research outcomes that are anchored in practice.

The book is divided into two parts: the first part is designed to provide an all-round appreciation of mini-grids and covers technical, practical design and non-technical aspects. The second part presents seven case studies covering a range of experience and analysis on mini-grids, all taken from South Asia.

The first part contains six chapters excluding the Introduction. A brief introduction to these chapters is given below.

In "Suite of Off-Grid Options in South Asia", Bhattacharyya, Palit and Kishore present an overview of off-grid electrification options in South Asia. This sets the tone of the book and highlights the multidimensional interactions involved in a mini-grid business.

Peter Boait presents a review of technical aspects relevant for rural mini-grids in "Technical Aspects of Mini-Grids for Rural Electrification". The electricity generation technologies relevant for mini-grids are first presented, followed by technical aspects of the downstream side. Boait highlights the importance of coordinated operation and management of such small systems to ensure reliable supply.

In "Smart Design of Stand-Alone Solar PV System for Off Grid Electrification Projects", Mohanty and Muneer present the practical design considerations for solar PV-based mini-grids. This chapter captures the standard design practices and suggests enhancements to incorporate smartness in the design for better cost-effectiveness, system reliability and optimal performance. System control and operation is also covered in this chapter. Written in a practical manual style, this chapter covers the entire gamut of mini-grid design and brings the most recent thinking in PV-based mini-grid system design.

Bhattacharyya, Mishra and Sarangi provide a review of alternative analytical frameworks that can be used to analyse mini-grid and decentralised off-grid systems in "Analytical Frameworks and an Integrated Approach for Mini-Grid-Based Electrification". Based on the review, the authors also formulate an integrated framework that covers the technical, economic, social and governance dimensions. The application of the suggested framework is presented in some of the case studies included in the second part of the book.

"Demand Management for Off-Grid Electricity Networks" touches on an important issue that is often neglected in the literature. Demand management as a concept is often related to large grid systems but Boait suggests that mini-grids face even a more challenging task because the resource is limited and the diversity of electricity use is less here than that in the grid. Accordingly, special care is required to ensure that available supply can meet the demand, without causing undue stress on the system or loss of reliability or waste of energy. He suggests various options for managing such issues.

In "Business Issues for Mini-Grid-Based Electrification in Developing Countries", Bhattacharyya discusses the business issues related to mini-grids. The chapter highlights the difficult local business environment and often non-supporting policy environment. It also presents the factors influencing the financial and economic viability of mini-grids. The chapter then explores alternative business models that are prevalent in the emerging area of mini-grids.

The second part contains seven case studies. In "Approach for Designing Solar Photovoltaic Based Mini-Grid Projects: A Case Study from India", Sharma and colleagues present the design approach and considerations used in the demonstration project carried out under the OASYS South Asia project in a cluster of five villages in Dhenkenal district in Orissa (India). The chapter provides a step-wise method to design a mini-grid project for remote areas, highlights the importance of livelihood assessments for integration of productive activities to enhance local income, stakeholder integration in the entire process and technical considerations in designing and implementing the system. The challenges that can affect such an activity are also covered.

"Renewable Energy-Based Mini-Grid for Rural Electrification: Case Study of an Indian Village" by Sen and Bhattacharyya reports a simulation exercise for an Indian village for which a hybrid renewable energy system is suggested. This applies the integrated framework of analysis presented in "Analytical Frameworks and an Integrated Approach for Mini-Grid-Based Electrification" and uses a computer model (HOMER) for the simulation, optimisation and sensitivity analysis. The study presents a hybrid system of electricity generation covering micro-hydro, solar PV, wind turbines and bio-diesel and identifies the appropriate technology for meeting the demand. The chapter goes beyond the standard simulation exercises by incorporating a pre-HOMER section to analyse the village electricity demand and a post-HOMER analysis where the business issues are explored.

In "From SHS to Mini-Grid-Based Off-Grid Electrification: A Case Study of Bangladesh", Bhattacharyya analyses the viability of mini-grids in Bangladesh and explores whether mini-grids offer a better solution than SHS for which the country has earned international recognition. Using census data at the village level and following the integrated analytical framework of "Analytical Frameworks and an Integrated Approach for Mini-Grid-Based Electrification", this study develops a set of alternative scenarios to investigate the feasibility of different scale of systems. The simulation exercise using HOMER followed by a detailed financial analysis of the alternative scenarios shows that the technical configuration may vary depending on the load, paying capacity and policy support.

"Application of Multi-Criteria Decision Aids for Selection of Off-Grid Renewable Energy Technology Solutions for Decentralized Electrification" makes an attempt to apply the multi-criteria decision approach to identify the appropriate decentralised electricity generation option in a mini-grid context in South Asia. Using stakeholder inputs, it reports the ranking of technology choices using PROMETHEE, a multi-criteria decision aid.

Palit, Sarangi and Krithika present a detailed analysis of solar PV mini-grids in Chhattisgarh state in India in "Energizing Rural India Using Distributed Generation: The Case of Solar Mini-Grids in Chhattisgarh State, India". The state renewable energy agency has successfully deployed a large number of mini-grids in remote areas and provides important lessons for enhancing electricity access, although this is underreported in the literature. This chapter provides a thorough investigation of the delivery model and its success factors, and fills the knowledge gap.

"Poverty Amidst Plenty: Renewable Energy Based Mini-Grid Electrification in Nepal" presents the mini-grid experience from Nepal. The country has experimented with mini-hydro-based mini-grids and offers significant experience. Sarangi et al. examine the experience and identify the road blocks that affect any scaling-up of this option for wider rural electrification in the country.

In "Viability of Husk-Based Mini-Grids in South Asia" Bhattacharyya investigates the possibility of exploiting the success of rice-husk-based electricity supply through mini-grids in South Asia, a region where rice cultivation is a major agricultural activity. The chapter presents the business models of Husk Power Systems and DESI Power and develops scenarios for rice-husk-based power generation considering the household demand, local commercial and industrial demand. The financial viability of various combinations is then analysed and an expansion of such activities in the rest of South Asia is explored.

References

1. ARE (2011) Hybrid mini-grids for rural electrification: lessons learned. Association for Rural Electrification, Brussels
2. Bazilian M, Nussbaumer P, Rogner HH, Brew-Hammond A, Foster V, Pachauri S, Williams E, Howells M, Niyongabo P, Musaba L, Gallachoir BO, Radka M, Kammen DM (2012) Energy access scenarios to 2030 for the power sector in Sub-Saharan Africa. Util Policy 20(1):1–16
3. China Statistical Yearbook (2008) National Bureau of Statistics of China, Beijing
4. Cook P (2011) Infrastructure, rural electrification and development. Energy Sustain Dev 15(3):304–313
5. DFID (2002) Energy for the poor: underpinning the millennium development goals, Department for International Development, UK
6. IEA (2002) World energy outlook. International Energy Agency, Paris (see Chapter 13, Energy and Poverty)
7. IEA (2011) Energy for all: financing access for the poor, special early excerpt of the world energy outlook 2011. International Energy Agency, Paris

8. IEA (2013) World energy outlook 2013. International Energy Agency, Paris
9. IED (2013) Identifying the gap and building the evidence base on low carbon mini-grids, support study on green mini-grid development. Innovation, Energy Development, Francheville, France
10. IFC (2007) Selling solar: lessons from more than a decade of IFC's experience. International Finance Corporation, Washington, DC
11. Szabo S, Bodis K, Huld T, Moner-Girona M (2011) Energy solutions in rural Africa: mapping electrification costs of distributed solar and diesel generation versus grid extension. Environ Res Lett 6:1–9 (IOP Publishing, UK)
12. UNDP (2005) The energy challenge for achieving the millennium development goals. United Nations Development Programme, New York
13. UNDP-WHO (2009) The energy access situation in developing countries: a review focusing on the least-developed countries and Sub-Saharan Africa. United Nations Development Programme, New York
14. WEC (2001) Living in one world. World Energy Council, London
15. World Bank (2008) Designing sustainable off-grid rural electrification projects: principles and practices. The World Bank, Washington, DC

Part I
Mini-Grid Concepts and Challenges

Suite of Off-Grid Options in South Asia

Subhes C. Bhattacharyya, Debajit Palit and V. V. N. Kishore

Abstract This chapter provides a review of alternative off-grid electrification options in South Asia. It covers four elements: the technical dimension, business models, regulatory governance and sustainability dimension of off-grid solutions. It concludes that in order to go beyond lighting applications, more careful consideration and investigation is required for electricity supply using local distribution networks (or mini-grids), particularly using hybrid technological options.

1 Introduction

As part of the research on off-grid electrification in South Asia, we have investigated the experience of rural electrification around the world and have identified various successful and not-so-successful experiences. These were reported in various academic publications (e.g. [1, 6, 18]). Simultaneously, a significant amount of review work was carried out to identify the technical options for off-grid electrification, participatory approaches are being used for delivery of such services

S. C. Bhattacharyya (✉)
Institute of Energy and Sustainable Development,
De Montfort University, Leicester, UK
e-mail: subhes_bhattacharyya@yahoo.com; subhesb@dmu.ac.uk

D. Palit
Social Transformation Division, The Energy and Resources Institute,
Lodhi Road, New Delhi, India
e-mail: debajitp@teri.res.in

V. V. N. Kishore
Independent Consultant in Renewable Energy,
Former Professor and Head of Department of Energy and Environment,
TERI University, Institutional Area, Vasant Kunj, New Delhi, India
e-mail: vvn_kishore@yahoo.com

S. C. Bhattacharyya and D. Palit (eds.), *Mini-Grids for Rural Electrification of Developing Countries*, Green Energy and Technology,
DOI: 10.1007/978-3-319-04816-1_2,

and financial arrangements for supporting off-grid electrification [5, 14, 15]. In addition, our research has also looked into the link between rural electrification and economic development [7], the importance of good regulatory governance [16] and alternative regulatory arrangements [3] for ensuring successful delivery of off-grid electrification for local development.

This chapter draws on the above research and provides a summary of alternative suite of off-grid options relevant for South Asia considering the multidimensional perspective involving techno-economics, social aspects, environmental issues and governance with a special emphasis on alternative participatory arrangements. In line with the objectives of the research project, we pay greater attention to options that meet the local needs for residential and productive uses. The purpose of this chapter is to take a regional perspective in enlisting potential suite of options that could be further investigated through project case studies (but not necessarily through this research project).

The chapter is organised as follows: Sect. 2 provides a summary of regional background, Sect. 3 presents the technological options; Sect. 4 summarises the delivery options; Sect. 5 presents the regulatory governance options, Sect. 6 analyses the sustainability of electrification options while Sect. 7 presents the concluding remarks.

2 Regional Background

South Asia consists of eight countries: Afghanistan, Bangladesh, Bhutan, India, Maldives, Nepal, Pakistan and Sri Lanka. The region covers an area of 5.13 million square kilometres (or 3.8 % of the world area) but holds 24 % of the global population (or 1.63 billion in 2010) [22]. The gross national income per person is reported as $1,176, which is about 13 % of the world's average of just above $9000 in 2010 [22]. However, the countries of the region show significant differences in terms of size, population density and economic development (see Table 1): the smallest country, the Maldives, has a very high population density and high income level; two other smaller countries, namely Bhutan and Sri Lanka also have significantly higher per capita income. However, the larger countries of the region dominate in terms of regional population and economic activities—in fact, three countries, namely Bangladesh, India and Pakistan account for 93 % of the regional population and 96 % of the economic output.

Although the region has recorded significant economic growth in the recent past, the region still suffers from a high incidence of poverty. More than 500 million (or about one-third of the region's population) is classified as poor who live on less than $1.25 a day. The region holds the world's largest concentration of poor people as a result.

Clearly, the high incidence of poverty manifests, among others, in terms of poor electrification rate of the region. More than 471 million people in the region were without electricity in 2010, implying an rural electrification rate of about 61 % on

Table 1 Basic information of South Asian countries in 2010

Country	Area, '000 (Sq. km)	Population, million	Population density (person/Sq. km)	GDP per capita, $	Poverty rate, % of population
Afganistan	652	34.4	52.8	410	36
Bangladesh	144	148.7	1032.6	700	31.5
Bhutan	38	0.7	19.1	1,870	23.2
India	3,287	1224.6	372.6	1,270	37.2
Maldives	0.3	0.3	1053.3	5,750	
Nepal	147	30.0	204.1	490	25.2
Pakistan	769	173.6	225.7	1,050	22.3
Sri Lanka	66	20.9	316.7	2,240	8.9
South Asia	5,131	1663.1	324.1	1,176	

Source World Bank [22]

average for the region [10]. There exists wide disparity in rural electrification at the country level however. Sri Lanka has a rural electrification rate higher than the global average while only 22 % of the rural population in Afghanistan is connected to the grid. India, Pakistan and Bangladesh alone constitute more than 90 % of the population that lack access to electricity in the region while the remaining 10 % is dispersed in the other smaller countries (see Table 2).

Limited access to clean energies, particularly in rural areas, results in significant differences in rural–urban energy use patterns. A recent survey in India, NSSO [17], provides a clear picture regarding the differences in the consumption pattern and the fuel mix of energy use by rural and urban consumers in India (see Figs. 1 and 2). While rural consumers still rely heavily on firewood and solid wastes for their energy needs, her urban counterpart relies more on modern energies. The fuel choice reflects the level of access to modern energies as well as consumers' ability to pay for a fuel and related government interventions. Rural households still rely heavily on firewood (87 % of rural households use firewood) for cooking with a per capita consumption of about 21 kg of firewood per month while the urban households have moved towards LPG (66 % urban households use LPG). The urban consumers are using about 5 kg of firewood and 1.8 kg of LPG per person per month. In terms of electricity use, there is a wide disparity as well: 94 % of urban households have reported electricity use while 69 % of rural households on average consumed electricity, indicating access constraints in rural areas. While an urban consumer has used about 24 kW h per month, a rural consumer has consumed only about 8 kW h per month. Limited affordability of rural consumers and limited ownership of electric appliances as well as electricity supply constraints explain the difference in rural–urban electricity consumption behaviour.

There is also a significant variation in consumption pattern by income class within urban and rural areas. Based on NSSO [17] data, a comparative picture is presented in Fig. 3. Although NSSO [17] provides consumption for each expenditure decile group, for convenience, we have aggregated the information into three categories—rich, medium income and poor—for urban and rural areas. This is based on the following assumption: The 10th decile is considered as the high

Table 2 Electricity access in 2010—South Asia

Country	Population without electricity (millions)	Electrification rate (%)		Per capita consumption (kWh)
		Total	Rural	
Afghanistan	22	30	22	35
Bangladesh	88	47	33	144
India	293	75	67	543
Nepal	7	76	72	81
Pakistan	56	67	55	475
Sri Lanka	5	77	75	418
South Asia	471	82.5	74.2	NA

Source IEA [9, 10]

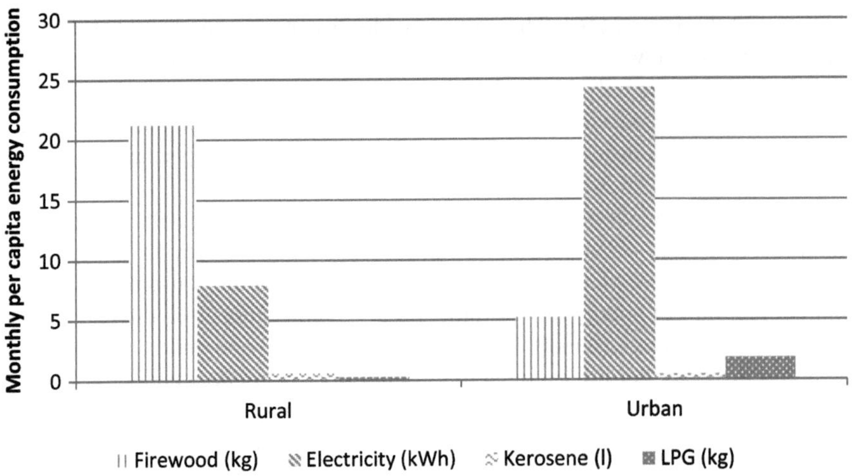

Fig. 1 Disparity in energy use between urban and rural India. *Data source* NSSO [17]

income group for both areas. For rural poor category, an average of first seven decile classes is considered while for urban poor, first six decile classes are considered. The remaining classes are considered as middle income groups. Figure 3 provides very interesting information:

(a) Fuelwood remains the main source of energy in rural areas irrespective of income class but this changes completely in the urban areas where fuel wood is only used by the urban poor;
(b) LPG remains essentially an urban fuel but the rural rich use some amount of this product; and
(c) Electricity use increases rapidly with income both in urban and rural areas but the scale is very different in urban areas compared to their rural counter parts. The rich in rural India approximately consume the same amount as the urban poor.

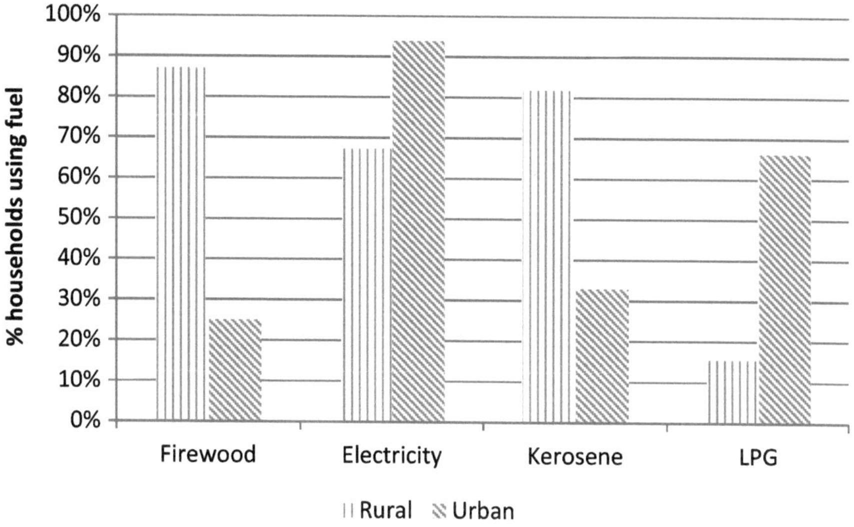

Fig. 2 Share of households using different energies. *Data source* NSSO [17]

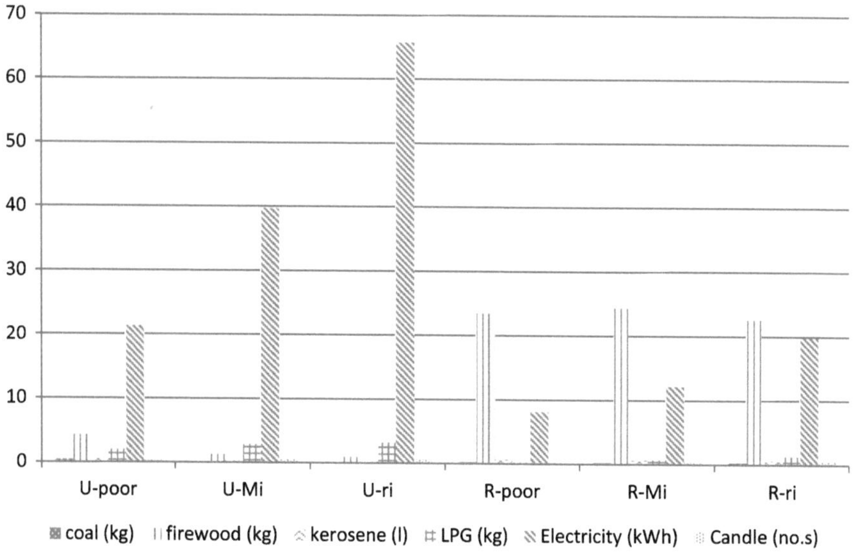

Fig. 3 Fuel consumption pattern by income class in India. *Note U-poor*—urban poor, *U-Mi*—urban middle class, *U-ri*—urban rich; *R-poor*—rural poor, *R-Mi*—rural middle class, *R-ri*—rural rich. *Source* Based on NSSO [17]

Although there will be some variation at the country level, similar information is not readily available for other countries. However, it is reasonable to believe that similar patterns are likely to prevail in other countries of the region. Clearly, lack of access to energies affects the development potential of countries in the region as

Table 3 Human development indicators for South Asia, 2011

Country	HDI	Life expectancy at birth, years	Mean years of schooling, years
Afghanistan	0.398	48.7	3.3
Bangladesh	0.5	68.9	4.8
Bhutan	0.522	67.2	2.3
India	0.547	65.4	4.4
Maldives	0.661	76.8	5.8
Nepal	0.458	68.8	3.2
Pakistan	0.504	65.4	4.9
Sri Lanka	0.691	73.4	8.2

Source HDI Database 2011

is evident from Table 3. Most of the countries of the region have a low Human Development Index (HDI) but those with better energy access have achieved higher HDI and per capita income. This observation is in line with the general trend of other developing countries (see [4] for a more detailed analysis).

More importantly, IEA [10] suggests that the number of people without electricity access in developing Asia will decline in its New Policies Scenario to 335 million in 2030 from 628 million in 2010. But India will remain the most important country with 150 million people without electricity access by 2030. This clearly suggests that additional concentrated efforts will be required to ensure universal electricity access by 2030. Given that the possibility of grid extension for universal electrification is limited in the region, the off-grid option is recognised as a feasible option.

3 Technologies for Off-Grid Electrification

Decentralised solutions have been promoted where the grid has not reached or is unlikely to reach in the near future. ESMAP [8] defines them as "an alternative approach to production of electricity and the undertaking and management of electrification project that may be grid-connected or not". Kaundinya et al. [13] indicate that the extent of decentralisation can exist at different levels: (1) village level where the focus is on providing electricity to meet the rural needs, (2) industry level where the demand of the industry is the main focus and any excess power is fed to the grid. Accordingly, the decentralisation can lead to grid-connected or off-grid (stand-alone) options. When a decentralised solution is not connected to the grid, it is known as an off-grid solution.

Off-grid systems are mostly used in areas where grid extension is difficult. These systems are demand-driven, small-scale operation for local needs; impose less pressure on resources due to smaller size; are often seasonal in supply due to technological characteristics and need storage systems that incur extra costs [13].

Off-grid options can be grouped into two categories—individual solutions and collective solutions. Individual solutions normally include small ready-to-use kit-based systems, such as Solar Home Systems (SHS), solar lamps and battery-operated systems. IFC [11] estimates that SHS has provided electricity access to between 0.5 and 1 million households in developing countries. Collective systems come in two modes of operation: stand-alone systems and local grid systems [8]. The local grid system serves a cluster of households or an entire village and provides electricity generally by generating from a diverse range of small local generators (such as solar PV, micro/mini hydropower, biomass-based technologies and diesel generators), with or without its own storage, and distributing it amongst the consumers. They are commonly called mini-grid and the service provider undertakes the business activities related to generation, distribution and sale (including billing, revenue collection and grievance redressal) of electricity. According to World Bank [21], portable 5–10 kW diesel generators are widely used as the conventional solutions. However, heavy reliance on diesel for small-scale power generation imposes cost burden on the utilities (more importantly on oil importing countries). The price fluctuations in the international market affect the overall cost of production and the viability of the business. This, in turn, imposes a heavy subsidy burden on the government.

Local grid system was also developed in hydro-dominated areas. For example, many small hydropower plants in China were initially developed using a local grid system and then connected to the main grid. In the stand-alone category, the solar photovoltaic systems for battery charging systems and charging solar lamps have emerged as a preferred off-grid solution for remote rural areas.

As elaborated in Kishore et al. [14], a number of energy sources can be used in off-grid applications, including potential energy of water, kinetic energy of wind, heat from solar energy and geothermal sources and a variety of fuels such as diesel, biodiesel, bio-methane and producer gas (chemical energy). Often a complex chain of processes is required to process the primary energy to deliver the usable form of secondary energy (see Fig. 4). Further details on the technological aspect are given in "Technical Aspects of Mini-Grids for Rural Electrification".

South Asia enjoys the benefits of both renewable and non-renewable energy resources. For example, the region has significant coal reserves and has some oil and gas resources. But generally, the region is import dependent to meet its fossil fuel needs. On the other hand, there is vast potential for renewable energies in the region. The potential for small/micro–hydro power exists in almost the entire HinduKush-Himalayan region, which includes Afghanistan, Pakistan, Nepal, Bhutan, northern India and Myanmar. Huge potential also exists in several locations in Sri Lanka and southern India due to their unique geo-climatic conditions [14]. Similarly, the average solar radiation incident over the South Asian countries varies from 4 to 7 kW h/day/m^2. With most of these countries having about 300 sunny days in a year [20], it is but natural for them to explore the possibilities of harnessing the energy of the sun. The region also relies heavily on biomass for its cooking energy needs and being an agrarian economy, there remains some potential for harnessing agro-wastes for energy purposes.

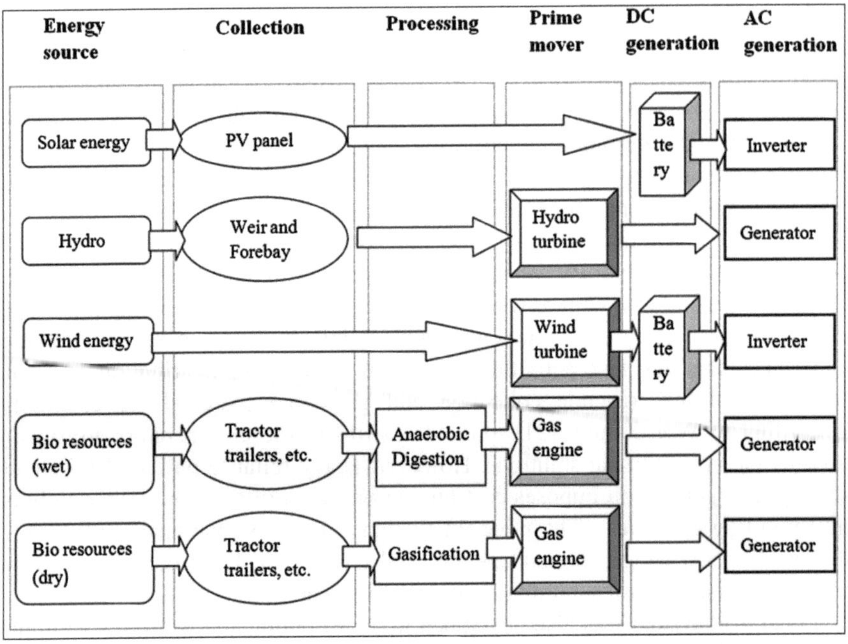

Fig. 4 Chain of operations involved in off-grid power generation. *Source* Kishore et al. [14]

The most common technologies used for off-grid electrification in the region are solar photovoltaic (PV) and mini/micro–hydro systems. The most common solar PV applications implemented in the region include both stand-alone solutions—solar home systems, solar lanterns and solar charging stations and local grid-based solutions such as solar mini-grids and solar DC micro-grids. While a typical SHS includes a 20–100-Wp (peak watt) PV array, a rechargeable battery for energy storage, one or more high efficiency lamps (either compact fluorescent or LED) and an outlet for a portable black and white television or other very low power consuming appliances,[1] the mini-grids are typically in the range of 2–150 kWp and provide AC electricity [18]. On the other hand, solar DC micro-grids are modular with capacity ranging from 75 Wp for connecting 10 households, using a direct current distribution grid, to around 1 kWp for connecting say around 200 households and usually provide only lighting services through LED lamps and facilities for charging mobile phones [19]. SHS has been the preferred option in many countries of the region for off-grid electrification since late 1980s and the development of LED lights has facilitated a reduction in the size of the systems, which in turn has helped the technology to reach the poorer sections of the population who could not afford the bigger systems even when they were

[1] Usually SHS with less than 40 Wp is used for lighting purpose whereas SHS above 40 Wp can be used for operating other electrical appliances such as TV, motor, fan, etc.

subsidised. However, the main weakness of the technology lies in its limited application potential and its inability to integrate productive uses of electricity directly. Thus SHS is often perceived as a pre-electrification option. In addition, India has also experimented with solar mini-grids (both AC and DC versions) especially in the Sunderbans (West Bengal), Chhattisgarh and Uttar Pradesh. Solar based AC mini-grid model has also been piloted in Bangladesh and its Ministry of Power is reportedly planning to expand the model under a programme called Remote Area Power Supply System (RAPSS) scheme, to be implemented by IDCOL.

The mini/micro–hydro systems (usually capacity in the range of 50 kW to 3 MW) have been used to create mini-grids to supply AC electricity locally. While Sri Lanka and Nepal have extensively used this technology to extend electrification to off-grid areas, such plants have also been installed in the hilly regions of India such as Arunachal Pradesh, Himachal Pradesh, Sikkim and Uttarakhand. Many mini/micro–hydro projects in the region have been driven by 'technology push', with micro–hydro now being a mature technology greatly improved by electronic load controllers, low-cost turbine designs, and the use of plastics in pipe work and penstocks. However, one of the key challenges faced by mini/micro–hydro systems, especially in India, is low utilisation factor due to unavailability of sufficient water discharge during dry season and very high discharges during monsoon in the Himalayan streams (when the plant has to be shut down to avoid damage to the penstock or turbine due to possibility of high quantity of silt coming with the water). The low load factors also result in high operating and maintenance (O&M) costs resulting in uneconomical operation in isolated mode in the hilly areas [18].

Biomass gasifiers have found use in India and to a limited extent in Sri Lanka for off-grid electrification. Biomass gasifier-based mini-grids are typically in the range of 10 to 500 kW. The technology, however, has found limited success for off-grid electrification. One of the key reasons for this is the absence of standardised performance-oriented technical specifications of the systems to ensure quality of the products and also due to non-creation of proper after-sales maintenance network to service the systems in the remote rural areas[2] [18].

In addition, diesel generators are widely used in the region for own electricity consumption (in homes and for agricultural purposes) but some enterprising villagers also work out an arrangement to provide power either to a cluster of houses or for some economic activity (e.g. village markets). The service depends on diesel availability and is generally a polluting activity but the skills for running such generators are often locally available.

[2] The performance of the biomass gasifier projects implemented under VESP (Village Energy Security Programme) or RVE (Remote Village Electrification) programme in remote rural areas is found to be unsatisfactory especially due to technology management and product quality issues On the other hand, biomass gasifiers implemented by private companies in some parts of India for electricity supply to 'not so remote' areas are reported to be working satisfactorily.

However, there has been limited experience with hybrid technologies particularly for off-grid applications with some exceptions. For instance, in the Sunderban region, the West Bengal Renewable Energy Development Agency had tried out hybrid systems (e.g. solar–biomass–wind) to improve the overall efficiency of their solar PV-based mini-grids. While PV was the dominant source of energy, wind systems—connected through the same battery–inverter systems—provided additional source of energy during the monsoon season while biomass gasifier systems meet this additional energy requirement during the dry season. The model not only helped to improve the reliability of the systems, but at the same time also addressed the incremental demand. In addition, the off-grid services have catered to limited needs by providing electricity for a limited period of time, mainly for lighting purposes in most cases. Thus, a round-the-clock service was not available and the demand growth over time was often difficult to meet.

4 Delivery Options

The business model for delivering off-grid electrification tends to vary depending on, among others, the mode of delivery (product sale or local grid-based supply), financing mechanism and ownership (see Fig. 5). For the stand-alone system delivery, four common delivery options are cash sales by retailers, rental fee-based service by an energy service company (ESCO), leasing arrangements by an ESCO and a micro-finance based scheme. The region has experience with all these options but some of the well-known cases include (see [15] for more details):

(a) Grameen Shakti, a non-profit company, has promoted SHS in Bangladesh through a micro-finance backed system. Following the example of Grameen Bank, Grameen Shakti has created a rural network of decentralised branches all over Bangladesh and has developed a sales strategy linked with consumers' access to finance. Consumers make an initial down payment and repay the rest over a period of time paying a low rate of interest on the loan. By 2012, GS has installed more than 1 million SHS in Bangladesh (the total SHS installation figure has increased to 2.6 million by November 2013), recording an eightfold growth in sales between 2007 and 2012.
(b) SELCO, a private social enterprise in India, has used 'lease-to-own' approach to promote SHS in South India. Although it does not provide finance to its customers, SELCO supports its customers in securing loans through third party finances (such as REEEP (Renewable Energy and Energy Efficiency Partnership) funding to cover the down payment requirement for securing a loan) and has been successful in installing 135,000 SHS since 1995.
(c) TERI's Lighting a Billion Lives campaign uses a fee-based rental model for promoting solar lighting in India and has covered around 100,000 households till end of 2013.

The frequency of transactions involved in the product delivery option can range from a single transaction (cash payment to the dealer and transfer of ownership) to

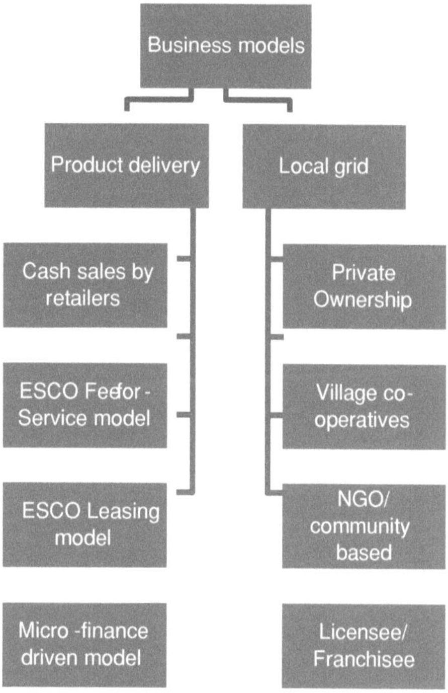

Fig. 5 Business models for off-grid electrification

regular interactions over a certain period of time (as in the case of fee-for-service or lease arrangements). This depends on the nature of the business arrangement chosen (see Table 4). However, the duration of the contract tends to be short- and long-term dependence of the parties to contract is not common.

In respect of local grid-based electricity supply in rural areas, the region has experimented with a number of alternative options as well. The private power generation and supply through diesel generators has already been mentioned. These use temporary networks and either provide the service for a specific use (fair, market, etc.) or offer the service to a limited cluster of households. These operate on a fully commercial basis and charge usually much higher rates for their supply, commensurate with the small size of their generators (therefore poor technical efficiency) and high cost of diesel fuel used for the supply. In addition to this category of service, renewable energy-based rural electricity supply has also emerged particularly in India, with the Husk Power System (HPS) being the most well-known case. HPS, a small start-up company based in Bihar, has electrified around 300 villages and hamlets since 2007 through the set-up of 80 plants, benefitting nearly 200,000 people. HPS builds village scale mini-grids using rice husk gasifiers, usually ranging between 30 and 200 kW systems. HPS works only in locations where at least 250 households agree to take connection and it charges a nominal installation charge as well as a regular fee for electricity, sometimes 45 INR per 15 W CFL. It charges a higher rate for commercial use than for residential

Table 4 Nature of transactions in the product delivery option

	Dealer—sale cash	Direct sale—credit	Fee-for-service	Lease arrangement
Frequency of interaction	One	Multiple—regular/irregular depending on the credit terms.	Multiple—regular transactions	Multiple regular transactions
Contract duration	Instantaneous	Defined by credit terms—normally 1–3 years	As agreed with the supplier, but with possibility for termination or non-payment	As defined by the lease arrangement, could be terminated due to non-payment
Ownership	Purchaser	Dealer until full payment, then purchaser	Supplier	Supplier
Maintenance responsibility	Purchaser	Purchaser	Supplier	Could be with supplier or user depending on the arrangement

use. Some of its plants have generated INR 40,000 monthly revenue from tariffs, considerably greater than average expenses of INR 20–25,000/month [15].

Community-managed systems have been widely used in the region. Micro hydro-based developments in Sri Lanka and Nepal, considered to be successful initiatives, are run by the local communities using the locally available water resource to meet their energy needs. On the other hand, India has implemented solar PV-based projects which were successful but the biomass gasifier systems implemented under Village Energy Security Programme (VESP) in India were unsuccessful. Generally, community involvement in a project brings a sense of ownership of the project and hence can enhance the acceptability of the project by the community. But a successful implementation and operation of the system requires appropriate skills, clarity of roles and responsibilities, social cohesion and strong commitment. Lack of technical skills in the community for system operation and financial management often impede with successful delivery of complex electrification projects.

Although the co-operative model has been widely used in Bangladesh, it was used for grid-based electricity supply and not for off-grid electrification as such. In India, the co-operative model has not been widely used but in the case of off-grid electrification in the Sunderbans, this model was used. As members of the co-operative consumers have a sense of ownership of the project, the management tends to take a formal approach than in the case of community-based systems. However, the issue of availability of skills and their retention remains a problem.

The franchisee model has been used in South America to engage private entities in rural electrification. In the case of grid extension in India, a similar approach was used but this has not been applied in the case of off-grid supply.

A comparative summary of these business models is presented in Table 5.

5 Regulatory Governance Options

Traditionally, when electricity generation and supply has been carried out by a vertically integrated utility, it was a regulated business activity and an electricity act (or a similar legal instrument) generally governed the operational and developmental activities of the industry. Even where the business has been unbundled and deregulated, the transmission and distribution business remains a regulated activity while generation and supply are subjected to lighter supervision and control. Investments in a network facility are largely sunk costs and hardly re-deployable, while the variable cost is relatively low. It is generally considered as a "classic natural monopoly" offering significant scale economies. In such a case, one entity can provide the service more economically than multiple entities. The main economic rationale behind the regulation arises from a dilemma involving natural monopoly and the possibility of consumer exploitation by monopolists. The economic logic would then require allowing one entity to provide the service, which ensures low-cost supply. But if a monopoly is allowed to operate it also has

Table 5 Comparison of alternative off-grid business models

Characteristics	Energy service delivery model				
	Co-operative	Fee-for-service/ ESCO model	Community managed	Franchisees	Private sector
Ownership	Members of the co-operative own and operate the model	Ownership vests with the ESCO	Can be of two types: • Owned by private/public entity and managed by communities • Owned and managed by communities	Ownership of assets vests with the discom; Franchisee is a custodian of the assets	Owned and operated by the private sector except in PPPs where the ownership of assets may remain with public entity
Management	Managed by a Board of directors or a governing body elected by the consumers	Managed by the ESCO	Managed either by an NGO or local self-governing institutions such as village committees or village councils etc	Managed by private sector NGO, SHG etc	Managed by private sector
Maintenance	Co-operative is responsible for O&M	Maintenance is undertaken by the ESCO	Maintenance is undertaken by the VEC village council etc	O&M undertaken by the franchise operator	O&M undertaken by the private sector
Pricing	Low upfront cost and monthly tariffs; usually regulated	Low to moderate tariffs (set-up by ESCO)	Low to moderate tariffs (mutually decided by the community and VEC)	Moderate electricity tariffs (regulated)	Moderate to high tariffs (set-up by service provider)
Community participation	Moderate to high participation. Communities are members of Co-operatives. Local youths may also be involved for bill collection, undertaking minor repairs etc	Limited participation	High participation. Communities are involved right from the planning stage till the end implementation stage. Several functions such as labour contribution for construction, management, maintenance, grievance redressal are performed by communities	Limited participation. Franchisee operators may involve locals for bill collection	Consumers are generally not involved in the planning or management of the business

(continued)

Table 5 (continued)

Characteristics	Energy service delivery model				
	Co-operative	Fee-for-service/ ESCO model	Community managed	Franchisees	Private sector
Risks	Amenable to political interference	Communities lack technical and managerial skills and this threatens the sustainability of the model	ESCO carries primary risk of theft. ESCO model is sensitive to uncertainty regarding grid extension	Franchisee depends fully on discom for power supply and so can't always meet the community's aspiration	Private operator can discriminate by charging high tariffs

Source Krithika and Palit [15]

the potential of abusing its power and charge excessive prices for its own profit motives. Forcing the natural monopolist to competition, on the other hand, is likely to lead to a situation of perpetual loss, which would not encourage any private provider to enter the market. The economic regulation tries to balance the dilemma by granting a monopoly status to the service provider but subjecting it to conditions that would protect the consumers as well.

The act normally specifies its area of application and does not generally distinguish between urban and rural areas. Generally, the service area of a supply provider includes a mix of urban and rural areas. Therefore, unless a specific exemption or waiver is granted or allowed, the rural electricity supply generally comes under the purview of the general provisions of the act and accordingly, the law of the country essentially decides whether rural electricity supply is a regulated activity or not.

However, the development of decentralised solutions in many countries around the world requires some attention. The poor state of (or even non-existent) rural electricity supply is a result of the failure of the existing delivery mechanisms. The emergence of the decentralised solutions thus can be viewed as a response to the existing deficiencies that are either arising as a consequence of modifications to regulatory arrangements or perhaps working outside the scope of electricity supply regulations. Two cases mentioned above, namely individual product delivery mode and collective service delivery mode, require specific attention in this respect.

5.1 Regulation of Individual Solutions

In the case of individual solutions which are delivered through sale or renting of products, electricity supply activity does not take place and hence it does not involve any distribution or transmission networks. The issue of natural monopoly does not arise accordingly. Some amount of electricity generation may take place at the consumer's premises in some of the options (e.g. SHS and mini generators). The law of the land may require consent/approval for such installations from competent authorities (such as local authorities and even electricity authorities or departments) but being a standard package solution that can be offered by multiple manufacturers, the basic need for economic regulation does not arise.

The responsibility for grid-based electricity distribution remains with the distribution utility but the product delivery mechanism provides a short-term relief until formal supply arrives. As the off-grid product delivery mode does not generally fall under the licenced (or regulated) activity, the electricity regulator does not control such activities. However, once the grid gets extended and the consumer tries to connect her decentralised system to the distribution network, the distribution utility and the electricity regulator become interested parties and the activity comes under regulator's jurisdiction.

Although the electricity regulator may not be involved in the product delivery mode, other rules/standards/regulations may apply to product/equipment delivery activities. The product or equipment will be subjected to technical standards (regulations) for quality, environmental standards and even consumer protection regulations/laws. Similarly, if credit-based systems are used, the financial intermediaries may be subjected to specific financial regulations to prevent cheating, rent seeking and exploitation. Absence or ineffectiveness in any of these governance mechanisms can reduce the benefits to the consumers. On the other hand, the absence of any proper market at present may justify the need for creating a protective environment for private entrepreneurs, which in turn may require allowing demarcated delivery zones even for product delivery option. This is likely to provide some monopoly rights over the area of delivery but depending on the authority used to grant such a protective environment, the regulatory control would be decided. In general, the competition commission or authority controlling monopoly and restrictive trade practices or a designated state agency would be responsible for monitoring and controlling such issues.

5.2 *Regulation of Collective Solutions*

In the case of a collective service provision, the decentralised service is provided as a substitute of grid extension that uses a distribution network and decentralised generators. This conforms to the commonly used definition of electricity supply and hence, unless specifically allowed by the country's law, the business activity will come under the supervision of the electricity regulator. The need for regulation arises for two reasons: (1) to ensure that the activity complies with the law of the land and (2) to protect the investors and the consumers following the standard principles of economic regulation indicated above.

The regulatory arrangement may depend on the mode of delivery chosen in this case and can take different forms.

(a) A generic waiver or exemption from the standard provisions applicable to the electricity supplier may constitute a simple solution. This is the approach followed in India where the Electricity Act 2003 allows the state government/ state commission to exempt certain types of organisations from the licence requirements for rural electricity supply either by notifying the rural areas to be covered by them or by the regulator specifying the terms and conditions for such exemption (see [2] for further details). However, unless the conditions of the waiver or exemption are clearly indicated, and the roles and responsibilities of the parties involved are clearly documented, this simple option can create confusion and may introduce uncertainties for the business. This can also create issues related to reporting and sharing of information and may prove to be an ineffective system.

(b) A simplified, standardised regulatory approach can be a more practical approach. Such a regulation should specify the role and duties of the provider, set the information filing requirements and ensure consumer protection mechanisms. The purpose of such a light-handed approach is to reduce the cost of regulation by imposing reduced burden on the regulatory agency. This is likely to be effective for local community-based organisations, non-profit organisations and private entities with socially driven motives. For-profit private organisations may try to take advantage of such light-handed systems to increase their profitability. Strong penalties and rule-enforcing mechanisms would be required as deterrents in such cases.
(c) A full-fledged regulatory arrangement constitutes the most formal regulatory approach. The existing electricity regulator can be entrusted with these duties or a separate rural electricity (or infrastructure service) regulator can be established. The regulatory powers are normally derived from a specific legislation (such as the Rural Energy Act) and the implementation and governance aspects follow the provisions of such legislation. However, such a regulatory arrangement is likely to be a costlier option and a careful cost-benefit analysis needs to be undertaken prior to the adoption of such an arrangement to ensure that the benefits of regulation would outweigh the costs.

5.3 Regulatory Supervision

It is evident from the above that the need for regulatory supervision is not the same for two types of delivery channels and for different types of delivery organisations. In the service mode of delivery, the regulatory supervision will depend on the ownership of the delivery system. For example, if a distribution franchisee model is chosen, the supervision need will perhaps be more extensive whereas in a co-operative model or a community-managed delivery system the threat of consumer exploitation may be limited. In general, the regulatory supervision covers the following aspects:

(a) Regulated business activities: The service provider is allowed to carry out specific tasks under the permission granted to it. In the off-grid electrification case, it would involve generation of electricity and supply using appropriate infrastructure. The sale of electricity is normally restricted to final users and reselling is not normally allowed. This would normally require a clear demarcation of the area of activity and a mechanism for avoiding overlaps with the incumbent utility's service area. Absence of clarity in this respect enhances business uncertainties.
(b) Activities requiring prior regulatory approval: The service provider is normally subjected to conditions requiring it to seek prior approval for a number of activities or transactions. These include sale of the business, engaging agents or transactions with affiliates, etc.

(c) Conditions of supply: Normally a condition of non-discriminatory supply to eligible consumers is imposed to ensure that all consumers meeting the supply criteria are connected. Similarly, any anti-competitive practices or practices leading to market abuse are also not permitted. The regulatory arrangement may provide specific conditions for connection and disconnection.
(d) Tariff related provisions: These constitute the most important element of the regulatory supervision. The cost of electricity supply using an off-grid system depends on the technology used, energy resources utilised, size of the system, demand pattern, infrastructure used, service quality and the cost of regulatory compliance. As the cost of supply tends to be high, full cost recovery may lead to limited access (due to limited affordability of consumers) while a limited cost recovery, either requires a well-defined subsidy scheme or leads to an unviable business proposition, thereby increasing the potential for under-achievement or failure of the system. The tariff issue can be the most contentious issue for the private sector involvement in the business while the challenge is somewhat mitigated in the co-operative or community-based service options.
(e) Consumer protection: Protecting the vulnerable consumers constitutes one of the main purposes of regulation. This can cover protection from abusive tariffs, poor supply quality and other customer grievances (related to billing, connection, disconnection, deposits, technical faults, etc.).
(f) Reporting requirements: All regulated entities are required to provide certain information to the regulator to indicate the level of activity, quality of service or for reporting incidents, disputes or grievances. A systematic flow of information allows the regulator to decide whether to intervene or not and whether any regulatory change is required.

The regulatory requirements for alternative delivery options are indicated in Table 6.

Evidently, regulation of the business activity will not be an easy process and would require significant amounts of training and capacity building both at the regulatory level and the service provider level.

6 Sustainability of Off-Grid Electrification Options

6.1 Methodology

For the sustainability analysis of energy access programmes, an adaptation of a framework suggested by Ilskog [12] is used. Ilskog [12] considered five sustainability dimensions—technical, economic, social/ethical, environmental and institutional sustainability. Each dimension was represented by a number of indicators and each indicator was scored on a scale of 1–7. The overall score obtained by simple averaging was used for final ranking of the programmes.

Table 6 Regulatory check for alternative delivery options

Regulatory conditions/requirements	Franchisee model	Co-operative model	ESCOs	State utility or community based
Electricity sale only for final consumption	Satisfied, as there is no intermediate transaction	Satisfied	Satisfied	Satisfied
Generation of electricity	Required as approved activity	Required as approved activity	Required as approved activity	Required as approved activity
Assignment of transfer of assets/business permission without prior approval	Required as a condition	Normally does not apply but may be required in case of merger or acquisition	Required as a condition	Normally does not apply
Engaging affiliates or subsidiaries	Condition required	May arise	May arise	Normally does not arise
Providing loans/guarantee on obligations	May arise and a suitable condition is required	May arise and a suitable condition is required	May arise and a suitable condition is required	May arise and a suitable condition is required
Undue preference	Suitable condition required	Unlikely to arise	Suitable condition required	Could arise and suitable condition required
Separate accounts for businesses	Required if franchisee operates different businesses	Required if different businesses are undertaken	Required if different businesses are undertaken	Required if different businesses are undertaken
Major incident reporting	Required as a condition	Required as a condition	Required as a condition	May per part of the overall utility reporting scheme
Seeking permission for disposing of or relinquishing assets or control	Required as a condition	Required as a condition	Required as a condition	May be part of the utility's overall regulatory obligation
Demand forecasting	Franchisee responsibility	Co-operative's responsibility	ESCO responsibility	Utility responsibility
Consumer protection	Franchisee responsibility	Co-operative responsibility	ESCO responsibility	Utility responsibility

(continued)

Table 6 (continued)

Regulatory conditions/requirements	Franchisee model	Co-operative model	ESCOs	State utility or community based
System planning	Franchisee responsibility	Co-operative responsibility	ESCO responsibility	Utility responsibility
Tariff regulation	Required to avoid exploitation	Not essential—no profit motive	Required to avoid exploitation	Could be part of overall utility regulation

We retain the five sustainability dimensions and identify relevant indicators for each dimension. We then apply this to alternative electrification programmes, namely grid extension, off-grid solar home systems, off-grid electrification through local mini-grids and apply a scoring on a scale of 1 (poorest) and 7 (highest). The score for each indicator was arrived at through a brainstorming session involving a number of energy specialists. We recognise that this is the weakest part of the methodology implementation which can be improved using a stakeholder survey at a future date.

More specifically, we consider the following (see Table 7 also):

(a) Technical sustainability is achieved if the system can meet the present and future needs reliably, efficiently and by using clean and renewable sources. This is captured by considering whether the programme can satisfy the present and future needs (both residential and productive), whether reliable, efficient and renewable energy-based supply can be delivered and whether supporting services for maintenance and running the systems are locally available or not.
(b) Economic sustainability is achieved if the system offers cost-effective and affordable supply at present and in the future. This is captured by considering the cost effectiveness and cost recovery potential of supply, capital and operating cost burden imposed on the users and financial support needs for the system.
(c) Social sustainability requires that the solutions should be widely acceptable and accessible to ensure reduction/removal of human drudgery and adverse effects on women and children.
(d) Environmental sustainability aims to reduce the environmental impacts on the users and the society. This is captured by considering contributions to local and global pollution, health damages and other environmental degradation.
(e) Institutional sustainability requires that the provision is locally manageable and controllable. This is represented by the degree of local ownership, availability of skilled staff, ability to protect consumers and investors and ability to monitor and control the systems.

While it is possible to include more factors in the sustainability analysis, the above provides a reasonable picture of the multidimensionality of the challenge. Although a more formal framework using multicriteria decision-making approach is also possible, the above provides a good starting point.

6.2 Analysis of the Results

The result of the analysis is presented in Fig. 6. Grid extension emerges as the preferred alternative for lighting option, although it received just about 50 % of the possible scores. This suggests its weakness as an all-round solution. The SHS received the weakest score of all because of its limited ability to meet the range of

Table 7 Indicators of sustainability of energy access programmes

Technical sustainability	Economic sustainability	Social/ethical sustainability	Environmental sustainability	Institutional sustainability
Ability to meet present and future domestic needs	Cost effectiveness	Wider usability amongst the poor	Contribution to reductions in carbon emissions	Degree of local ownership
Ability to meet present and future productive needs	Cost recovery potential	Need for micro-credit or financial support systems	Contribution to reduction in indoor pollution	Need for skilled staff
Reliability of supply	Capital cost burden on the user	Potential to reduce human drudgery	Contribution to reduction land degradation	Ability to protect consumers
Reliance on clean energy sources	Running cost burden on the user	Potential to reduce effects on women and children	Contribution to reduction in water pollution	Ability to protect investors
Technical efficiency	Financial support needs			Ability to monitor systems
Reliance on local resources	Contribution to income generating opportunities			
Availability of support services				

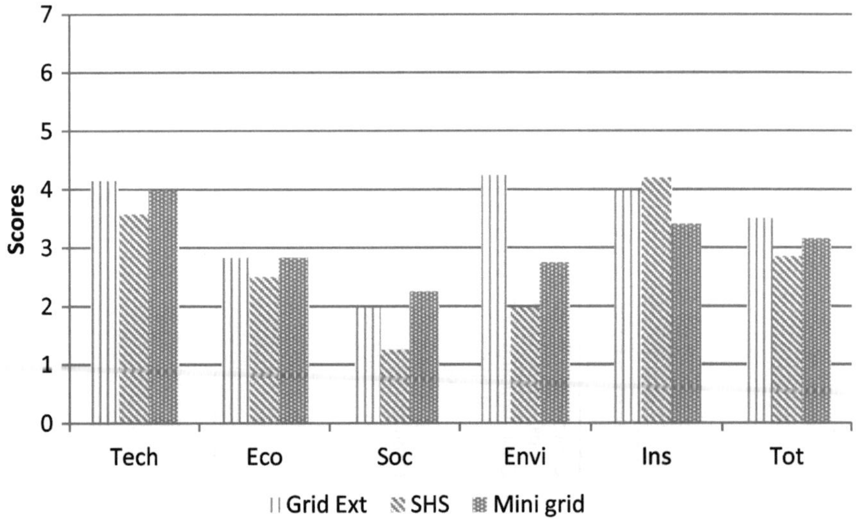

Fig. 6 Sustainability comparison of alternative electricity access programmes. *Source* Bhattacharyya [4]

needs. However, no option reaches the highest score of 7 in the overall assessment, indicating their weakness in certain areas.

The scores related to the economic sustainability are relatively low, reflecting the problem of sustaining such solutions without some support mechanisms. Local resources fare better in this respect but cost issue remains a main problem in all options. Moreover, some sustainability issues exist for each dimension. The inability to meet future demand, particularly productive needs and poor technical efficiency and reliability of the systems are serious technical constraints faced by most of the access options considered in this exercise. The inability of lighting-oriented access options to reduce drudgery and to promote wider clean energy use amongst the poor affects the social sustainability of such options. The lighting options also score relatively low in terms of environmental benefits as these options miss the most important energy needs of the people.

The above analysis suggests that the existing practices of providing electricity access are generally unsustainable from a number of perspectives although this has received limited attention. Over-emphasis on limited impact options needs to be avoided and a rebalancing of provision options is required to ensure a more sustainable approach to resolve the problem. Hybrid options are likely to perform better in this respect, although our analysis did not focus on these solutions. Also, further analysis of country-specific experiences is required to identify sustainability challenges in specific cases.

7 Conclusion

One of the main challenges of our time is to ensure universal access to energy but IEA [10] indicates that unless concerted efforts are directed to address the challenge, South Asia will remain a problem area even by 2030. As grid extension is unlikely to materialise in the entire region, off-grid options will play an important role. Although the region has gained some experience in off-grid electrification over the years, the focus has been mostly on solar home systems but given its limitations in terms of service SHS can only be considered as a pre-electrification or a temporary solution until grid comes or electricity is supplied through a mini-grid from a local generator. Moreover, the initial investment being significantly high relative to the ability to pay by the poor, this option has hardly reached the poorest section of the population, even where microfinance has been organised. Thus, a step change is required where the local supply will include productive use of electricity that can positively catalyse the economic development at the local level.

The local grid-based systems thus require more attention. Although the region has experimented with this option as well, particularly using micro–hydro technologies, solar PV and agricultural wastes (e.g. rice husk), the result remains mixed. While a properly designed and operated micro–hydro system can provide the cheapest electricity, the reliability of supply can be a major challenge due to seasonal water availability, poor maintenance and lack of local skills in managing the system. Similarly, although there is now a reasonably long-lasting experience of dealing with solar PV-based mini-grids in the region, long-term issues such as meeting demand in the future, ensuring a reliable and cost-effective supply and supporting productive uses for rural development remain. Further, there is very limited experience with hybrid systems and clearly, there is a significant vacuum in the regulatory sphere.

The suite of technological options, business models, regulatory and governance options enlisted in this chapter can provide useful information for developing further research particularly using hybrid options for promoting reliable and cost-effective solutions at the local level.

References

1. Bhattacharyya SC (ed) (2013) Rural electrification through decentralised off-grid systems in developing countries. Springer, London
2. Bhattacharyya SC, Srivastava L (2009) Emerging regulatory challenges facing the Indian rural electrification programme. Energy Policy 37(1):68–79
3. Bhattacharyya SC, Dow SR (2013) Regulatory issues related to off-grid electricity access, chapter 11. In: Bhattacharyya SC (ed) Rural electrification through decentralised off-grid systems in developing countries. Springer, London
4. Bhattacharyya SC (2012) Energy access programmes and sustainable development: a critical review and analysis. Energy Sustain Dev 16(3):260–271

5. Bhattacharyya SC (2013) Financing energy access and off-grid electrification: a review of status, options and challenges. Renew Sustain Energy Rev 20:462–472
6. Bhattacharyya SC, Ohiare S (2012) The chinese electricity access model for rural electrification: approach, experience and lessons. Energy Policy 49:676–687
7. Cook P (2011) Infrastructure, rural electrification and development. Energy Sustain Dev 15(3):304–313
8. ESMAP (2001) Best practice manual: promoting decentralized electrification investment. World Bank, Washington, DC
9. IEA (2009) World Energy Outlook 2009. International Energy Agency, Paris
10. IEA (2012) World Energy Outlook 2012, Electricity Access Database. International Energy Agency, Paris
11. IFC (2007) Selling solar: lessons from more than a decade of IFC's experience. International Finance Corporation, Washington, DC. http://www.ifc.org/ifcext/sustainability.nsf/Content/Publications_LOE_SellingSolar
12. Ilskog E (2008) Indicators for assessment of rural electrification an approach for comparison apples and pears. Energy Policy 36:2665–2673
13. Kaundinya DP, Balachandra P, Ravindranath NH (2009) Grid-connected versus stand-along energy systems for decentralized power—a review of literature. Renew Sustain Energy Rev 13:2041–2050
14. Kishore VVN, Dattakiran J, Nand Gopal E (2013) Technology choices for off-grid electrification, chapter 3. In: Bhattacharyya SC (ed) Rural electrification through decentralised off-grid systems in developing countries. Springer, London
15. Krithika PR, Palit D (2013) Participatory business models for off-grid electrification, chapter 8. In: Bhattacharyya SC (ed) Rural electrification through decentralised off-grid systems in developing countries. Springer, London
16. Minogue M (2013) Regulatory governance of off-grid electrification, chapter 10. In: Bhattacharyya SC (ed) Rural electrification through decentralised off-grid systems in developing countries. Springer, London
17. NSSO (2012) Household consumption of various goods and services in India. NSS 66th round, National Sample Survey Office, National Statistical Organisation, Government of India, New Delhi
18. Palit D, Chaurey A (2011) Off-grid rural electrification experiences from South Asia: status and best practices. Energy Sustain Dev 15:266–276
19. Palit D (2013) Solar energy programs for rural electrification: experiences and lessons from South Asia. Energy Sustain Dev 17(3):270–279
20. Raman P, Murali J, Sakthivadivel D, Vigneswaran VS (2012) Opportunities and challenges in setting up solar photo voltaic based micro grids for electrification in rural areas of India. Renew Sustain Energy Rev 16:3320–3325
21. World Bank (2008) The welfare impact of rural electrification: a reassessment of the costs and benefits, an IEG impact evaluation. World Bank, Washington, DC
22. World Bank (2012) The little data book 2012. World Bank, Washington, DC

Technical Aspects of Mini-Grids for Rural Electrification

P. J. Boait

Abstract This chapter aims to provide an overview of the main technical options and decision points that arise when developing a plan for a rural mini-grid. It begins by considering the four potential sources of renewable energy—water, sun, wind and biomass. For each the relevant technologies and methods for quantifying the available resource are considered. Key options for the system architecture are then covered including the choice of AC or DC distribution, network topology, hybrid or single generator systems, and battery technology. Finally, the critical technical factors in system management are summarised.

1 Introduction

The challenge of avoiding catastrophic climate change has motivated governments in many countries to introduce policies aimed at incentivising a transition to low or zero carbon forms of electricity generation. This has stimulated investment in research and development of all forms of renewable generation, whether powered by water, wind, waves, sun or biomass. Because these resources are inherently distributed widely dependent on geography, and are available at every scale from watts to gigawatts, the technologies that have emerged are equally capable of operating at all scales. Increasing manufacturing volumes are also driving down costs making a wider range of technologies accessible to mini-grid designers and investors.

This chapter therefore aims to review the technical options for mini-grid system components and architectures including innovations that have reached, or are approaching, viability for deployment in rural areas of the developing world. This requires proven reliability and ease of installation and support, as well as acceptable

P. J. Boait (✉)
Institute of Energy and Sustainable Development, De Montfort University, Leicester, UK
e-mail: p.boait@dmu.ac.uk

S. C. Bhattacharyya and D. Palit (eds.), *Mini-Grids for Rural Electrification of Developing Countries*, Green Energy and Technology,
DOI: 10.1007/978-3-319-04816-1_3, © Springer International Publishing Switzerland 2014

cost. The discussion of generation options in the next section focuses on methods for assessing the available energy resource using elementary physics and mathematical methods that are amenable to spreadsheet programming. Often these methods are embedded in software tools such as HOMER (covered in "Analytical Frameworks and an IntegratedApproach for Mini-Grid-BasedElectrification"), but an understanding of the underlying physics provides insights into the limitations and data sensitivities of these tools. Later sections cover system architecture and management technologies.

2 Generation Technologies

2.1 Micro-Hydro

Most societies have exploited the power of running water as an energy source for centuries, so generation of electricity forms a natural successor technology to mechanical water mills. Hydro generation is by far the largest source of renewable electricity generation globally reflecting the number of large- and medium-scale plants that have been operating since the first half of the twentieth century. But there remains much potential—the World Small Hydropower Development Report published by UNIDO [1] indicates that in South Asia only 19 % of potential capacity is currently exploited (compared to 95 % in Northern Europe).

An illustration of the key components of a mini hydro scheme is shown in Fig. 1, from Practical Action [2], which also provides videos of a mini hydro in operation. A weir on the stream allows a controlled fraction of the flow to be diverted to the forebay where the water is slowed sufficiently for suspended material to settle. Floating debris is filtered out with a comb of bars (a trash rack) to prevent damage to the turbine. Screening measures to prevent fish entering the system may also be required. The potential energy of the water is converted into a combination of head pressure and kinetic energy in the penstock and delivered to the turbine and generator.

The instantaneous power P kW produced by an ideal hydro generator is given as

$$P = gHQ \tag{1}$$

where H is the head in metres, Q is the flow rate in m^3/s and g is the acceleration due to gravity (9.81 m^2/s). The head is the vertical height between the forebay and the turbine. In practice, P is significantly less due to losses in the flow path, turbine and generator. However, about 70 % is an achievable water-to-wire efficiency [3], so for assessing a potential resource Eq. (1) can be reduced to:

$$P = 7HQ \tag{2}$$

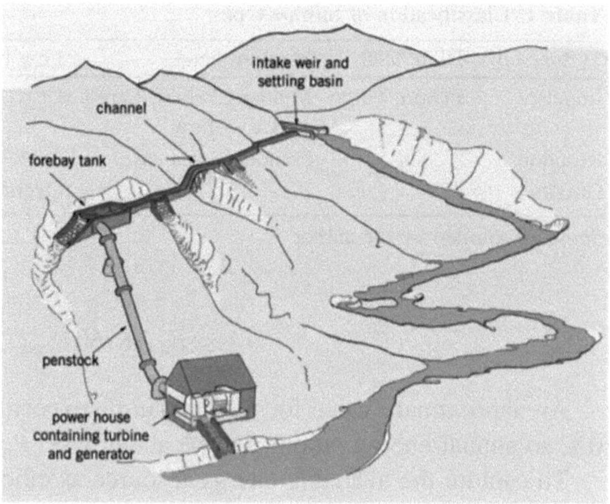

Fig. 1 Components of a mini hydro scheme [2]

The energy that can be produced then depends on the variation in the available flow rate over the year. A simple way of measuring the approximate flow rate is to set up a weir with a rectangular gap in it through which all the water must flow. The flow rate can then be estimated from the depth of water flowing through the gap. It is given as

$$Q = 1.8(w - 0.2d)d^{1.5} \tag{3}$$

where w is the width of the rectangular gap in the weir and d the depth of water, both in metres. If Q is surveyed over a year then this will provide the basis for several key decisions. The first is the proportion of the flow that can be diverted into the hydro system. This will depend on many factors such as the ecosystems to be maintained in the main stream, any local regulatory requirements and the variability of the flow. A typical judgment would be to set the flow rate that is exceeded for 85 % of the year as a minimum flow above which abstraction for hydro generation can take place.

Having determined the system flow rate over a year, the maximum power P_{max} (capacity) of the turbine and generator can be determined. Clearly there is a trade-off between the cost of a larger turbine and generator, and the proportion of the year for which its full capacity will be used. A useful benchmark is the annual mean flow Q_{mean}, which is often a reasonable level to choose to set P_{max} using Eq. (2). For schemes with a head above 10 m, or high flow variability, it may be appropriate to set P_{max} for a flow up to 1.5 Q_{mean}. A factor in this decision will be the ability of the mini-grid consumers to make economic use of seasonal higher levels of power output. If peak river flows correspond with a crop processing task that can profitably use the power, then P_{max} can be chosen accordingly. The relationship between P_{max} and Q_{mean} determines the annual energy output E_a that can be obtained via the capacity factor C_f of the generator. This is a useful metric for all types of generator—it is given as

Table 1 Classification of turbine types

Turbine type	High head	Medium head	Low head
Impulse	Pelton, Turgo	Multi-jet Pelton, Turgo, Cross-flow	Cross-flow, undershot waterwheel
Reaction		Francis (spiral case)	Francis (open flume), propeller, Kaplan
Gravity			Overshot water-wheel, Archimedes screw

Source Compiled by the author

$$C_f = E_a/(8760\,P_{\max}) \tag{4}$$

An approximate value for C_f that can be expected if P_{max} is set from Q_{mean} is 0.4, so annual energy output will be about 3500 P_{max} kWh.

To capture the available energy resource as efficiently as possible, the type of turbine employed must be matched to the characteristics of the site. The dominant factor is the available head, which may be classified as high (>50 m), medium (10–50 m) or low (<10 m). The turbine types usable within these head ranges are shown in Table 1. Impulse turbines are driven by the kinetic energy of water jets and have the advantage of being compact and achieving a high rotation rate, which is desirable for efficient generation. Reaction turbines operate completely immersed like the propeller of a ship and can exploit the suction of water falling away from the turbine. Gravity turbines exploit the mass of water falling over the entire head. The choice of device for a given head is a complex trade-off between cost, efficiency at reduced rates of flow and the physical constraints of the site. Okot [4] provides a useful review of these issues, while Williamson et al. [5] examine the options for very low power (<5 kW) installations.

The use of an Archimedes screw for generation is a relatively recent innovation arising from improvements in the cost and efficiency of the gearbox needed to convert its slow rotation rate into a speed suitable for a generator. They are particularly attractive for micro hydro because of their ability to use a low head and tolerate debris and particulate matter in the water. The civil engineering requirements for installation are also relatively simple, and fish can usually pass through safely.

As an energy source for a mini-grid, a micro hydro system on a favourable site has the considerable advantage of producing electricity continuously all year round, though usually with variations in output. This eliminates or minimises the need for backup plant such as batteries and diesel generators. However, they do require sustained maintenance and supervision particularly of the screens and trash rack that filter the water entering the system. These can easily become blocked by seasonal debris and require regular manual clearance.

2.2 Solar Photovoltaic

The universal availability of solar energy makes this an attractive resource for a mini-grid, which may be in a remote location where there are few or no other options. Its seasonal and daily variations lead to the need for energy storage, usually batteries—the system design issues around battery sizing and configuration are covered in a later section. Here, we consider the assessment of the solar resource available at a given site and the technologies for photovoltaic (PV) generation.

The average energy flux S (i.e. power per unit area) reaching the top of the Earth's atmosphere from the sun is 1.367 kW/m^2. When it reaches the Earth this conveniently corresponds to about 1 kW/m^2, which is taken as the benchmark expected value for a clear day on a plane surface with the sun vertically overhead. To turn this figure into the energy resource available in a practical case is a complex trigonometric calculation. There are many free software tools available on the Web to perform this analysis such as the US National Renewable Energy Laboratory's PVWatts [6]. However, these tools do not necessarily have weather data available for a desired location, or contain oversimplifying approximations, or are unable to represent a particular PV system configuration. So this section summarises the calculation from first principles so that readers can cross-check tool outputs. For a comprehensive presentation of the mathematics Duffie and Beckmann [7] is recommended (Table 2).

The sun's angle of elevation α determines the clear sky energy flux on the ground plane—it is simply sin α kW/m^2. But because of the need to separate out beam (i.e. direct) radiation and radiation that is diffused by clouds, it is easier to start with the energy flux impinging on the top of the atmosphere that is directed at the point of interest—this is $S \sin \alpha$. The calculation proceeds by obtaining α for a given time of day, date and location and integrating the energy flux over a day to obtain the extra-terrestrial energy input. This input must be divided into beam and diffuse components at ground level depending on the sunshine hours in the day. Each component must also be modified to take account of the angle of the solar panels, and the sum of the two components provides the energy captured by the panels. From there the efficiency of the panels and subsequent power conversions can be used to find the energy that would actually be presented to a mini-grid.

The first step is to calculate the sun's declination δ for a given day n where $n = 1$ on 1st January:

$$\delta = 23.45 \sin\{360(284 + n)/365\} \tag{5}$$

Then sin α can be found for a latitude ϕ and time of day ω:

$$\sin \alpha = \sin \delta \sin \phi + \cos \delta \cos \phi \cos \omega \tag{6}$$

Table 2 Nomenclature for estimation of solar energy resource

Symbol	Represents
α	Solar elevation
β	PV panel angle of elevation to horizontal
δ	Solar declination
ϕ	Latitude of location for PV system
ω	Solar hour angle—noon is zero
ω_s	Solar hour angle of sunrise, sunset is $-\omega_s$
A	PV panel area m^2
B	Solar beam energy on unit horizontal area over given day and location
D	Solar diffuse energy on unit horizontal area over given day and location
e_m	PV module efficiency
e_p	PV output conditioning plant efficiency
G	Total solar energy on unit solar panel area over given day and location and panel angle of elevation
H_o	Extra-terrestrial energy directed at a unit horizontal area over a given day and location
K	Clearness index for day
n	Day number in year, n = 1 on 1st January
N	Number of integer hours of daylight for given day and location
r_b	Factor expressing the effect of panel angle β on incident beam radiation
r_d	Factor expressing the effect of panel angle β on incident diffuse radiation
s	Number of sunshine hours in the day
T_O	Reference temperature for PV panel performance, usually 25 °C
T_a	Average ambient temperature during daylight
t_{cp}	Coefficient of PV panel power variation with temperature
S	Solar radiation power constant 1.367 kW/m^2
w	Weighting factor when calculating daily average r_b

Sunrise ω_s and sunset ($-\omega_s$) times are also needed for daylight duration and are given as

$$\omega_s = \arccos(-\tan\phi\tan\delta) \tag{7}$$

H_o can now be found with adequate accuracy by summing the energy in each of the integer daylight hours (1:N hours):

$$H_0 = \sum_{i=1}^{i=N} S\sin\alpha_i \tag{8}$$

Suehrcke's method [8] as validated by Driesse and Thevenard [9] can be used to divide this energy into beam and diffuse components depending on the amount of cloud cover. This begins by calculating the clearness index K from sunshine hours data using the empirical relationship:

$$K = 0.65\sqrt{[3]s/N} \tag{9}$$

where s is the number of sunshine hours in the day and N the number of integer hours of daylight. If s is small giving a K value below 0.2, Eq. (9) is less accurate as shown by Driesse and Thevenard, so a better approximation is obtained setting $K = 0.2$. The beam and diffuse components B and D of the radiation at horizontal ground level are then found using two more empirical relationships:

$$B = 1.11 H_0 K^2 \tag{10}$$

$$D = H_0 K - B \tag{11}$$

The empirical constant 0.65 in Eq. (9) provides a good fit for a relatively cloudy climate similar to that of northern Europe. Increasing it to 0.75 provides a better result for consistently sunny climates.

In general the amount of solar energy collected over a year can be optimised with respect to the location, climate, battery capacity and service reliability by placing the panels at an angle to the horizontal. A common decision is to orient the panels for maximum output during the month containing the winter solstice. The modification factor r_b (an improvement for most of the day) on beam energy flux caused by positioning the PV panel south facing with an elevation angle β can be calculated as for a given date, location and time of day as

$$r_b = \frac{\cos(\phi - \beta)\cos\delta\cos\omega + \sin(\phi - \beta)}{(\cos\phi\cos\delta\cos\omega + \sin\phi\sin\delta)} \tag{12}$$

To obtain the effect of this modification on a day's energy collection an average value of r_b is taken, with the value for each hour weighted by the normalised fraction of H_o expected in the hour. Because of the ill-conditioned nature of (12) it is best to constrain the upper limit of daylight duration N to 12 h, i.e. 6 h either side of noon. The weighting factor w_i is given as

$$w_i = \sin\alpha_i / \sum_{j=1}^{j=N} \sin\alpha_j \tag{13}$$

Then

$$\overline{r_b} = \frac{1}{N}\sum_{i=1}^{i=N} r_{bi} w_i \tag{14}$$

The effect of the panel angle β is however to reduce diffuse energy flux by a factor r_d:

$$r_d = \frac{1}{2}(1 + \cos\beta) \tag{15}$$

The day's total solar energy incident on the panel G for unit area is thus equal to

$$G = \overline{r}_b B + r_d D \tag{16}$$

Table 3 Commercial photovoltaic technologies

Technology	Typical efficiency e_m (%)	Typical temperature coefficient of power t_{cp} (%)	Comments
Cadmium telluride thin film	11	−0.25	Good performance in hot climate
Copper-indium-gallium-selenide (CIGS) thin film	12	−0.45	Potentially low cost
Silicon amorphous	8	−0.2	Low cost per unit area
Silicon polycrystalline	14	−0.44	Widely used for small scale systems
Silicon monocrystalline	15	−0.5	Also widely used for small scale systems
Silicon heterojunction with intrinsic thin layer amorphous (HIT)	18	−0.3	Highest output per unit area

Source Compiled by the author

The conversion of this energy into useful electricity then depends on the PV module efficiency, the operating temperature and the efficiency of the electronic plant matching the PV output to the grid. Typically, this is a maximum power point tracker (for a DC grid) or an inverter (AC grid). The PV module efficiency e_m and temperature coefficient of power output t_{cp} are usually given on manufacturer's data sheets, while the plant efficiency e_p will be dependent on plant loading and so must be selected as a reasonable approximation off the manufacturer's efficiency curve. The electricity output E kW h for the day will then be given as

$$E = e_m e_p AG\{1 + (T_a - T_0)t_{cp}\} \tag{17}$$

where A is the panel area, T_a is the average ambient temperature during daylight, and T_o is the panel reference temperature (usually 25 °C). If temperature varies considerably during a day the average should be weighted similarly to that performed in Eqs. (13) and (14). By repeating this calculation over a year using local meteorological records providing sunshine hours and temperature an annual profile of the expected output can be obtained.

Table 3 provides a summary of the photovoltaic module technologies that are readily available commercially with typical efficiencies and temperature power coefficients with respect to a T_o of 25 °C.

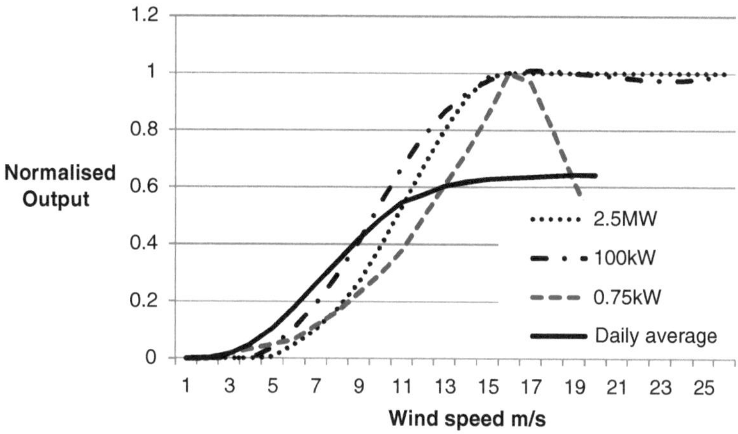

Fig. 2 Normalised wind turbine power curves

2.3 Wind

Like solar energy, some level of wind energy can be found everywhere and it has the merit of not being limited to daylight hours. However, the highly nonlinear relationship between wind speed and the energy that can be collected implies that coastal and hilly locations will always be preferred. Because wind generation is usually higher in winter or during a monsoon season, it can partly complement PV generation in a hybrid system so that the battery capacity required for a given level of reliability can be reduced.

The power P that can be generated by a wind turbine with swept area A in a wind speed v is given as

$$P = e_t \frac{1}{2} A \sigma v^3 \quad (18)$$

where e_t is the efficiency of the turbine, which has a theoretical limit of 0.59 [10] and σ is the density of air (about 1.2 kg per m^3). In practice, the efficiency of a wind turbine varies depending on the wind speed and the manufacturer's optimisation, and is typically between 0.35 and 0.5 including the conversion from mechanical to electrical energy. In Fig. 2 three manufacturer's power curves are shown for turbines with widely differing peak power ratings of 2.5 MW, 100 kW and 0.75 kW. The output is normalised with respect to the peak power to allow comparison. All exhibit to some extent the cubic relationship in Eq. (18). Output also flattens or falls away over 16 m/s for all three devices.

Meteorological records typically provide average wind speeds which do not directly provide a reliable indication of average power output because of this nonlinear relationship between wind speed and output. However, the statistical properties of wind can be used to estimate the expected output of a wind generator.

The Weibull distribution provides the most general model of wind speed variation (Justus et al. [11]), but when the Weibull shape factor (which controls the evolution of the distribution from an exponential form to a bell curve) is set to 2, it simplifies to a Rayleigh distribution. This shape factor is appropriate for a wind turbine that is employed for a rural mini-grid because it is intermediate between a mainly exponential function applicable to sheltered situations given by shape factor 1 and a bell curve distribution given by shape factor 3, which is more applicable to locations such as offshore with a sustained high average speed [12].

A Rayleigh distribution of wind speed v with mean μ is given as

$$f(v) = \frac{\pi v}{2\mu^2} e^{-\frac{\pi v^2}{4\mu^2}} \tag{19}$$

If this distribution of wind speeds during a day with average μ is applied to the appropriate manufacturer's power curve then the energy output for the day can be calculated. The 'Daily average' curve in Fig. 2 takes the power curve for the 2.5 MW turbine (chosen because its shape is intermediate between the other two) and plots the resulting average normalised daily power obtained by applying to it the Rayleigh distribution of wind speeds from a given daily average wind speed. It can be seen that the higher wind speeds that occur at low average wind speeds push up the energy output over the day, while at higher averages the capping of power at 16 m/s limits the daily energy output. Carillo et al. [13] provide a useful database of the key parameters for this calculation for a wide range of turbines. Noumi et al. [14] use this method to evaluate the annual output expected from a specific model of 1 kW wind generator (Unitron H80) at 19 locations in India with results ranging from 297 kWh at Malwan to 1769 kWh at Muppandal with an average of 839 kWh.

Depending on the size of the wind turbine envisaged for a given project, it may be desirable to adjust wind speeds from historic data, which are likely to be collected at about 10 m above ground level, to give wind speeds at the actual hub height (i.e., height of the centre of rotation of the turbine). Wind speed rises with height, following approximately a power law relationship:

$$\frac{v_h}{v_r} = \left(\frac{z_h}{z_r}\right)^{\alpha} \tag{20}$$

where z_r is the reference height at which data were collected, z_h is hub height, v_r is the observed wind speed and v_h is velocity at hub height. The exponent α is an empirical constant depending on the local topography and surface roughness from vegetation and dwellings. A value of 0.14 for α is often used and is suitable for a level open field. Because of the sensitivity of wind turbine output to wind speed, it is desirable to make the hub height as high as is reasonably practical.

Technological innovation in wind generator design has generally been concerned with refinement of methods for driving the generator from the turbine and within the generator itself. The horizontal axis turbine, with 2 or 3 blades, remains

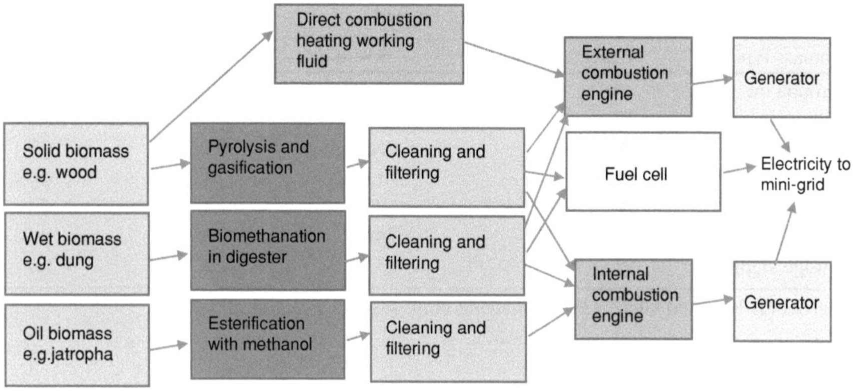

Fig. 3 Processes for conversion of biomass into electricity (*source* this study)

dominant. There are manufacturers producing other designs such as vertical-axis turbines, which allow the generator to be conveniently located at ground level, but the design trade-offs have not resulted in radical improvements. Kjellin et al. [15] provide a useful analysis of performance from a simple 12 kW H-rotor vertical axis wind turbine.

A potentially useful aspect of wind as a resource is that it is possible to build a wind generator from scrap materials. This usually relies on finding a suitable low speed permanent magnet motor that can be recycled, but given access to good quality magnets it is possible to wind alternator coils by hand and build the generator. There are many Internet sites giving details for this; a book reference is Bartmann and Fink [16].

2.4 Biomass

In this section, we consider all forms of biological energy resource, even though they may be liquid or gaseous at some or all stages of processing. This covers a wide range of potential sources and processes to arrive at generated electricity. The considerable potential for biomass generation is illustrated by the case study given in "Viability of Husk-Based Mini-Grids in South Asia" which describes the rice husk fuelled mini-grid systems deployed in South Asia. Figure 3 summarises the process options, which derive from the three main forms of biomass:

- Solid material, such as wood and crop residues, which can be subjected to direct combustion, or converted into a gas via a pyrolysis process.
- Wet material, such as food waste, animal dung and urine, which provides a good basis for anaerobic digestion to produce biomethane.

Table 4 Example data on energy content of biomass

Biomass type	Water content (%)	kW h/m^3	kW h/kg	Source
Jatropha oil	<5	9,460	11	(a)
Willow wood	12	3,816	4.4	(b)
Eucalyptus	15	Not given	4.3	(c)
Bamboo	11	1,062	4.2	(b)
Rice husks	10	729	4.2	(b)
Rice straw	11	448	3.7	(b)
Sewage sludge	57	319	1.2	(b)

Sources (a) Koh and Ghazi [38] (b) Chiang et al. [39] (c) Pérez et al. [40]

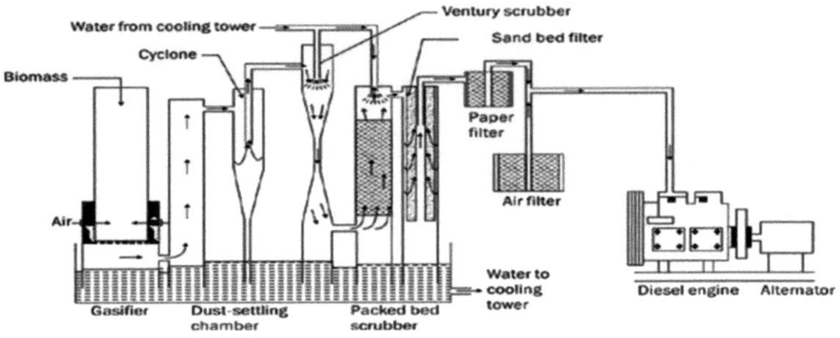

Fig. 4 Schematic of a gasifier system for power generation (*Source* www.teriin.org)

- Oils obtained from crops, such as rice bran and jatropha, by pressing. This is readily converted into an oil suitable for a diesel engine through a chemical process (transesterification).

The process steps to convert each of these biomass forms into a usable form of chemical energy are described in the next three sections, followed by consideration of the technologies required to transform that chemical energy into electricity. Table 4 provides example figures for the energy content of various forms of biomass. These are highly dependent on the particular plant varieties and handling methods assessed in the data sources indicated. The process efficiency then determines the proportion of the biomass energy content that can be converted into electricity.

2.4.1 Solid Biomass Processing

Where solid biomass will be used for direct combustion (e.g. in a log-fired boiler) processing need only consist of air drying to a suitably low moisture level (below 20 %) and cutting or pressing the material to the required size for the fuel feedstock. Otherwise, gasification by pyrolysis is the preferred process. Figure 4 illustrates the components of a mini-grid scale gasifier system.

This involves heating the biomass with a limited amount of air in a reactor vessel, which causes the volatile organic compounds to be driven off and a series of complex reactions to take place. The carbon content remains as char creating a reducing environment that converts water vapour to hydrogen and carbon dioxide to carbon monoxide. The resulting gas (known as producer gas or syngas), which also contains methane and other hydrocarbons, must then be cooled and cleaned to remove dust and heavier organic compounds that condense to form tar. This process has been employed for many decades to convert wood into a gas suitable for internal combustion engines, so it is a mature and low cost technology for which plant is available at scales from small gasifiers that can feed a tractor engine to industrial plant capable of generating several MW.

2.4.2 Wet Biomass Processing

This process employs a tank or chamber containing bacteria in a wet and warm environment from which air is excluded, hence the name anaerobic digestion. The feedstock comprises the biomass mixed with water to form a slurry with about 8–11 % of solid material by weight. This slurry is consumed in a batch process with each batch taking 40–100 days, at the end of which the spent slurry must be removed and can be used as a fertiliser. The breakdown of the chemical chains of the input biomass material to produce methane is complex, involving four different bacterial processes. Hydrolysis is the first stage, which breaks down the chemical bonds of fats, carbohydrates and proteins to form sugars, fatty acids and amino acids. Acidogenesis breaks down the products of hydrolysis, degrading them into carbon dioxide, hydrogen, alcohols and organic acids including acetic acid. Some of these products (e.g. acetate, hydrogen and carbon dioxide) can be directly used by the methane-producing bacteria in the fourth stage of anaerobic digestion (methanogenesis). Other products, such as some alcohols and organic acids, need to be degraded into acetates by acetate-forming bacteria (acetogenesis) before they can be used as a substrate for the methane-forming bacteria.

For sustained production of methane, the digestion vessel must be kept at a constant temperature between 32 and 45 °C and the slurry should be stirred so that all four bacterial processes take place simultaneously. The output gas typically contains about 50–65 % methane, 30–40 % carbon dioxide and a variable balance comprising water vapour, nitrogen, hydrogen and hydrogen sulpfhide. For use as a fuel the moisture content is reduced by condensing the water vapour, and the hydrogen sulphide is absorbed by chemical 'scrubbing'.

2.4.3 Oil Biomass Processing

The first step for oil-bearing biomass is to mechanically press the material, with application of heat in some instances, to extract the oil. From then on the objective is to prepare the oil for use in a diesel engine. This requires the viscosity to be reduced so that it can be vaporised by the engine injectors and will ignite under compression. There are a variety of methods possible for achieving this, including blending with solvents and forms of pyrolysis, but the preferred method for producing a reliable fuel is transesterification. This is a chemical reaction, which typically employs methanol and sodium hydroxide as a catalyst to convert the input oil into a mixture of esters with the desired properties and glycerol as a by-product. The process is easy to operate on a small scale; detailed instructions for plant construction and operation are available on the Internet such as Addison [17]. A comprehensive review of transesterification process techniques and issues is provided by Demirbas [18].

2.4.4 Generation of Electricity from a Biofuel

To convert the chemical energy from a biomass-derived fuel into electricity, there are two routes. The first and potentially the simplest and most efficient is a fuel cell, which oxidises the fuel with oxygen from the air in a controlled catalytic process similar to a battery that generates electricity directly. In current practical and relatively low-cost implementations such as the BlueGen offered by Ceramic Fuel Cells [19], this operates with a gas fuel, such as hydrogen or methane, which could in principle be obtained from a biological source. However, there are two related issues, which make this technology unlikely to be viable for rural electrification in the immediate future. The first is the need for the fuel gas to achieve a high level of purity and stability in its chemical makeup so that the operating conditions and catalysts of the fuel cell can be chosen accordingly. The second is the long-term reliability of the catalysts, which has proved difficult to achieve even with gas with a tightly controlled specification from a main gas grid. But given the high energy conversion efficiency of fuel cells (about 60 %) and because hydrogen fuel cells are seen as a possible approach to de-carbonisation of transport (in effect replacing diesel and petrol engines) a considerable research and development effort is in progress, which in time should deliver devices with the required durability and low cost. Biomethane from anaerobic digestion is the most suitable biomass-based feedstock for a fuel cell because of its relative purity. NREL [20] provides a good summary of the state of the art in biogas-powered fuel cell systems including descriptions of demonstration plant.

The commonplace alternative to the fuel cell is some form of thermodynamic heat engine, which uses heat from combustion of the fuel to create mechanical motion that can drive a generator. A primary choice with respect to the heat engine is between internal and external combustion. An internal combustion engine is just a conventional gas or diesel engine. These are available often with integrated generator for a wide range of capacities and fuels. The constraint they apply is that the quality of the fuel must be reasonably well controlled to ensure reliable operation. Even then, they tend to require quite a lot of maintenance but at least the required skills are widely available.

External combustion has the big advantage that the fuel can take any form and its chemical composition will not be critical, so extensive preprocessing such as gasification is not essential. Mechanical reliability is also considerably improved because the internal components of the engine do not suffer from the corrosive effect of combustion by-products. Historically, the most common form of external combustion engine has been the steam engine. These were often used to drive mechanical agricultural processes, such as sugar cane crushing or palm oil milling. The waste biomass from the process (bagasse or nut shells) provided fuel for the boiler and the exhaust steam from the engine could be used for process heat. Because steam engines have largely been abandoned in the developed world their merit for rural development has possibly been overlooked. Sookkumnerd et al. [21] found steam engines a valid solution for electricity generation in Thailand from rice husks, while Teixera et al. [22] evaluated a range of engine options for electricity generation from biomass in the Amazon region, and found a steam-driven option the most cost-effective. Small steam engines complete with generator sets are still manufactured, for example in India, by Aadhunik Global Energy [23]. For electricity generation, the disadvantage of solid-fuelled small-scale steam power is the relatively low efficiency, possibly in the range 10–20 %, and the manual labour associated with stoking the boiler, which will be difficult to sustain continuously. These issues are a trade-off from the simplicity of fuel preparation.

Other forms of external combustion engine are organic Rankine and Stirling cycle devices. A Rankine engine is similar to a steam engine in principle, but uses a volatile organic compound as the working fluid in a similar way to refrigeration plant. The Stirling engine operates with a stable and sealed volume of gas such as nitrogen as the working fluid. Advantages of both are being able to operate with a low temperature differential and without the high pressure boiler needed for steam, which can be hazardous. Corria et al. [24] examine the merits of a Stirling engine for rural electricity generation in Brazil. From a purely technical perspective, these external combustion technologies are more suitable than internal combustion engines as prime movers driving generation for a remote rural mini-grid because of the simplicity and reliability of the complete process chain. But because they are not manufactured on the same scale as internal combustion engines their cost tends to be higher.

3 System Architecture and Performance

In this section, we consider the main system design issues and decisions that arise once the energy resources to be exploited have been determined. These may be summarised as:

- The selection of direct current (DC) or alternating current (AC) for distribution, the operating voltage and the voltage tolerance at supply points.

- The central system architecture that can deliver the planned service capacity, availability and reliability. This defines the requirements for hybrid generator configurations and batteries.
- Distribution network cable sizing and topology.
- Technical options around balance-of-plant items, such as maximum power point trackers and inverters.

3.1 Selecting DC or AC Distribution

This choice may be simple if the primary source of electricity is predetermined to be a conventional diesel generating set equipped with an AC generator. If this is not the case, then DC operation should be considered first to see if the constraints it imposes are acceptable. The reason for evaluating DC distribution first is the proliferation of electricity-consuming devices and appliances that are designed to operate on DC and require an internal or external power conversion module to use AC. The shift to DC within appliances is not confined to light-emitting diode (LED) lighting and electronic devices such as computers and peripherals. It is also taking place wherever electric motors are required because DC operation allows precise control over motor speed and torque resulting in energy savings and improved reliability. Converting AC to DC has both an initial and continuing cost—Garbesi et al. [25] estimate that eliminating these conversions reduces energy consumption by about 14 %.

There is therefore an emerging initiative in developed countries to migrate electricity distribution in homes and offices to DC. At the time of publication this is still at the stage of pilot and demonstration projects. However, standards are being drafted, published and implemented, in particular, by the Emerge Alliance [26]. It seems likely that the 24 and 380 V standards promoted by this body will be widely adopted, initially by warehouses, supermarkets and data centres where the savings are most significant. The 24 V standard will also be taken up in homes and small offices where eliminating the clutter of 'black plug' power converters will be attractive alongside the energy and cost savings. Once large-scale adoption occurs, then lower level components, such as connectors and circuit breakers, will be manufactured in the volumes needed to be competitive with AC components.

For a mini-grid that is primarily powered by PV or wind energy there is the additional advantage in DC distribution of eliminating the cost and losses of AC inverters. The present critical limitations of DC distribution are:

- Operation at 24 V limits the area covered by the mini-grid before voltage losses, or the cost of suitably low resistance cable, becomes unacceptable. However, the availability of useful light from a 3 W LED lamp combined with a 2 W mobile phone charger allows a household to benefit from service as low as 5 W. Table 5 shows the possible distribution cable distance for different service loads

Table 5 Distribution cable length for 1 V drop on a 24 V DC system

Total consumer load (W)	Length (m) 1 mm^2 cable	Length (m) 1.5 mm^2 cable	Length (m) 2.5 mm^2 cable	Length (m) 4 mm^2 cable
5	109	166	267	436
10	55	83	133	218
15	36	55	89	145
20	27	41	67	109

Source Author

Table 6 Distribution cable length for 10 V drop on a 230 V AC system

Total consumer load (W)	Length (m) 1 mm^2 cable	Length (m) 1.5 mm^2 cable	Length (m) 2.5 mm^2 cable	Length (m) 4 mm^2 cable
10	5227	7931	12,778	20,909
20	2614	3966	6389	10,455
50	1045	1586	2556	4182
100	523	793	1278	2091
500	105	159	256	418

Source Author

and cable conductor sizes, assuming an acceptable voltage drop of 1 V, which corresponds to distribution losses of about 4 %.

- Some types of electrical plant important to a rural mini-grid, for example, refrigeration appliances and small industrial equipment, such as grinders and mills, are currently more expensive or not readily available in DC-powered form.

Once the components and operational safety practices associated with 380 V DC distribution become well known it will be possible to overcome the distance limitation. DC plant, and the cost savings associated with it, will also become accessible to rural mini-grid projects as developed world migration to DC proceeds. But Table 5 (based on IET Wiring Regulations) [27] shows that 24 V operation does permit safe distribution for a compact village if household provision is mainly lighting; and higher power requirements, such as computers, televisions and agricultural process plant, are co-located with the generator and balance of plant such as batteries.

If a DC system cannot provide the coverage or choice of load devices needed, then a 220–240 V AC system is the logical alternative. The exact operating voltage will depend on the plant chosen and any national regulations for the location. Table 6 provides distribution of cable lengths for a 10 V drop on a 230 V system, against the higher consumer loads likely to be offered. For a system with industrial plant in the kW region and generation of 20 kW and above, a three-phase AC system will be preferred. This extends possible distribution distances

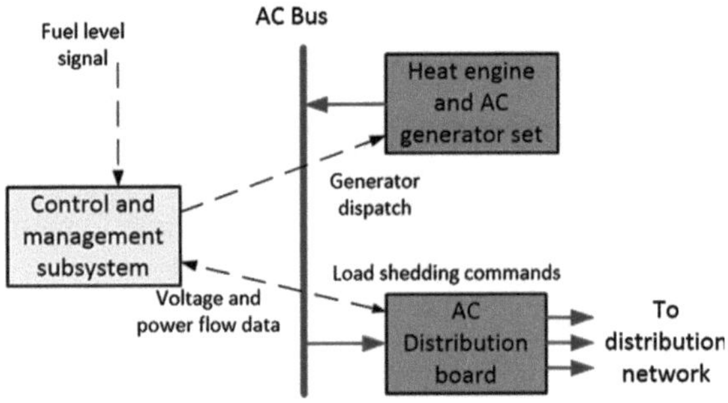

Fig. 5 Simple central system with heat engine and AC generator

further but requires greater attention to safety because of the 400 V interphase voltage.

For other voltages and consumer loads, a useful formula for the length l m of cable that will cause a 1 V drop is

$$l = 22.7AV/W \tag{21}$$

where A is the conductor area in mm^2, V is the distribution voltage and W is the consumer load in Watts. This is an approximate equation for high quality copper cable and assumes 2-wire distribution with a margin of about 30 % to allow for cable heating. Longer lengths may be achievable in cool climates with overhead cable distribution.

3.2 Central System Architecture

A mini-grid central system can be as simple as a diesel generator set supplying a distribution network via a distribution board that provides current limiting for safety on the various network feeder lines and central voltage and current metering for system management purposes. This is a practical option where fuel can be stored to buffer variations in the supply chain allowing the generator to be dispatched on a schedule or operated continuously. It also applies for mini-hydro systems where the water flow is reliable or there is an upstream reservoir that can be used to regulate the availability of the water resource. But where the energy resource is uncontrollably variable or intermittent, more complex configurations are needed to ensure a supply commitment can be given to the electricity consumers.

Figure 5 shows a simple central system as used, for example, by the rice husk-fuelled systems described in "Viability of Husk-Based Mini-Grids in South Asia". Some form of control and management subsystem is needed even at this level to avoid the need for continuous operator attention and to collect data allowing performance to be monitored. This could comprise just a time clock scheduling the dispatch of the generator with an interlock ensuring enough fuel is available. The

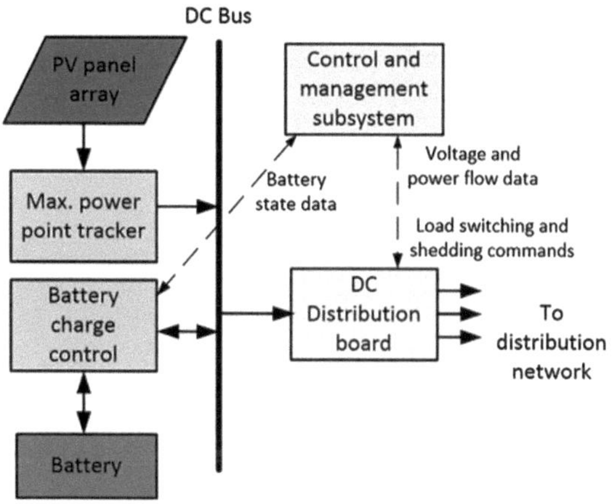

Fig. 6 Central system architecture for PV generator with DC distribution

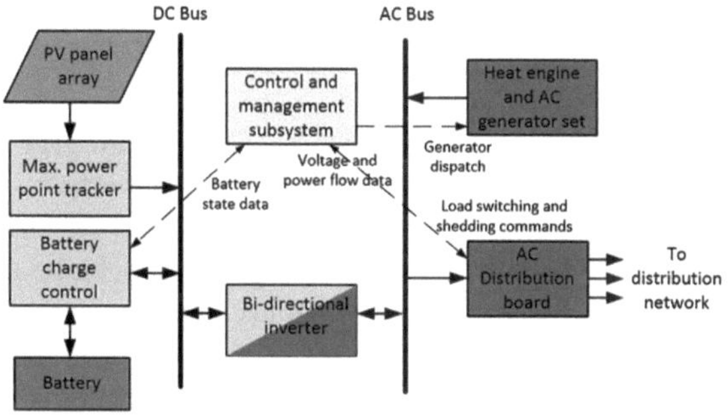

Fig. 7 Central architecture for a dual bus hybrid system

data collected could be simply fuel consumed, voltage levels and metered energy outgoing to the network. Fuel consumption data will allow the efficiency of the heat engine to be monitored, while aggregate output metering in addition to household level metering will allow electricity theft to be detected. There should preferably also be some ability for the control subsystem to respond automatically to overload conditions by shedding load at the distribution board in a predetermined priority order.

The simplest possible central system for a mini-grid with PV generation and DC distribution is shown in Fig. 6. Key additional components are the maximum power point tracker (MPPT) and battery with charge controller. The MPPT has the function of adjusting the load seen by the PV panel so that the voltage and current it produces are optimised and the collected power flow is converted to the DC bus voltage, which will be set by the state of charge of the battery. The role of the control subsystem is similar to that in Fig. 5, but because battery capacity will be the dominant constraint on the service that can be provided, the construction of an operating schedule by which different segments of the distribution network are energised at certain times becomes critical. The management subsystem should ideally provide software tools to assist this task. The issues and techniques around

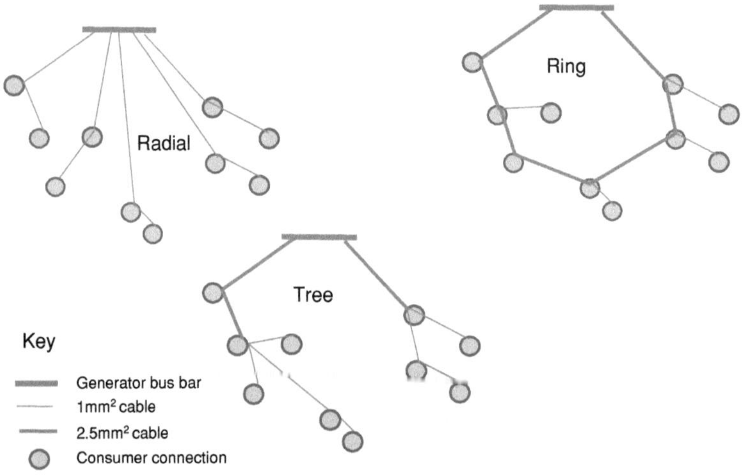

Fig. 8 Mini-grid distribution network topologies

matching supply and demand for a mini-grid are considered in "Demand Management for Off-Grid Electricity Networks".

An illustrative architecture for a dual-bus system designed to provide a 24-h AC electricity supply is shown in Fig. 7. To achieve this level of service without an extremely costly amount of battery capacity, a fuel powered electricity generator, such as a diesel generator set or a fuel cell, is needed. Such systems are usually referred to as hybrid. The bi-directional inverter is able to convert DC from the PV panels and/or battery to AC to supply the loads and can also accept any surplus AC generator capacity to recharge the battery. The control subsystem ensures the generator is dispatched whenever it predicts that the battery charge will not sustain the service until PV production resumes, or when PV production is inadequate. These dispatch decisions must take account of the need to operate the generator over a sufficient period and with a sufficient load so that its reliability is not compromised. In a suitable location, PV hybrid systems of this type are capable of scaling-up to meet the needs of a small town—Léna [28] provides a good review of the scaling trade-offs.

There are many possible variations on the dual-bus architecture of Fig. 7. A common simplification for small systems is to omit the fuel-powered generator and supply AC power from the inverter alone. Where fuel supplies are ample, it is possible to build a system with multiple generators that can be dispatched according to the level of demand, and reduce the battery capacity to the minimum level required for bridging between PV and generator operation. If the location has a useful wind resource, then combining wind generation and PV can work well as a hybrid system since wind speeds are often higher in winter months when PV output is low.

3.3 Network Topology

In general, it is always possible to connect the consumers of a mini-grid to the central plant using different network topologies. Figure 8 illustrates radial, tree and ring options for a simple case of a 24 V system with 5 W loads in each dwelling (a total of 10 dwellings) and a maximum cable distance of 55 m to the most remote dwelling. In many cases, the options are restricted by the geography of the settlement or location allowing a selection from the remaining alternatives by inspection, taking account of factors such as the need to separate out loads of different priority on different feeders to facilitate load shedding. It is also important where three-phase AC distribution is adopted to design the network so that the load is balanced across the three phases. In some situations a ring solution is preferred because the available capacity at any point on the ring is doubled potentially giving scope for expansion, while reliability of service is enhanced because a single break in the ring will not deprive any consumer of supply. Sebitosi et al. [29] show that in the absence of other constraints a tree structure is most efficient for an even distribution of consumers and provide a simple design method.

For larger networks with hundreds or thousands of consumers, it may be preferable to use more automated methods to find a preferred topology that minimises cable cost, constrains voltage drop and satisfies other requirements that arise. This is a classic problem for which a range of discrete optimisation techniques as applied to graphs can be employed. Genetic algorithms are probably the most suitable for mini-grid topology optimisation because they generate a range of promising solutions from which the best can be chosen taking account of human factors that are not easily encoded in a fitness function. Coley [30] and Silva et al. [31] describe how to apply genetic optimisation methods to this type of problem. Genetic algorithm add-on programmes are available for established mathematical software, such as Matlab and Excel.

3.4 Batteries

A battery will be required in every mini-grid, if only for starting an engine driving the generator. The sizing of its capacity will depend on the system architecture and reliability as discussed in Sect. 3.2. Here we consider the performance options arising from the battery chemistry. Lead-acid batteries are the established technology, but for mini-grid use they must withstand regular deep discharge cycles to 80 % of capacity. The sealed valve regulated type (VRLA) can provide this, with wet, gel and AGM (absorbed glass matt) variations in the medium holding the electrolyte. Gel types generally have the best performance with charge efficiency (i.e. discharge energy as a proportion of charge) of 90–95 % and a life of up to 1,200 deep discharge cycles for individual 2 V cells with tubular construction.

Flat plate batteries similar to those for automotive use have a shorter life of around 500 cycles. Hence, these batteries receive less preference in mini-grid applications.

In wider applications where weight and energy density are critical factors, batteries based on lithium-ion chemistry have emerged as the preferred solution. These are not important for mini-grid applications, but the lifetime and deep discharge tolerance of lithium batteries have also benefited from the extensive research and manufacturing volume now deployed for this technology. Krieger et al. [32] performed a thorough comparative evaluation of four different battery types when charged by a wind turbine. This is a challenging test because of the bursty nature of wind generation. The battery types are flat plate lead-acid (VRLA), lithium cobalt oxide (LCO), LCO-lithium nickel manganese composite (LCO-NMC), and lithium iron phosphate (LFP). Lithium iron phosphate is clearly the most durable technology to the extent that the authors believe it is a cost-effective alternative to VRLA even though the cost per Ampere-hour is about 4 times higher. This is justified because the LFP battery can be taken close to complete discharge, whereas the VRLA would normally be sized for 50 % discharge in normal operation. The LFP will also last more than twice as long at 2,000 discharge cycles. If costs of shipment, maintenance and recycling are taken into account, LFP is clearly the preferred technology if the capital cost can be supported.

4 System Management

The management challenge for an operational mini-grid of matching demand to available electricity supply has already been mentioned. "Demand Management for Off-Grid Electricity Networks" examines this issue in detail and reviews the technologies and devices that may be installed at the consumer's connection point to manage demand. In this section, we briefly consider the information systems that may be deployed to assist operators. These are generally closely tied to payment mechanisms, which are not covered here. However, two technical factors have a strong influence on practical options. The first is the availability of mobile phone coverage at the mini-grid location. This allows data collection and control to be performed remotely. Bit-harvester [33] uses text messaging for this purpose, while Gram Power [34] provide a smart meter and switch that is enabled via the mobile phone network, and used for load prioritisation and monitoring as well as revenue collection.[1] The ability to supervise a mini-grid remotely allows more complex solutions in all aspects of mini-grid design than is possible for systems that have to be totally self-sufficient.

The second key technical factor is whether a mini-grid is power or energy limited. Biomass-fuelled or mini hydro systems are typically power limited in that

[1] These examples are given for illustration purposes only. This is not, in any way, an endorsement of these products.

the maximum generator power provides the dominant limit on demand. Where PV or wind generation is the main resource, battery energy capacity becomes the critical limit on demand as the instantaneous power available from batteries and inverters will be higher than the level that can be sustained. For power limited systems, careful consideration has to be given to the population of electricity-consuming appliances that are likely to be connected and active at any given time. To manage the service, the types of load operating at different times of day need to be categorised and prioritised, so the information system needed to support these decisions must capture data on the loads in use and their timing.

For energy-limited systems, the constraint that must be applied to consumers is a daily energy budget. This limit will vary from day to day and seasonally because of the weather dependence of PV or wind generation. The technology for equitable allocation of variable energy budgets is still emerging at the time of writing. Devices such as the Urja Bandu (CAT [35]) and the Dispensador [36] can enforce a daily energy budget for each consumer and provide feedback on its status. The information system then needed for management must be able to monitor and predict the state of battery charge and facilitate decisions on energy budget allocation by the operator.

5 Conclusions

With the levelised cost of PV-generated electricity at or near 'grid parity' in many locations including India [37] it is likely that innovation in mini-grid technology will be more rapid for system components that make use of the solar energy resource. Combined with the evolution to highly efficient DC-powered lighting, entertainment and information systems this should enable PV powered systems, augmented where necessary with biomass-fuelled generators, to have enough capacity to support small-scale industrial and commercial uses of electricity. Development of information systems that are easy to operate and allow flexible allocation of daily energy budgets would complete the portfolio of technology needed to bring electricity to every rural area.

References

1. UNIDO (2013) World small hydropower development report
2. Practical Action (2014) Micro-hydro power. http://practicalaction.org/simple-ideas-micro-hydro. Accessed 22 Jan 2014
3. BHA (2012) A guide to UK mini-hydro developments. British Hydropower Association, Wimborne
4. Okot DK (2013) Review of small hydropower technology. Renew Sustain Energy Rev 26:515–520

5. Williamson SJ, Stark BH, Booker JD (2014) Low head pico hydro turbine selection using a multi-criteria analysis Renew Energy 61:S
6. NREL (2013a) PVWatts calculator. http://www.nrel.gov/rredc/pvwatts/. Accessed 05 Jan 2013
7. Duffie JA, Beckman (1991) Solar engineering of thermal processes, 2nd edn. Wiley, New York
8. Suehrcke H (2000) On the relationship between duration of sunshine and solar radiation on the earth's surface: Angstrom's equation revisited. Sol Energy 68(5):417–425
9. Driesse A, Thevenard D (2002) A test of Suehrcke's sunshine-radation relationship using a global data set. Solar Energy 72(2):167–175
10. MacKay D (2009) Sustainable energy without the hot air. UIT, Cambridge
11. Justus CG, Hargraves WR, Mikhail Amir, Graber Denise (1978) Methods for estimating wind speed frequency distributions. J Appl Meteor 17:350–353
12. Archer CL, Jacobson MZ (2003) Spatial and temporal distributions of U.S. winds and wind power at 80 m derived from measurements. J Geophys Res 108(D9):4289–4300
13. Carrillo C, Obando Montano AF, Cidras J, Diaz-Dorado E (2013) Review of power curve modelling for wind turbines. Renew Sustain Energy Rev 21:572–581
14. Nouni MR, Mullick SC, Kandpal TC (2007) Techno-economics of small wind generator projects for decentralized power supply in India. Energy Policy 35:2491–2506
15. Kjellin J, Bulow F, Eriksson S, Deglaire P, Leijon M, Bernhoff H (2011) Power coefficient measurement on a 12 kW straight bladed vertical axis wind turbine. Renew Energy 36:3050–3053
16. Bartmann D, Fink D (2009) Homebrew wind power. Buckville Publications, Masonville
17. Addison K (2013) The Biodiesel Bible. http://journeytoforever.org/books/biodiesel-bible.cgi
18. Demirbas A (2009) Progress and recent trends in biodiesel fuels. Energy Convers Manag 50:14–34
19. Ceramic Fuel Cells (2013) BlueGen. http://www.bluegen.info/. Accessed 13 Dec 2013
20. NREL (2013b) Biogas and fuel cells workshop summary report. In: Proceedings from the biogas and fuel cells workshop golden, Colorado, 11–13 June, 2012
21. Sookkumnerd C, Nobutaka I, Koji K (2007) Feasibility of husk-fuelled steam engines as prime mover of grid-connected generators under the Thai very small renewable energy power producer (VSPP) program. J Cleaner Prod 15(3):266–274
22. Teixeira MA, José C, Escobar P, César RS, Electo E, Silva L, Osvaldo JV, Dirk A (2013) Assaí—an energy view on an Amazon residue. Biomass Bioenergy 58:76–86
23. Aadhunik Global Energy (2013) Tinytech steam power plants. http://www.tinytechindia.com/steampowerplan.htm
24. Corria ME, Vladimir MC, Electo SL (2006) Perspectives of stirling engines use for distributed generation in Brazil. Energy Policy 34(18):3402–3408. ISSN 0301-4215
25. Garbesi K, Vossos V, Shen H (2011) LBNL-5364E catalog of DC appliances and power systems. Lawrence Berkeley National Laboratory, Berkeley
26. Emerge Alliance (2013). http://www.emergealliance.org/. Accessed 23 Jan 2014
27. IET (2008) IET wiring regulations: BS7671, 17th edn. Institution of Engineering and Technology, London
28. Léna G (2013) Rural electrification with PV hybrid systems. International Energy Agency Photovoltaic Power Systems Programme, St. Ursen
29. Sebitosi AB, Pillay P, Khan MA (2006) An approach to rural distribution network design for sub-Saharan Africa. Energy Convers Manag 47:1101–1112
30. Coley DA (1999) An introduction to genetic algorithms for scientists and engineers. World Scientific, London
31. Silva HA, Britto AS, Oliveira LES, Koerich AL (2013) Network infrastructure design with a multilevel algorithm. Expert Syst Appl 40:3471–3480
32. Krieger MA, Cannarella J, Arnold CB (2013) A comparison of lead acid and lithium based battery behavior and capacity fade in off-grid renewable charging applications. Energy 60:492–500

33. University of Nairobi (2013) Bit-Harvester. http://sciencepark.uonbi.or.ke/index.php?title=bit-harvester&article=achivements. Accessed 29 Dec 2013
34. Gram Power (2013). Gram power smart meter. http://www.grampower.com/solutions/. Accessed 29 Jan 2013
35. CAT Projects (2011) Bushlight India project. http://www.catprojects.com.au/bushlight-india-hardware#.UsAaLmfuPcs. Accessed 29 Dec 2013
36. Graillot A, Briganti M, Solano-Peralta M, Vallvé X (2012) 15 Years of field experience with the "daily energy allowance" concept as the basis for load control and guide for social behaviour in rural micro grids". In: Proceedings of 6th European PV-hybrid and mini-grid conference
37. World Energy Council (2013) World energy perspective—cost of energy technologies. World Energy Council, London
38. Koh MY, Ghazi TI (2011) A review of biodiesel production from Jatropha curcas L. oil. Renew Sustain Energy Rev 15:2240–2251
39. Chiang K-Y, Chien K-L, Lu C-H (2012) Characterisation and comparison of biomass produced from various sources: suggestions for selection of pre-treatment technologies in biomass-to-energy. Appl Energy 100:164–171
40. Pérez S, Renedo CJ, Ortiz A, Manana M, Delgado F, Tejedor C (2011) Energetic density of different forest species of energy crops in Cantabria (Spain). Biomass Bioenergy 35:4657–4664

Smart Design of Stand-Alone Solar PV System for Off Grid Electrification Projects

Parimita Mohanty and Tariq Muneer

Abstract Solar PV has seen a remarkable growth across the globe with the total PV installation now exceeding several gigawatts (GW). The technology has the versatility and flexibility for designing systems in different regions, especially in remote rural areas. While conventionally straightforward designs were used to set up off-grid PV-based systems in many areas for a wide range of applications, it is now possible to adopt a smart design approach for the off-grid stand-alone solar PV system. A range of off-grid system configurations are possible, from the more straightforward design to the relatively complex, depending upon power and energy requirements, electrical properties of the load, as well as on site specificity, and available energy resources. The overall goal of the off-gird system design should be such that it can give maximum efficiency, reliability and flexibility of the system at an affordable price. In this chapter, the three basic PV systems, i.e. stand-alone, grid connected- and hybrid-systems are briefly described. Assessment of energy requirement including load profiling, load distribution and load categorisation is provided in a lucid way. That is then matched with a section on solar resource availability. Furthermore, a systematic approach has then been presented regarding sizing and designing of the above systems. Guidelines for selection of PV components and system optimisation are provided. Owing to the natural variance of the solar resource there is a critical need for storage batteries to satisfy energy demand during nocturnal and overcast periods. Furthermore, under a highly variable cloud regime there is a frequent demand for charging and discharging of battery in a PV system. So, the type of battery used in a PV system is not the same as in an automobile application. Detailed guidelines for selection of battery are therefore also provided.

P. Mohanty (✉)
The Energy and Resources Institute (TERI), New Delhi, India
e-mail: parimita@teri.res.in

T. Muneer
School of Engineering and the Built Environment, Edinburgh Napier University, Edinburgh, Scotland
e-mail: T.Muneer@napier.ac.uk

S. C. Bhattacharyya and D. Palit (eds.), *Mini-Grids for Rural Electrification of Developing Countries*, Green Energy and Technology,
DOI: 10.1007/978-3-319-04816-1_4,

1 Introduction

One of the primary concerns in designing a solar photovoltaic (PV) system is the determination of the optimum relationship between the PV array and the other Balance of System to supply a required unit of energy at a certain level of reliability. In order to get higher reliability of the PV system, where the weather pattern and load profile of a site can hardly be accurately predicted, the system is generally oversized, making it highly expensive. However, there are several smart approaches of designing the solar PV system, which would not only help improve the reliability of the system, but would also assist in improving the efficiency of the overall system.

This chapter is an introduction to such a smart design approach for the off-grid stand-alone solar PV system. A range of off-grid system configurations are possible, from the more straightforward design to the relatively complex, depending upon its power and energy requirements, electrical properties of the load as well as on the site specificity and available energy resources. However, the overall goal of the off-gird system design should be such that it can give maximum efficiency, reliability and flexibility of the system at an affordable price. The experiences clearly show that clean energy systems in rural and remote locations need to be designed, taking into account that access to frequent service, repair and maintenance is difficult, that is, the system performance should be reliable. Also, since the capital cost of clean energy systems is usually high, it is imperative that the efficiency of such systems is optimised so that cost per unit of electricity generated is minimised. Lastly, the system should have a significant level of flexibility such that both the grid (when it is eventually extended to an off-grid site) and other sources of energy can be coupled with the system as and when required. Therefore, the option to easily connect to the grid and earn revenue from sale of electricity must exist. The above requirement can be met not through huge investment in capital cost but mainly through proper planning, optimum design and reconfiguration of the system. The incremental cost involved here can be easily recovered through the kind of efficiency, reliability and flexibility such type of system gives in the long run. While considering the above mentioned points, the following sections cover the designing of stand-alone solar PV system used in off-grid electrification projects.

1.1 Types of Solar PV Systems

PV systems are broadly classified into three distinct types:

1. *Stand-alone systems* where the power is generated and consumed in the same place and which does not interact with the main grid. Normally the electricity consuming/utilising device is part of the system, i.e. solar home systems, solar street lighting system, solar lanterns, and solar power plants.

2. *Grid-connected systems* where the solar PV system is connected to the grid. The grid connected system can either be a grid tied system, which can only feed power into the grid and such system cannot deliver power locally during blackouts and emergencies because these systems have to be completely disconnected from the grid and have to be shut down as per national and international electrical safety standards for both safety reasons. On the contrary, grid-interactive PV systems with their bi-directional energy transfer capability can not only connect to the public utility grid but can also provide power locally in case of an islanding condition. In this way, grid interactive system provides commercial and other users highly reliable and cost-effective energy while staying environmentally aware and responsible.
3. *Solar PV-hybrid system* In a hybrid system another source of energy, such as wind, biomass or diesel, can be hybridised to the solar PV system to provide the required power. Here, the objective is to bring more reliability into the overall system at an affordable way by adding one or more energy source(s).

Although this chapter focuses more on design approaches adopted for stand-alone solar PV system, it also mentions how certain approaches in design and component selection should be taken care of from the beginning in order to upgrade the stand-alone system into a hybrid or grid interactive or grid tied system in future with minimum interventions and incremental cost involvement.

2 Guidelines for Smart Designing of Stand-Alone Solar PV Systems

A systematic approach is important and required when sizing and designing stand-alone solar PV systems. The following procedures are generally followed

A. Planning and site survey
B. Assessment of energy requirement and solar resource availability
C. Sizing of main component of the PV systems
D. Selection of main components of PV system.

2.1 Planning and Site Survey

The output of PV array is site specific. For the generation of maximum power, it is important to choose the best site for the array. The maximum power will be obtained if the module area is exposed to direct sunlight for a maximum period of a day. Therefore, the first thing that needs to be done in a PV system design is site survey and planning in order to avoid site-related confusion and contradiction at a later stage. The following points need to be covered during the site survey to check the suitability of the site [1]

- site orientation, total land area/surface area of roof available;
- structure and type of roof;
- possible routes for cables, battery and inverter location;

The proposed site for installation can be surveyed with a solar path-finder to check whether there is any possibility of seasonal shading problem. In addition to this, the most essential input parameters like incident solar radiation, the ambient temperature, and wind velocity, which are likely to vary widely from site to site, need to be collected for the specific site. The air temperature significantly affects the cell operating temperature (COT) and hence the output energy. So it is necessary to have access to the data and information on all these parameters in order to carry out an optimum PV system design exercise.

2.2 Assessment of Energy Requirement and Solar Resource Availability

In a stand-alone solar PV system, estimating the energy requirement and assessing the realistic solar resource availability are the most important tasks, which have to be done properly. This is also critical from the point of view of adding any smart load management and resource management features (see Box-1).

Following steps are adopted for carrying out this exercise

Box-1 Guidelines for system optimization

Efficient design lowers the initial system expenses. For example: reducing the electric lighting load by 75 % by shifting from incandescent to fluorescent lights or by reducing the load by 40 % by shifting from CFL to LED will reduce the modules and batteries needed for load. Eliminating module shading by relocating the mounting system does not cost any additional money and can increase the system's efficiency by a large percentage. Inefficiency caused by excessive voltage drop in the system's wiring can also be inexpensively eliminated. Intelligent advance planning doesn't have high incremental cost, but can drastically reduce system's initial cost.

In general the designer should consider some important points while trying to optimise the system [4]

- Siting a system correctly so that it is not shaded;
- Orientation of the system is a critical element in maximising annual PV output based on local climatic conditions;
- Mounting options can maximise or even increase insolation gain;
- Modules should be selected according to a system's parameter;

- Wiring should be designed to minimise voltage drop;
- Controllers must operate a system efficiently while meeting the needs of the system;
- Battery storage must be sized to the specific installation (For how long the system should work when the sun is not shining).
- Loads determine the size of the system and should be minimised by intelligent planning.

2.2.1 Load Profiling, Load Distribution and Load Categorisation

The load of a particular site may vary from season to season and event to event. Sometimes, the load may be as low as few watts (W) and sometimes it may be as high as few kilowatt (kW). Therefore, it is important to assess the type and utilization of such load and profiling of that load. The first step is to identify and enumerate the typical load, which can vary from simple lighting load to heavy motor load, and its characteristics as per Table 1. The power rating of the appliances can be obtained from catalogued information. If details are not provided, the power rating can be assumed to be the same as other similar appliances.

Once the type of the load is identified, the next step is to map the distribution of these loads and create the load distribution map; this is critical from design point of view as it helps in estimating the number of feeders to be used, routing of the distribution feeders, length of the distribution cable/feeder and sub-feeder, voltage drops, total power losses in cables, etc. It will also help in identifying the actual location of the street lights and the solar PV power plant to be installed. Typical load distribution maps from site (as shown in Fig. 1) can be created.

Once the load distribution is made, load at different feeder as well as the total load is calculated and the information is filled as per Table 2.

A graph between time of the day (*X*-axis) and total load at solar PV power plant end (*Y*-Axis), i.e. load profile can be plotted as shown in Fig. 2.

Box-2 Calculate load and energy-related requirement Calculate

- Total connected Watts
- Average daily daytime energy requirement
- Average daily nocturnal energy requirement.

Table 1 Load details

Load appliances	Number of household (No.)	AC or DC load	Number of load (No.)	Power rating of each load (W)	Single phase (1Φ) or three phase (3Φ)	Total power required (W)
Lights						
Fans						
Street lighting						
Water pump						
Others						
Load 1						
Load 2						
Load 3						
Total						

Source This study

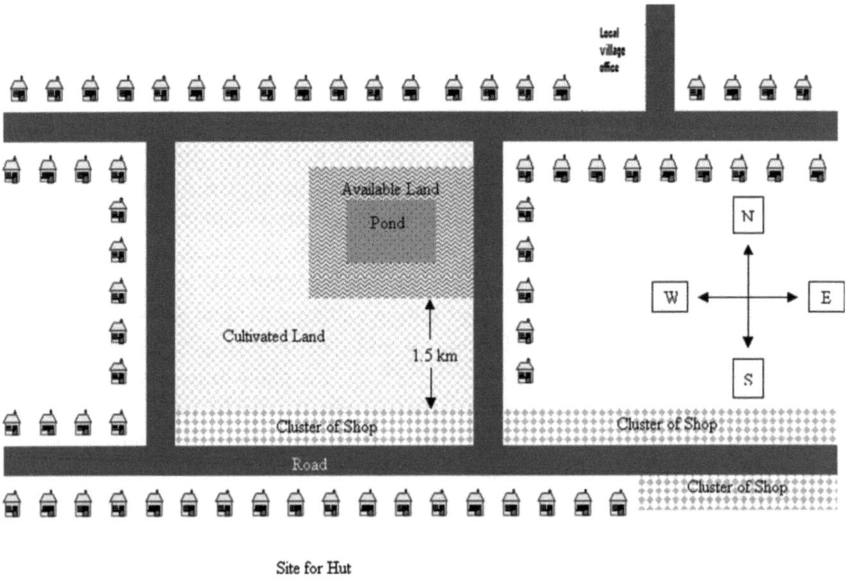

Fig. 1 Typical load distribution maps (*Source* Authors' compilation)

Based on the load profile, the total connected load (Watt ac) and "Average daily daytime energy requirement" (Watt-hour) and "average daily nighttime energy requirement" (Wh) may be calculated as indicated in Box-2.

Subsequently the load can be categorised based on their priority and load profile. There are 2–3 ways of categorising the load and each of the load categorisation is done in order to meet certain objectives. Table 3 shows the type of load and its category

Table 2 Load at different feeder

Time (Hrs)	Load at feeder 1(W)	Load at feeder 2 (W)	------	Load at feeder n (W)	Total load at solar PV power plant end (W)

Source: this study.

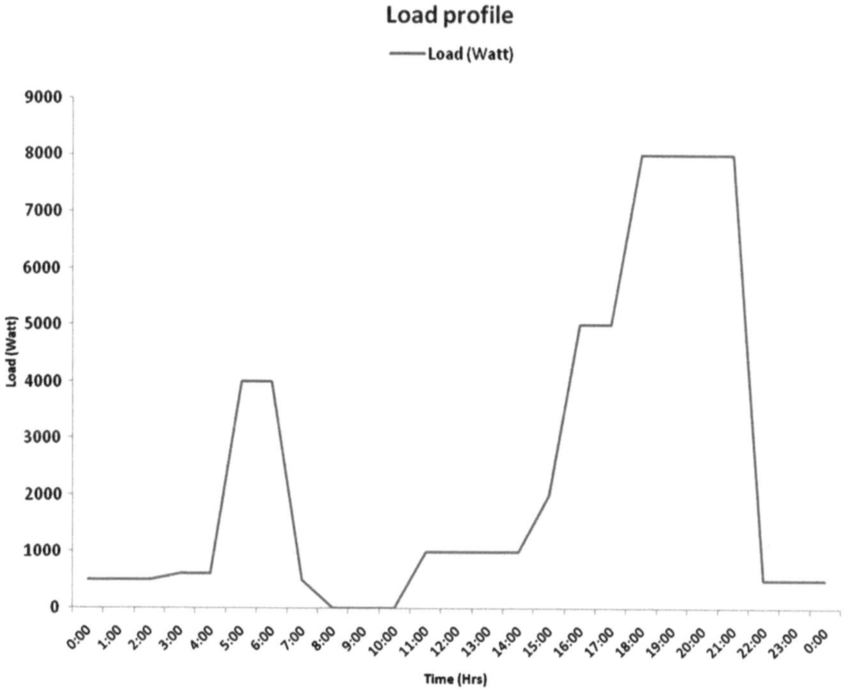

Fig. 2 Load profile of the proposed site (*Source* Authors' compilation)

- In order to design the configuration and optimum capacity of the energy storage system as well as the entire mini-grid system optimally and efficiently, the total load of each of the site is categorised into (i) day load and (ii) night load.

Table 3 Type and category of load

Category of load	Type of load
Daytime load	
Nocturnal load	
Critical load/Essential load	
Non-essential load	

Source Author's compilation

- Furthermore, in order to give preference to certain loads over others in case of limited availability of energy (for which only certain loads can be chosen to which energy can be supplied), the loads are also categorised based on their priorities, i.e. critical or essential load and (ii) non-essential load.
- From the tariff payment point of view, the loads can also be categorised as (i) minimum load, and (ii) desirable load based on the willingness to pay.

The advantage of categorising the load during the design stage is that it will allow automatic connects/disconnects of certain loads based on their priorities. For example, let us assume that there are certain non-essential loads, which are not as critical as the lighting load of the household. In that case, if adequate power is not available from the solar system and the battery is not amply charged, then the non-essential loads can be disconnected automatically, without affecting the smooth operation of the critical or priority load.

2.2.2 Assessment of Solar Energy Resource

The assessment of solar energy resource is very important from the sizing point of view because that helps in estimating the output of the solar module or array and also assist in identifying the appropriate PV technology to be used for the PV power system. Hence, in this exercise, solar radiation falling on the plane of the solar array is to be assessed. Solar radiation consists of direct radiation and diffuse radiation and the value for both of those components should be known. The total of direct and diffuse radiation is known as the global solar radiation.

Solar radiation is measured in units of solar energy falling on a horizontal surface over a period of time—the standard unit is kilowatt-hour per square metre (kWh/m^2). This is usually available for an average year or for an average day in a given month.

While assessing the solar energy resource, the following information is to be collected [4]

- Average annual global solar radiation and average daily global solar radiation (kWh/m^2) on horizontal surface;
- Monthly-averaged daily global and diffuse solar radiation on horizontal surface (as shown in Fig. 3);
- How many sunny days and overcast days occur in a year;

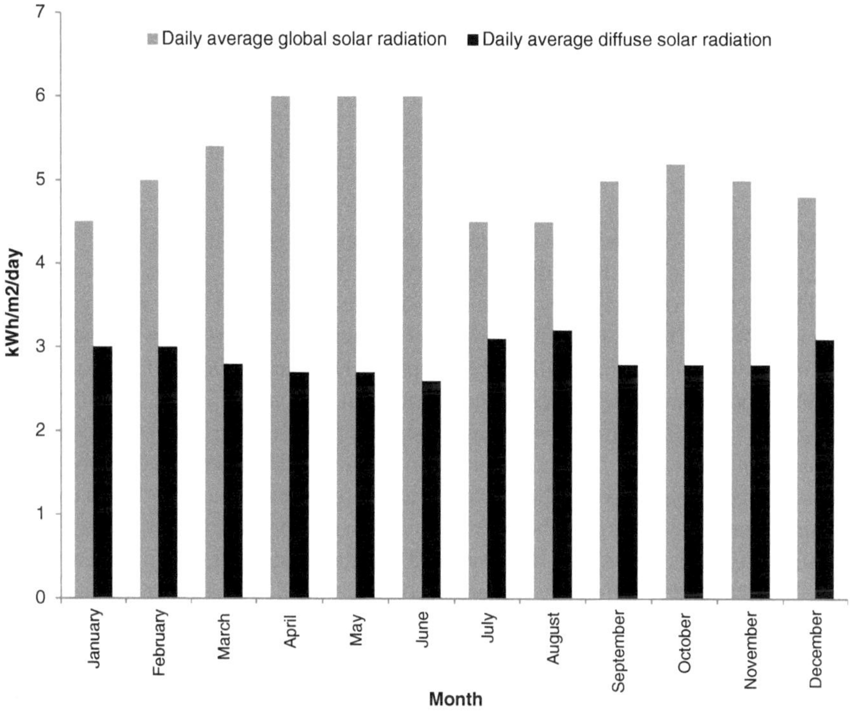

Fig. 3 Monthly-averaged daily global direct and diffused solar radiation on a horizontal surface

The above monthly-averaged data can be obtained from NASA surface meteorology website or similar authorised national solar energy resource centres. However, since the power output of the solar array depends upon the incident solar radiation, the solar radiation falling on the horizontal surface needs to be adjusted with a *modification factor* (*which is generally in the range of* 1.03–1.1) in order to take the surface tilt and orientation into consideration. This information is used in the subsequent section for estimating the sizing of the solar PV array.

2.2.3 Sizing of Main Component of the PV System

One point that needs to be decided at this stage is the *System Voltage*, i.e. whether a 12, 24, 48 V or higher voltage is chosen. This is critical from the point of view of determining the capacity of the components. Once the system voltage is chosen, the following steps can be followed for sizing the power system.

Step I: Specifying the inverter rating

i. Calculate “total connected AC Wattage”;
ii. Capacity or rating of the inverter should be the nearest higher value to the “total connected AC Watt”;

iii. Define the efficiency of the inverter (η_{inv})—This value can be taken from the technical brochure or specification chart of any of the reputed inverter manufactures;

Capacity or rating of inverter $\geq$ total connected AC Wattage

More details on how to select the inverter are provided in Sect. 5.

Step II: Specifying the battery capacity

The capacity of the battery depends upon three things: (a) daily energy requirement (E, W h); (b) days of autonomy (A) (i.e. number of days for which storage is required) and (c) maximum allowable Depth of Discharge (DoD) of the battery.

This approach for determining the capacity of the battery is different in a smart design concept than from a standard design approach. In a standard design, the capacity of the battery is decided based on the total daily energy requirement E. It does not segregate between the daytime energy requirement and night-time energy requirement. The difference between these design approaches does not seem substantial if there is no or little daytime load and daytime energy requirement. But it makes considerable difference if there is considerable daytime energy requirement.

For a daytime load, there is no need of storing a large proportion of solar generated DC electricity in battery. Rather, the DC electricity can be directly converted to AC electricity through the inverter. This would result not only improve the efficiency of the entire system (by around 5–7 %), but would also lead to smaller capacity of the battery and thus reduce the capital as well as replacement cost of the battery. However, the battery capacity should be such designed, that it can take care of the sudden fluctuation in the solar radiation (such as swift movement of clouds would result in a sudden drop of solar irradiation). So, in order to cater to the above points, on an average, 5–8 % of the daytime energy requirement can be added to the nocturnal energy requirement to find out the ideal *daily energy requirement,* which is used to design the battery capacity.

Daily energy requirement (for battery sizing) E = nocturnal energy requirement + 0.08 $\times$ day-time energy requirement

Secondly, to find out the days of autonomy, we need to take a decision on the following:

(a) whether to increase the capacity of the solar PV array in such a way that there would not be any situation when load requirement is more than the generated solar electricity and thus there is no extra day for storing energy in battery

or

(b) certain number of extra days (generally 2–3 days) is considered for storing energy in battery

or

The value of global solar radiation at horizontal surface (kWh) is equivalent to the peak sun-shine hours or Equivalent Hours of Sunshine (EHSS). The relationship is as follows

kWh = Peak power of the module (Wp) × Number of peak sunshine hours
Number of peak sunshine hours is the total number of hours when the module is exposed to the standard test condition of 1,000 W/m^2, 25 °C and AM of 1.5.

(c) solar PV array is sized in such a way that it can cater to the required energy in most of the months in a year and in the remaining month battery can be charged through another energy resource (such as diesel or wind generator) and thus there is no extra day (or maximum one day) for storing energy in battery.

From the smart design point of view, option (a) and (b) will not be recommended as it will permit an unnecessary increase in the designed size of solar PV array or battery. Instead, option (c) can be considered.

Box-3

In a *standard design* the battery capacity is done based on the total daily energy requirement, whereas in *a smart design concept* the battery sizing is done based on the total nocturnal energy requirement and 5–8 % of the daytime energy requirement.

Once the above-mentioned information is collected, the minimum capacity of the battery, Q required in Ampere-hour (Ah) can be calculated as per the formula given in Eq. 1 [1]

$$\boxed{Q = \frac{(E \times A)}{V \times T \times \eta_{\text{inv}} \times \eta_{\text{cable}}}} \qquad (1)$$

where,

E	Daily energy requirement for battery sizing (Wh)
A	Days of autonomy, i.e. number of days of storage required
V	System voltage
T	Maximum allowable depth of discharge of the battery (0.3–0.9)
η_{inv}	Inverter efficiency (0.8–0.95)
η_{cable}	efficiency of the cables delivering the power from the battery to the load.

The maximum allowable depth of discharge of the battery depends upon the type as well as characteristic of the battery (*whether lead acid battery—flooded type*, *Sealed Maintenance Free* (*SMF*), *Valve Regulated Lead Acid* (*VRLA*) *or gel type*; *lithium* (Li) *based—lithium ion, lithium phosphate*, etc.). More details about the selection of the batteries are given in Sect. 4.

Step III: Sizing of solar PV array

In a standard design, the sizing of the solar PV array in a stand-alone solar PV system can be done by using the Eq. 2

$$\boxed{W_{PV} = E_{total} \div ESSH \div \eta_{sys,\,overall}} \tag{2}$$

where,

W_{PV}	Peak wattage of the solar PV array
E_{Total}	Total daily energy requirement in Watt-hour (Wh)
ESSH	Equivalent Hours of Sunshine (i.e. Peak Sunshine hours)
$\eta_{sys,overall}$	Total system efficiency which includes the losses on account of PV array mismatch, cable losses, losses within inverter and charge controller, battery Ah-efficiency, etc. Here the value is around 0.6.

In general, overall efficiency is calculated as given in Eq. 3

$$\eta_{sys,\,overall} = \eta_{PV} \times \eta_{cable(PV,\,bat)} \times \eta_{cc} \times \eta_{batt} \times \eta_{inv} \times \eta_{distb\ cable} \tag{3}$$

where,

$\eta_{sys,overall}$	Total system efficiency
η_{PV}	Losses due to array mismatch, dirt, shading, etc. the efficiency ~90 %
$\eta_{cable\ (PV,bat)}$	Losses due to voltage drops in cable from PV array to battery (value ~98 %)
η_{cc}	Charge controller efficiency (value ~98 %)
η_{inv}	inverter efficiency (value ~90 %)
η_{batt}	Battery Ah efficiency (value ~85 %)
$\eta_{distb\ cable}$	losses in distribution cables (value = 98 %).

In a smart design, the sizing of the solar PV array can be achieved by estimating and adding the following

a. Solar PV array sizing for supplying the daily energy required for the battery; and
b. Solar PV array sizing based on daily daytime energy required.

Solar PV array capacity for supplying the daily energy required for the battery can be obtained as per the formula given below (Eq. 4),

Table 4 Total capacity of PV array required with different combination of energy requirement

Total energy required (Wh)	Daytime energy required (Wh)	Night time energy required (Wh)	Total size of PV array required (Wp)
4,000	0	4,000	1,350
4,000	1,000	3,000	1,250
4,000	2,000	2,000	1,170
4,000	3,000	1,000	1,090
4,000	4,000	0	1,000

Source This study
Note ESSH of 5 h is considered

$$\boxed{W_{\mathrm{PV-batt}} = E \div \mathrm{ESSH} \div \eta_{\mathrm{sys,overall}}} \tag{4}$$

where,

$W_{\text{PV-batt}}$ Peak wattage of the array
E Total daily energy requirement for battery in Watt-hour (Wh)
ESSH Equivalent Hours of Sunshine (i.e. Peak Sunshine hours)
$\eta_{\text{sys,overall}}$ Total system efficiency ($\sim$65 %) which includes the losses on account of PV array mismatch, cable losses, losses due to inverter and charge controller, battery Ah-efficiency etc.

Solar PV array capacity for supplying the daily daytime energy required can be obtained as per Eq. 5

$$\boxed{W_{\mathrm{PV-daytime}} = E_{\mathrm{day}} \div ESSH \div \eta_{\mathrm{sys,\,daytime}}} \tag{5}$$

where,

$W_{\text{PV-batt,daytime}}$ Peak wattage of the array
E_{day} Total daily daytime energy requirement in Watt-hour (Wh)
ESSH Equivalent Hours of Sunshine (i.e. Peak Sunshine hours)
$\eta_{\text{sys,daytime}}$ System efficiency which includes the losses on account of PV array mismatch, cable losses, losses due to inverter and charge controller (value is around 80 %).

The total solar PV capacity requirement can be calculated by adding $W_{\text{PV-batt}}$ (Eq. 4) and $W_{\text{PV-daytime}}$ (Eq. 5)

$$\boxed{W_{\mathrm{PV}} = W_{\mathrm{PV-daytime}} + W_{\mathrm{PV-batt}}.} \tag{6}$$

The table below (Table 4) shows how the total capacity of the PV array varies with different combination of daytime energy requirement and night-time energy requirement although the total energy requirement is kept fixed.

As per Table 4, in case of standard design, the total capacity of the PV array required for meeting the daily energy requirement of 4,000 Wh is 1,350 Wp irrespective of whether it is used in daytime or night-time. However, if smart design concept is followed, total capacity of the solar PV array can be reduced based on their daytime and night-time energy requirement.

Caution

Although system design is made based on certain assumptions, it is difficult to predict how a system will be used in practice, particularly in an off-grid location. Therefore, it is also difficult to predict the realistic energy requirement. Again there is every possibility that although the loads are designed for daytime but it might be used in night-time or vice versa. Therefore, in a smart design concept, these uncertainties are also addressed by having additional protections for the load sensitive components, specifically for the batteries. Here, in a worst case scenario, if energy withdrawn from the battery bank is very high compared to the available energy, then some of the loads, which are not very critical (*non-essential load*), are automatically disconnected depending upon the State of Charge of the battery.

Step IV: Sizing system wiring

The following steps can be followed to size the system wiring [4]

i. Divide the "AC total connected Watts" by the "DC system voltage". This will provide "Maximum DC amps continuous" which the system will have to provide;
ii. The above value will be used to size the cable from battery to inverter. That is the "maximum array amps" the controller would encounter under a short circuit condition.

Power losses and voltage drop in a PV system

Assessing the power loss and voltage drop is very critical in a PV system particularly if it is a small power system of few kW and the power is distributed to a large distance. If it is not assessed properly and realistically, this may lead to severe problems in the system, such as

- Battery might not get charged properly due to large voltage drop in the cable between the solar module and the battery;
- Some appliances may not work properly or get damaged if the voltage reaching them is quite low as compared to its operating voltage level.

So here, the voltage drop at each section (solar *array to battery*, *battery to inverter*, *inverter to load distribution box*, *in the local distribution network*, etc.) of the PV system should be calculated.

Voltage drop within the PV system can be calculated as follows

$$V_D = 1 \times R_c$$

where,

V_D the voltage drop in the cable

I the current in the cable and

R_c the cable resistance, which depends upon cable length and cross sectional area.

Generally, the cable length and the cross sectional area is chosen in a such a way that voltage drop between any two sections of the PV system is within the permissible voltage level. Most national electric code provides a method to calculate the voltage drop and cable size for the range of cables approved in the code either through the formula or by using the guidance tables. The designer or installer should make them familiar with the guidance table and the concerned formula in order to identify the appropriate cable size and type for a defined distance. The load distribution map developed during the initial stages of the design would be used here to carry out the power loss/voltage drop assessment exercise.

3 Guidelines for the Selection of PV Components

The following sections discuss the critical points that are to be considered while selecting the critical components of the PV system such as PV modules, batteries, inverter and charge controller.

3.1 Guidelines for Selection of PV Module

In principle, while procuring a module or an array, the following parameters have to be considered.

- PV module electrical characteristics and its specification;
- PV module temperature tolerance, hail impact resistance;
- PV module efficiency, dimension and weights;
- PV module with system compatibility;
- PV Quality requirement, Quality marks, standard and specifications; and
- PV module warranties and guarantees.

3.1.1 PV Module Electrical Characteristics

The PV module electrical characteristics cover

- Open circuit voltage (V_{oc});
- Short circuit current (I_{sc});
- Voltage at the maximum power point (V_m);
- Current at the maximum power point (I_m);
- Peak power (P_m); and
- Fill Factor (FF).

For a solar power plant, modules with higher wattage should be used because it will require less number of modules thus covering less roof area, fewer mounting structure, fewer inter-module electrical connections, etc. Power tolerances are also very critical and these need to be checked before the purchase of the module as the module output varies significantly on the negative side and can badly reduce the total output of the array due to array mismatching. Generally power tolerance of PV module is ±3 to ±5 %.

3.1.2 PV Module Temperature Tolerance, Hail Impact Resistance

The open circuit voltage of an individual silicon solar cell reduces by 2.3 mV per degree rise of cell temperature. Therefore, the temperature coefficient of the voltage is negative and expressed as shown in Eq. 7

$$ {}^{\mathrm{d}V_{oc}}/_{\mathrm{d}T} = -2.3\,\mathrm{mV}/^{\circ}\mathrm{C}. \tag{7} $$

As a module consists of large number of cells in series, the voltage coefficient is very large, which is expressed as shown in Eq. 8

$$ {}^{\mathrm{d}V_{oc}}/_{\mathrm{d}T} = -2.3n_c\,\mathrm{mV}/^{\circ}\mathrm{C} \tag{8} $$

where n_c is the number of cells in series.

The temperature coefficient of the current is slightly higher and the current increases very slightly with increase in cell temperature. However, the temperature coefficient of the power is negative for the silicon solar cell, whereas that of the thin film it is positive. That means, with the rise in temperature, the power output of the thin film solar cell increases slightly as compared to decrease in power output for silicon solar cell. This temperature coefficient effect needs to be taken care of in the system design and in estimating the yield of a PV array.

3.1.3 PV Module Efficiency, Dimension and Weight

PV module efficiency has a practical relevance because module with lower efficiency will simply cover a larger roof space (to provide the same power output), which is not a problem if there is a considerable surface area available. So, if the per Wp cost of the solar module is substantially lower for the lower efficiency module, then it makes sense to install relatively lower efficiency modules. So, here module with same Wp but of different efficiencies can be compared with their cost and the available area. The most optimum module can thus be chosen.

3.1.4 PV Module with System Compatibility

PV module or array should be compatible with the rest of the electrical components in the system. Particularly, the PV module or array's electrical characteristics should be compatible with the inverter, charge controller and the batteries. For example, the voltage of the PV string should match with the battery voltage in order to charge the battery appropriately or it should match with the input voltage required for the inverter in order to provide the desired AC output.

3.1.5 PV Quality Requirement, Quality Marks, Standard and Specifications

Modules which comply with appropriate national and international standards developed by recognised institutions, such as International Electrotechnical Commission (IEC), American Society for Testing and Material (ASTM), Sandia National Laboratory, etc. can be considered as reliable and likely to have longer life. The compliance of the standard will also take care of the minimum degradation of power output (%) from solar PV module in a defined time period.

3.1.6 PV Module Warranties

PV module warranties are extremely critical and the modules where the output will be guaranteed to be a minimum of 80 % of its original rating after 20–25 years should be purchased.

3.2 Performance Comparison of Different PV Technologies

Based on the site-specific conditions and requirements, Table 5 helps in identifying the most appropriate PV technology that is to be used.

Table 5 Performance of different PV technologies

	Mono-crystalline	Poly-crystalline	Thin-film		
			Amorphous	Cadmium Telluride (CdTe)	Copper Indium Gallium di-selenide (CIGS)
Typical module efficiency	15–20 %	13–16 %	6–8 %	9–11 %	10–12 %
Best research cell efficiency	25.0 %	20.4 %	13.4 %	18.7 %	20.4 %
Area required for 1 kWp (in m^2)	6–9	8–9	13–20	11–13	9–11
Typical length of warranty	25 years	25 years	10–25 years		
Temperature resistance	Performance drops 10–15 % at high temperatures	Less temperature resistant than mono-crystalline	Tolerates extreme heat	Relatively low impact on performance	
Additional details	Oldest cell technology and most widely used	Less silicon waste in the production process	Tend to degrade faster than crystalline-based solar panels		
I–V curve fill factor (idealised PV cell is 100 %)	73–82 %		60–68 %		
Module construction	With anodized aluminium		Frameless, sandwiched between glass; lower cost, lower weight		
Inverter compatibility and sizing	Lower temperature coefficient is beneficial		System designer has to consider factor such as temperature coefficients, $V_{oc} - V_{mp}$ difference, isolation resistance due to external factors		
Mounting systems	Industry standard		Special clips and structures may be needed. In some cases labour cost is significantly saved		

NB Authors' compilation, 2014

Table 6 Comparisons of two cases of solar PV system

Details	Solar system 1	Solar system 2
Manufacturer	A	B
Model	M1	M2
Efficiency	17 %	16 %
Number of solar panels	40	40
Output (year one)	8,000 kWh	9,000 kWh
Total cost	$40,000	$36,000
Area	34.8 m^2	45.1 m^2
Value	$5/kWh	$4/kWh

Let us now apply the above guidelines for selection of PV system assuming the following cases provided in Table 6. From a financial point of view, solar PV system 2 makes more sense for the consumer to buy and install. However, solar PV system 1 may make more sense if space for system installation is a constraint.

4 Guidelines for the Selection of Battery

Since solar modules can generate power only when exposed to sunlight, there is a need for storage batteries to store energy collected during periods of high irradiance and make it available at night and also during overcast period. There is a frequent demand for charging and discharging of batteries in a PV system. So, the type of battery used here is not the same as an automobile battery. Figure 4 shows the different categories of batteries.

The following types of batteries are available for PV application

- Lead acid battery (most common)
- NiCd (Nickel Cadmium)
- NiMH (Nickel Magnesium Hydroxide)
- Lithium based

Lead acid batteries are most widely used in solar PV application. Currently, lithium ion-based batteries are also being used in solar PV applications, particularly for small scale applications. While the useful life of the battery increases, the installation cost becomes very high in such cases. Other battery types, such as NiCd or NiMH, are used in portable devices. Typical solar system batteries' lifetime spans from 3 to 5 years. We can, however, achieve life of around 6–8 years with proper battery management and regular preventive maintenance of the battery bank. There are a large numbers of batteries available in the market and each of these is designed for a specific application.

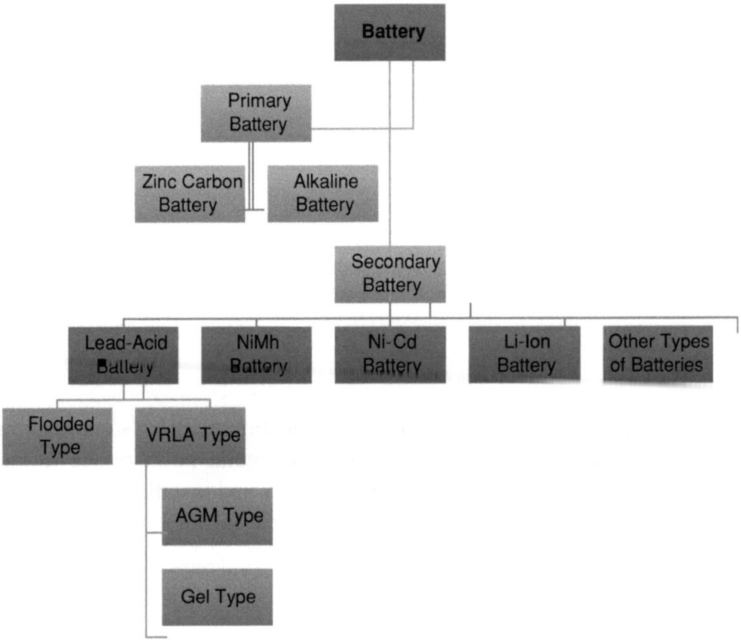

Fig. 4 Categorization of batteries (*Source* Author's compilation)

For product designers, an understanding of the factors affecting the battery capacity and life is vitally important for managing both product performance and warranty particularly with high cost batteries. Following characteristics are very important while selecting a battery used for solar application.

- Battery capacity at what rate
- Cycle life and temperature
- Battery Ah efficiency
- High charging current capability
- Good reliability under cyclic, deep discharge conditions
- Good power density, high recharge efficiency, rapid rechargeability
- Maintenance free, wide operating temperature
- Low cost per Ah, low self-discharge rate.

4.1 Battery Capacity

The capacity of a battery is measured in Ampere-hour (Ah). It provides information on the number of hours a specific current can be delivered by a fully charged battery before it gets completely discharged.

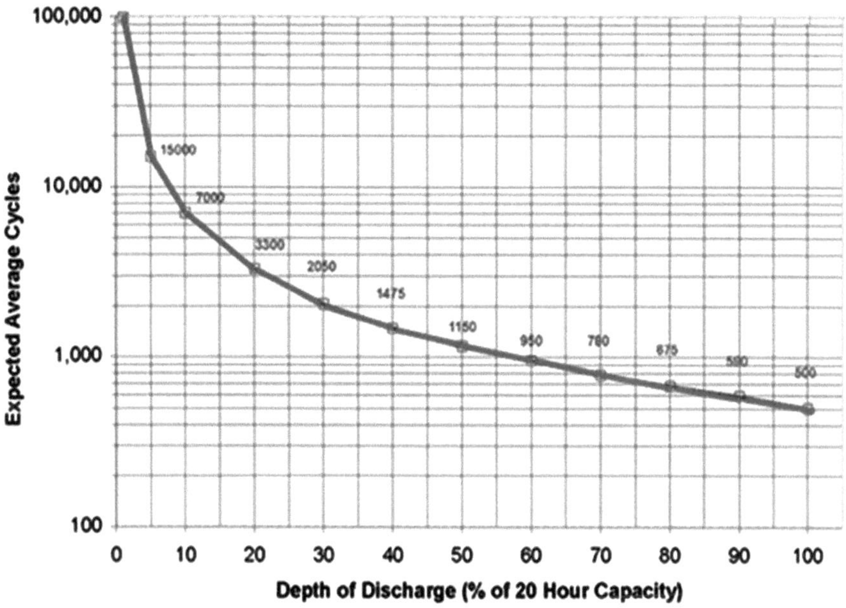

Fig. 5 Depth of discharge versus cycles of battery discharge (*Source* Author's compilation)

Battery capacity depends on

- *Discharging current*—the higher the discharging current the lower the capacity, and vice versa. For example: a battery delivering 1 A current for 100 h has a capacity of 100 Ah. However, if the same battery is discharged (delivering a current) with 8 A current it may provide that current for 10 h. That means the capacity is no more 100 Ah and instead reduces to 80 Ah (see Fig. 5). *Therefore, from the designer's perspective, the selection of the battery based on its capacity is important but more importantly at what rate* (*C*/10 *or C/*20 *or C*/100) *is most critical* [3]. *Because that decides what would be the actual capacity of the battery if it is not discharged at the designed rate.*
- *Temperature*. The capacity of the battery available in the market is specified at 25–27 °C. When the temperature exceeds, the capacity of the battery increases with temperature, whereas the life of the battery decreases (see Fig. 6). The performance of all batteries drops drastically at low temperatures. At −20 °C (−4 °F), most nickel-, lead- and lithium-based batteries stop functioning. Specially-built lithium ion brings the operating temperature down to −40 °C, but only on a reduced discharge. So, for lower temperature lithium ion battery can be chosen.

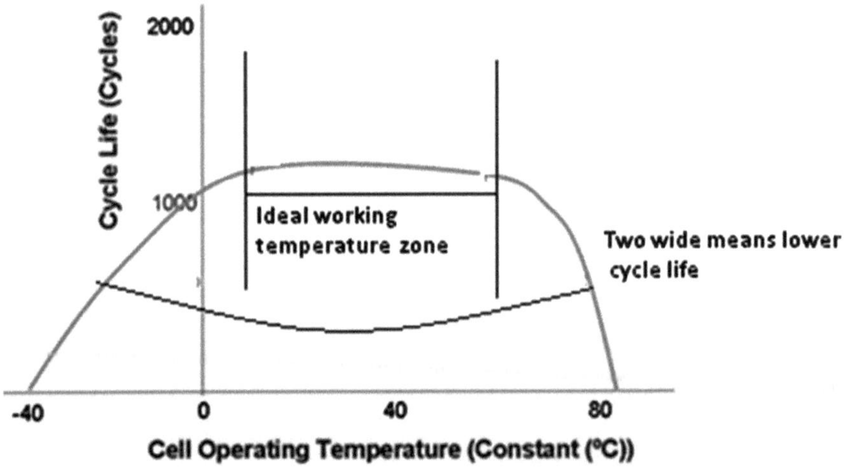

Fig. 6 Operating temperature of the battery versus cycle life

Figure 6 shows the ideal working temperature of the battery in order to get the optimum life cycle. Operating above 55 °C damages the battery permanently. One of the main functions of the battery management system is to keep the cells operating within their designed operating temperature window.

4.2 Life of a Battery

The life of the lead acid batteries is quoted as 5, 10, and 20 years. However this is a conditional statement and generally that condition is overlooked. The condition is the life of the battery will be *X* years, if the battery is kept in a specific temperature band, is kept at a specific float voltage. If these guidelines are not followed, the life of the battery will reduce. The rate of reduction of life depends upon how far its operation deviates from the designed value.

To supply a given load of *X* Watt, a bigger battery will be drained by a lower percentage, and would last for more cycles. That is to supply a 50 Ah load, a 100 Ah battery would be drained 50 %, whereas a 200 Ah battery would only be drained 25 %. The bigger battery will last longer. From Fig. 7, it can be seen that if the DoD changes from 50 to 25 %, then the life cycle of the battery would be more than doubled. Thus, a bigger battery is always recommended. But a bigger battery would also increase the cost. Thus, a balance between the cost of the battery and its capacity can be made to select the most appropriate capacity and type of battery.

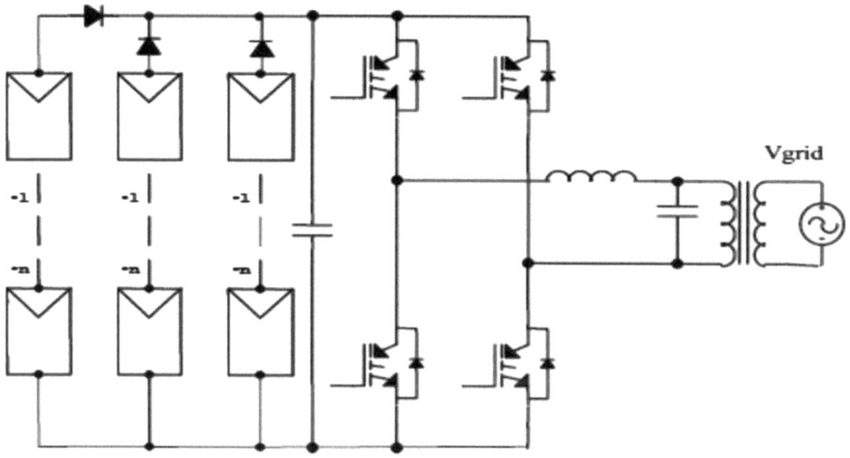

Fig. 7 Connections of solar PV strings

4.3 Self Discharge of Battery

This parameter is typically critical for a remote electrification project where the transportation of the material takes more time than standard time [1]. The typical self-discharge rates for common rechargeable cells are as follows:

- Lead Acid 4–6 % per month
- Nickel Cadmium 15–20 % per month
- Nickel Metal Hydride 30 % per month
- Lithium 2–3 % per month.

4.4 Comparison of Batteries Available in the Market

Table 7 compares the characteristics of four commonly used rechargeable battery used for solar applications, showing average performance ratings at the time of publication.

Based on the above-mentioned parameters, the appropriate battery for particular applications can be selected.

Table 7 Comparison of characteristics of different batteries

Specification	Lead acid	NiCd	NiMH	Lithium based		
				Cobalt	Manganese	Phosphate
Specific energy density (Wh/kg)	30–50	45–80	60–80	150–190	100–135	90–120
Cycle life	200–300	1,000	300–500	500–1,000	500–1,000	1,000–2,000
Fast charge	8–16 h	1 h	2–4 h	2–4 h	1 h or less	1 h or less
Overcharge tolerance	High	Moderate	Low	Cannot tolerate		
Self-discharge (month) from start	5 %	20 %	30 %	<10 %		
Maintenance requirement	3–6 months (topping)	30–60 days deep discharge	60–90 days deep discharge	Not required		
Cell voltage	2 V	1.2 V	1.2 V	3.6 V	3.6 V	3.3 V
Charge temperature	−20 to 50 °C	0–45 °C				
Charge temperature	−20 to 50 °C	−20 to 60 °C				
Safety requirement	Thermally stable	Thermal protection mandatory				

5 Guidelines for the Selection of Inverter

The size and capacity of the distributed solar PV system varies widely from few kW to MW scale and thus it is important to know whether selection on the type of inverters is different for different capacity of the PV system. The section below describes the different configurations that exist and its significance for PV system.

5.1 Single Stage/Central Inverter

The single stage inverter (central inverter) is widely used for large scale power applications. Here, the single power processing stage takes care of all the tasks of Maximum Power Point Tracking (MPPT), voltage amplification and grid side current control. In this configuration, the solar modules are connected in series to create strings with output voltage high enough to avoid an additional voltage boost stage [7]. In order to obtain the desired power level, the strings are connected in parallel through interconnection diodes (string diodes) as shown in Fig. 7.

Although this configuration is widely used, the global efficiency of the generation system is effectively reduced. The main reason of reduced performance is due to the centralised MPPT control that fixes a common operating point for all PV modules (shaded as well as un-shaded), whereas different operating points should be adopted for each module in order to extract the maximum power from the source. Because of these limitations, more advanced inverter topology is used based on the use of PV fields arranged in strings rather than arrays.

5.2 Double or Multi-stage Inverter

Here each string is connected to a double- or a single-stage inverter. If a large number of modules are connected in series to obtain an open circuit voltage higher than 360 V, the DC/DC converter can be eliminated. On the other hand, if a few number of PV modules are connected in series; a DC/DC boost converter is used. The DC–DC converter is responsible for the MPPT and the DC–AC inverter controls the grid current.

5.3 Multi-string Multi-stage Inverters with High Frequency Transformer

Another topology adopted is multi-string multi stage inverter. The multi string inverter has been developed to combine the advantage of higher energy yield of a string inverter with the lower costs of a central inverter. Lower power DC/DC converters are connected to individual PV strings. Each PV string has its own MPPT, which independently optimises the energy output from each PV string. All DC/DC converters are connected via a DC bus through a central inverter to the grid. Depending on the size of the string the input voltage ranges between 125 and 750 V. Here, system efficiency is higher due to the application of MPPT control on each string and higher flexibility comes from the ease of extensions for the photovoltaic field. This topology is more convenient for power levels below 10 kW. The multi-string inverters provide a very wide input voltage range (due to the additional DC/DC-stage), which gives the user better freedom in the design of the PV-system. However, the disadvantages are that it requires two power conversion stages to allow individual tracking at the inputs.

5.4 Selection of Inverter Based on Switching Devices

To effectively perform Pulse Width Modulation (PWM) control for the inverter, Insulated Gate Bipolar Transistor (IGBT) and Metal Oxide Semiconductor Field Effect Transistor (MOSFET) are mainly used for switching devices. *IGBT is used in more than* 70 % *of the surveyed inverter products and MOSFET is used in around remaining* 30 % *of the inverter*. As far as differences in characteristics between IGBT and MOSFET are concerned, the switching frequency of IGBT is around 20 kHz and it can be used for large power capacity inverters of exceeding 100 kW [6]. On the other hand, although the switching frequency of MOSFET can go up to 800 kHz, its power capacity is reduced at higher frequencies and MOSFET are used for output power range between 1 and 10 kW. So, in a nutshell, both IGBT and MOSFETs are used for small to medium range PV system with power capacity of 1–10 kW, whereas IGBTs are used for large scale power plant with power rating equal to or more than 100 kW. High frequency switching can reduce harmonics in output current, size, and weight of an inverter and thus nowadays High Frequency (HF) inverter with a compact size is available and widely used.

5.5 Selection of Inverter Based on Operational Perspective

In order to assess the inverter's performance in terms of operational perspective, a literature review and collection of secondary information was carried out by the authors. For the analysis purpose, information of about 200 models of different inverters with different capacity and types were collected. The following sections bring the findings from that survey.

5.5.1 Features of Grid Connectivity

The distributed or off-grid inverter should have the feature for grid connectivity (both incoming and outgoing) so that these solar PV systems would not be completely obsolete when grid extension reaches. With the massive plan of conventional rural electrification, it is always wise to select an inverter having grid connectivity features from the beginning with some incremental cost than completely changing the inverters later on with a high replacement cost.

5.5.2 AC Voltage and Frequency Range

For the standard values, the inverter can be operated substantially without any problems within the tolerance of $+10\ \%$ and $-15\ \%$ for the voltage, and ± 0.4 to $1\ \%$ for the frequency, specified by the grid standards of any country. For example, in India, where the single phase AC line is specified as 230 V, 50 Hz, the inverter should work at any voltage value between 253 and 198 V and any frequency value between 49.5 and 50.5 Hz without any problem. Any inverter which does not have this wide range, might not be considered, particularly for distributed power system, which are installed in relatively remote and rural locations, where wide fluctuation of voltage and frequency is prevalent [2].

5.5.3 Operational DC Voltage Range

The operable range of the DC voltage differs according to rated power of the inverter, rated voltage of the AC utility grid system, and design policy. In this survey, the operable range of the DC voltage for a capacity in the range of 180–500 W includes 14–35 V, 30–60 V. Similarly, the operable DC voltage range for a capacity of 10 kW or over includes 330–1,000 V. Hence, depending upon the operational range of the voltage range of the inverter, the capacity and configuration of the solar PV modules should be decided. So, while designing or designing the capacity of the solar PV system, this is one criterion that decides how many modules need to be connected in series or parallel to get the required DC operating voltage.

5.5.4 AC Harmonic Current from Inverter

Minimisation of harmonic current production is required as harmonic current adversely affects load appliances connected to the distribution system and can impair load appliances when the harmonic current is increased. The results of this survey show that Total Harmonic Distortion (THD), the total distortion factor of the current normalised by the rated fundamental current of many of the inverter, is

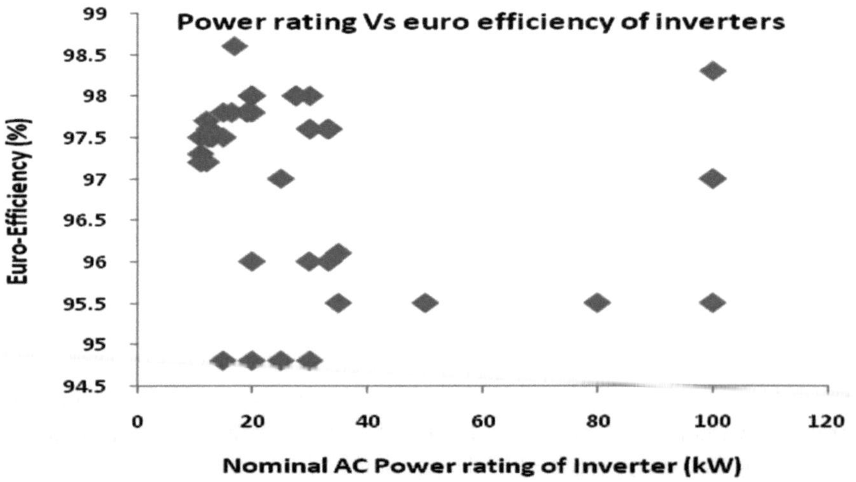

Fig. 8 Power rating versus euro efficiency of inverters

3–5 %. However, there are certain inverters in the power rating of 10–100 kW, that have THD in the range of 1–5 %.

5.5.5 Inverter Conversion Efficiency

Figure 8 presents the performance of a range of inverters.

Figure 8 shows that the efficiency range for all the inverters varies from 94.5 to 98.7 % [5]. In the medium scale range (10–20 kW), there are several inverters available with the euro efficiency range of 97–98 % and the incremental cost of these inverters are not much more than that of the low efficiency inverters (generally USD 50/kW). Thus, the project designer can evaluate the cost versus benefit of the inverter in terms of the enhanced efficiency.

5.5.6 Operational Environment

The installation conditions of the inverter (the indoor installation specification or the outdoor installation specification), the ambient temperature, the requirements for water- and dust-proofing, actual audible noise level of the inverter, and applicable regulations for EMC (electro-magnetic compatibility) needs to be examined carefully. As per the survey (Fig. 9), the maximum acceptable ambient temperature at nominal AC power is in the range of 40–75 °C. This range is relatively wider for 10–40 kW inverter, whereas it is narrower in the operational range of 40–50 °C for larger inverters.

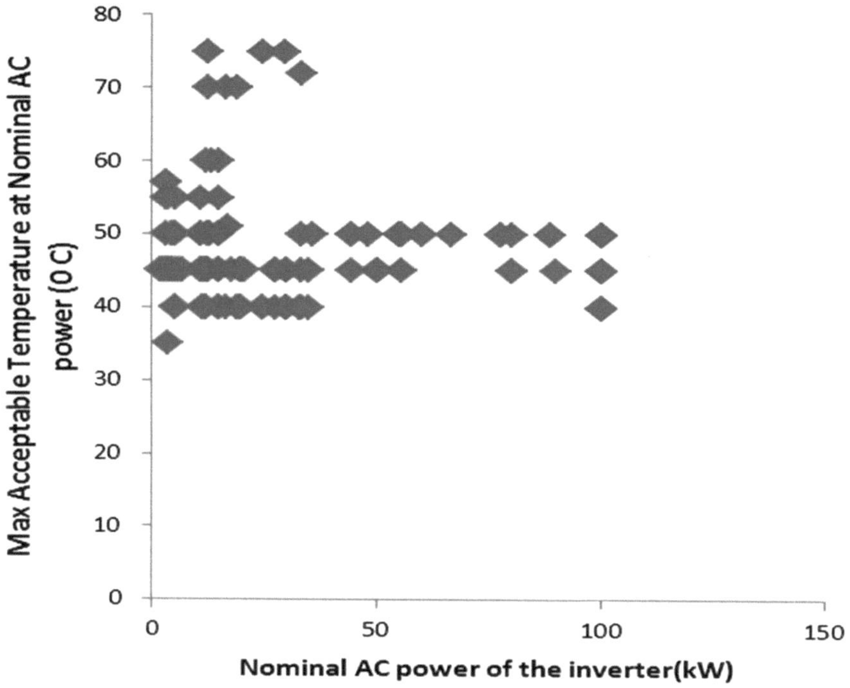

Fig. 9 Maximum acceptable temperature at nominal AC power

5.5.7 Required Protection Devices or Functions

Protective functions include protection for the DC- and the AC-sides. The protective functions for the DC side include those for DC overpower, DC overvoltage, DC undervoltage, DC overcurrent, and detection of DC grounding faults. Protective functions for the AC side include AC overvoltage, AC undervoltage, AC overcurrent, frequency increase, frequency drop, and detection of AC grounding, etc. Most of the inverters include these basic protections.

5.5.8 Standby Power Consumption

The standby power consumption of the inverter is a very important parameter that needs to be checked, particularly for off-grid PV applications where the only source for providing power is PV. So, by practice the lesser the standby power consumption is, the better it is for distributed PV applications. As per Fig. 10, it is observed that there is a wide range of products that are available with normalised standby power consumption for inverter capacity of less than 20 kW. So, the

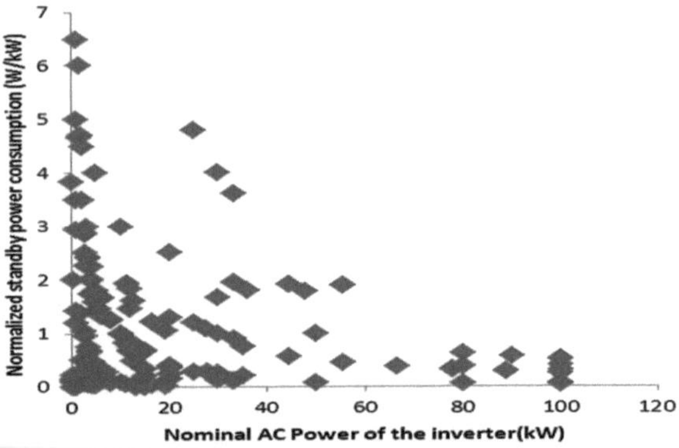

Fig. 10 Normalized standby power consumption of a selection of inverters

inverter can be judiciously selected so that self-consumption of these devices would not be significant.

5.5.9 Inverter System Cost, Size and Weight

The cost of the inverter system is very crucial when considering the economy of a photovoltaic power system. Although the cost of the inverter varies from country to country as well as with make and model, the average cost range of the inverter is found to be USD 600–1,000 per kW [5]. The weight of the inverter system differs considerably according to presence/absence of the isolating transformer (shown in Fig. 11). However, from the project developer's point of view it is a very critical parameter to judge as the weight of system affects the total transportation, handling and installation costs. For an inverter for a household PV power system, weight reduction is important when the inverter is installed indoors or it is to be mounted and thus an appropriate inverter with lower weight should be preferred.

5.5.10 System Guarantee

System guarantee plays a crucial role as it influences the entire system economics. Although the cost data for each of the inverter is not available, it is noted that there are some inverters that are available with an extended guarantee of 25 years.

6 Conclusion

Although PV systems are not technologically new, they have received increased attention because of their ability to meet peak power demand, provide backup power, offer improved power quality and reliability to micro-grids and rural

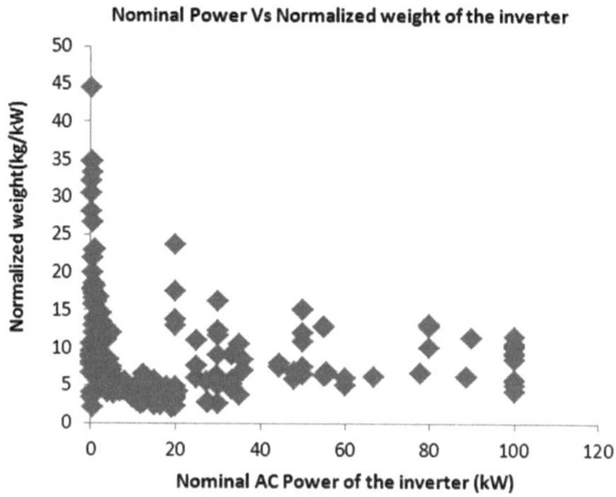

Fig. 11 Normalized weight of different inverters

electrification projects and substantial cost reduction of the PV systems in the last few years. The PV systems require specific power electronic converters to convert the power generated into useful power that can be directly interconnected with the utility grid and/or can be used for specific consumer applications locally. The roles of these power converters become very critical, particularly when it is used in one of the relatively expensive energy generating sources such as PV. With the advent of few state-of-the-art power converters and inverters certain smart design approaches has been possible, which is explained in this chapter. This chapter shows that if the smart design approaches, such as load profiling, load categorisation, and resource profiling, are carried out properly then the total capacity and thus the cost of the system can be reduced substantially. The chapter also shows if the battery and inverter are chosen properly, substantial cost savings in terms of reduced but appropriate PV system sizing can be achieved.

References

1. Antony F, Durschner C, Remmers K-H (2007) Photovoltaic for professionals—solar electric systems marketing, design and installation. Solarpraxis AG Publishers, Earthscan, London
2. Kim H, Kim J, Min B, Yoo D, Kim H (2009) A highly efficient PV system using a series connection of DC–DC converter output with a photovoltaic panel. Renewable Energy 34(11):2432–2436
3. Lasnier F, Ang TG (1990) Photovoltaic engineering handbook. IOP Publishing, Pennsylvania
4. Mohanty P (2008) Solar photovoltaic technology-renewable energy engineering and technology. TERI Publication, India
5. Photon International (2012) July and August issue 2012
6. Kjaer S, Pedersen J, BlaaBjerg F (2005) A review of single phase grid connected inverters for photovoltaic modules. IEEE Trans Indus Appl 41(5):1292–1306
7. Tang Y, Zhao L (2010) Maximum power point tracking techniques for photovoltaic array. New Energy, June 2010, pp 48–51

Analytical Frameworks and an Integrated Approach for Mini-Grid-Based Electrification

Subhes C. Bhattacharyya, Arabinda Mishra and Gopal K. Sarangi

Abstract Although rural electrification using mini-grids has attracted recent global attention, the concept has been there for quite some time. Consequently, a number of analytical approaches exist to support the decision-making process. This chapter first provides a review of literature dealing with analytical frameworks for off-grid and mini-grid based electrification projects. The range of analytical options includes simple worksheet-based tools to more sophisticated optimisation tools for technology selection as well as assessments based on multi-criteria analysis. This is followed by an evaluation of mini-grid based off-grid electrification projects in India that allows the identification of critical factors for the success of such projects. Finally, the chapter proposes an integrated approach for analysing decentralised mini-grid projects in a holistic manner.

1 Introduction

Decentralised electricity supply using local grids is a complex undertaking that requires all the activities of a conventional electricity supply system at a smaller scale in a remote area. Being a small, self-contained system, any such project requires careful preparation and planning, operation, control and maintenance as well as appropriate organisational arrangement, although it may not enjoy the

S. C. Bhattacharyya (✉)
Institute of Energy and Sustainable Development, De Montfort University, Leicester, UK
e-mail: subhesb@dmu.ac.uk; subhes_bhattacharyya@yahoo.com

A. Mishra
Faculty of Policy and Planning, TERI University, New Delhi, India
e-mail: amishra@teri.res.in

G. K. Sarangi
TERI University, New Delhi, India
e-mail: gopal.sarangi@teri.res.in

S. C. Bhattacharyya and D. Palit (eds.), *Mini-Grids for Rural Electrification of Developing Countries*, Green Energy and Technology,
DOI: 10.1007/978-3-319-04816-1_5,

benefits of accumulated expertise, resources of a large organisation, or the support of a large resource and consumer base to offer stability and diversity. Accordingly, it is important to undertake a careful study prior to any investment decision. This is increasingly gaining importance with private sector gradually venturing into the decentralised energy space as a potential actor.

Developing countries are increasingly looking forward to off-grid solutions like mini-grids not as an isolated form of energy provision, but in the overall context of community life and as an integral component of other rural improvement efforts [50]. For project developers of decentralised mini-grid systems, some inherent constraints arise because of high upfront costs, 'thin' rural markets due to small size of buyers and sellers, difficulty in arranging institutional finance and ambiguous government policies. More specifically on the policy front, the existing responses are limited to pilot scale implementation models and do not have a clear framework to mainstream decentralised systems in the national planning process [46]. A greater challenge to promote and accelerate the off-grid process in developing economies is linked to inadequate capacities at the community or village level to effectively operate and maintain the infrastructure. Studies point that capacity building efforts are central to wider replication of these systems [19, 114].

A review of literature reveals that there is a dearth of studies with integrated frameworks of analysis. Most of the studies either focus on techno-economic assessments or present policy narratives without adequate attention to local contexts and key determinants shaping the development trajectory of these projects. To bridge the gap, this chapter reviews alternative methodological options and presents an integrated framework for mini-grid analysis. We shall rely on the relevant literature to perform the above tasks. The chapter is organised as follows: Sect. 2 provides a review of literature covering the wide variety of approaches that have been used in the past and includes techno-economic feasibility studies, various analytical approaches such as indicators, optimisation, multi-criteria decision-making and systems approach and project-based literature (including project reports, manuals and best practices). Section 3 then considers the field-level experience from India while Sect. 4 presents an integrated approach for mini-grid interventions at the country level. Finally, a concluding section captures the main findings of this chapter.

2 Literature Review

The literature[1] on renewable energy and electricity supply is very well developed, and a number of strands can be identified from this body of knowledge.

(a) First, the focus of most of the literature is on the technical design of the system and its cost effectiveness analysis using some economic indicators (see [9] for a review of literature on hybrid renewable energy systems). Given the

[1] This section is based on Bhattacharyya [11].

technological diversity of renewable energies, this set of literature has often relied on the case study approach where the application of individual technologies or a combination of technologies has been considered to meet the energy demand. Some studies provide a review of the technological and economic readiness of alternative energies as well.

(b) Second, a number of tools have been used by various authors. For example, Hybrid Optimisation Model for Electric Renewables (HOMER), developed by National Renewable Energy Laboratory, USA (NREL), appears repeatedly in the literature as a preferred tool. It can handle a large set of technologies (PV, wind, hydro, fuel cells, boilers, etc.), loads (AC/DC, thermal and hydrogen), and can perform hourly simulations. Table 1 provides a list of some examples of HOMER application. HOMER is an optimisation tool that is used to decide the system configuration for decentralised systems. Other software tools include HYBRID2 developed by the Renewable Energy Research Laboratory (RERL) and Hybird Optimisation by Genetic Algorithms developed by the University of Zaragoza, Spain (HOGA), which are available freely.[2]

The use of optimisation approach has a long tradition in rural energy supply analysis. The most common application relies on linear programming due to its ease of use but more advanced applications have also been reported. In addition, simulations, multi-objective evolutionary programmes, and multi-criteria decision-making approach have been used. Given that the decision-making often involves trade-offs amongst various competing objectives and because such decisions may change depending on the stakeholder preferences, MCDM provides an effective alternative for reconciling conflicting viewpoints.

(c) Third, the practice-oriented literature (such as manuals and best-practice experiences) also provides some guidelines on decentralised electricity supply and in some cases recommends steps or critical factors for such projects.

In this section, we shall review some literature covering all three areas. Given the vastness of the literature and scarcity of resources, the review will necessarily be a partial one but we shall try to capture the essential points from each of the above sets.

2.1 Techno-Economic Feasibility Studies

There is a large volume of literature available of this sort that focuses on various technologies and country cases. These studies generally follow a common approach—assessment of technological appropriateness, evaluation of economic viability and determination of financial or other incentives required to make the

[2] For a list of such tools and their characteristics see [9].

Table 1 Some examples of HOMER application

Reference	Technology application	Country of application
Sen and Bhattacharyya [92]	Mini-hydro, solar, bio-diesel, wind	India
Demiroren and Yilmaz [23]	Wind, PV, wind-PV hybrid, and diesel generator	Turkey—Island example
Lau et al. [62]	PV-diesel hybrid	Malaysia
Weis and Ilinca [112]	PV-diesel	Canada
Setiawan et al. [93]	PV-diesel, wind-diesel, PV-wind-diesel	Maldives
Himri et al. [45]	Wind-diesel hybrid	Algeria
Nandi and Ghosh [69]	Wind-PV-battery	Bangladesh
Turkey and Telli [106]	Wind, solar, hydrogen	Turkey
Dalton et al. [21, 22]	PV, diesel, wind	Australia—energy supply options for a large hotel
Khan and Iqbal [58]	Wind, diesel, battery, fuel cell	Newfoundland, Canada
Shaahid and Elhadidy [94–98]	PV-diesel-battery	Saudi Arabia
Nfah et al. [70]	PV, micro-hydro, LPG generator, battery	Cameroon
Bekele and Palm [7]	PV-wind hybrid	Ethiopia
Prodromidis and Coutelieris [82]	PV, wind, battery	Greek Islands

Source Based on Bhattacharyya [11]

project viable at a given location [57]. Most of these studies compare the grid and off-grid investment options using a financial or economic cost-benefit evaluation.

Reddy et al. [86] provided a comparative costing of grid-connected, off-grid and energy conservation systems in the Indian context. This study highlighted the importance of improved technology for rural energy services and provided a detailed evaluation of grid-connected, off-grid and energy conservation options using the life cycle costing approach. Sinha and Kandpal [97] compared the cost of electricity supply through grid extension against the cost of supply from decentralised sources for rural India. This study considered the cost of extending the grid in terms of investments for the distribution network. It also considers the cost of grid electricity at various distances from the grid as well as for different levels of load factors and transmission-distribution losses. The cost of electricity supply from alternative sources is then considered and compared to see how the cost effectiveness changes as the distance from the main grid increases, the load factor changes and peak demand increases. The study using data from late 1980s found that the decentralised solutions are viable in isolated, small villages with low load factors. Similarly, for villages located 25 km away from the 33 kV grid, decentralised options become viable. It might be useful to undertake such a study using more recent data and taking other country examples.

Bernal-Agustin and Dufo-Lopez [8] have performed an evaluation of the grid-connected solar PV system in the Spanish case. The study analysed the economic and environmental benefits of a PV project considering the net present value and pay-back periods. The paper considered the initial cost of the grid-connected PV system (i.e. the cost of the generator, the cost of the inverter, and the cost related to installation) along with any subsidy that is available to the investor. The net cash flow generated by the project per year is estimated taking income and expenses into consideration. Income is generated either by selling electricity to the grid or by reducing electricity purchase through auto-consumption of electricity generated through the PV system. Expenses include costs related to operations and maintenance of the system, insurance and financing the project. Given that cash flow occurs over the lifetime of the project and that the investment is an initial cost, the net present value of the investment is considered assuming a life of 25 years for the project. The paper then applies this to a case in Zaragoza (Spain) and analyses the NPV and payback period for different electricity prices, interest rates, and subsidy sizes. The paper also considered the environmental benefits of PV electricity by considering the amount of emissions avoided and CO_2 emission mitigated. It then estimated the avoided costs of externality in monetary terms considering different scenarios of electricity substitution by the PV system (e.g. avoidance of thermoelectric power, power from sub-bituminous coal, etc.).

Chakraborti and Chakraborti [17] analysed the case of solar PV for an island use in Sagar Dweep, India. The study considered PV and grid extensions as alternative systems and evaluated the options from economic and environmental perspectives. The study shows that grid extension over long distances is not cost-effective.

2.2 Analytical Approaches

In this section, we shall review a number of alternative approaches that are found in the literature that have been used to analyse and decide the appropriate energy systems for rural areas.

2.2.1 Levelised Cost of Supply

The levelised cost of supply is a common indicator used for comparing cost of electricity supply options. The levelised cost is the real, constant cost of supplying electricity that if recovered from consumers over the lifetime of the plant would meet all costs associated with construction, operation and decommissioning of a generating plant. This generally considers capital expenditures, operating and maintenance costs, fuel costs, and any costs involved in dismantling and decommissioning the plant. It can also consider the external costs and other relevant costs such as costs of back-up power in the case of intermittent energies. This indicator has been routinely used to analyse the cost effectiveness of renewable energy options compared to other conventional energies. Examples include IEA [16, 48, 49, 88, 102].

However, care has to be taken in using this method due to a number of factors:

(1) First, the levelised cost is calculated based on a specific rate of utilisation (capacity factor) of a technology. Technologies with similar utilisation rate can be easily compared using this method but technologies with different load profiles or loading patterns can give misleading results. For example, if a technology is used for base load and the other for peaking purposes, the levelised cost for the base load plant will always be favourable due to higher level of utilisation.
(2) Second, in many cases, the variability of fuel costs is inadequately captured (or underestimated), making fossil fuel-based plants more cost effective.
(3) Third, for non-firm supply technologies, the cost of back-up or standby power could be inappropriately considered.
(4) Finally, this often ignores the external costs related to fossil fuel use, thereby putting the renewable energies at a disadvantage.

For decentralised electricity supply in developing countries, the levelised cost approach has been used in a number of studies. Banerjee [6] has presented a detailed study of cost estimations in the Indian context. Similarly, Nouni et al. [72–74] presented cost estimations for specific technologies in the Indian context.

Nouni et al. [71] has used this approach to identify the potential areas for decentralised electricity supply in India. They considered the delivered cost of electricity supply for different load factors and for villages located within a radius of 5–25 km from an existing 11 kV substation for two cases: plain terrain and hilly terrain, where the cost of local distribution tends to be higher. They also considered the cost of supply from decentralised renewable energy options. Considering typical village load data from 1991 Census statistics, they estimated that the average peak load of a remote rural household to be 0.675 kW. Considering the population of villages, they suggested that villages with <50 kW peak load could be considered for decentralised electricity supply through renewable energy technologies. The authors then considered the trade-off between grid extension and off-grid supply to find the cost effective electricity supply option for remote villages. While this provides a framework of analysis from the cost of supply perspective, the analysis does not consider the external costs related to fossil fuel use, cost of security of supply, cost of stand-by power for renewable energies. Accordingly, the study is likely to favour fossil fuels and undermines the potentials of renewable energies.

Kolhe et al. [59] presented a life cycle cost comparison between a stand-alone PV system and a diesel power plant in India. The study followed an approach similar to levelised costs but derived these using specific parameters for diesel plants and PV systems.

Perhaps the most comprehensive study of alternative generating technologies suitable for energy access projects is found in ESMAP [33]. This study presents a review of a range of technologies covering a wider spectrum of capacities—50 W to 500 MW. The review is presented for 37 technologies (renewable, conventional and emerging) under three categories—off-grid, mini-grid and grid-connected electricity supply. The report provides the technical features of alternative technologies, presents alternative configurations that are used in practice and discusses the cost and performance assumptions of each technology used in the analysis.

The study has assessed the economics of the above technologies using the levelised cost method and presented the results for three different time horizons—2005, 2010 and 2015, to reflect the effect cost reduction in some technologies. The study also considers the effect of sensitivity of key variables on the economic viability of technological options. The costs of local distribution as well as long distance transmission are also considered where applicable. However, the study does not include the external costs and security of supply concerns for fossil fuels and stand-by power costs for renewable energies.

The levelised cost of electricity supply for renewable energy options is presented in Table 2, while the levelised cost of conventional/emerging electricity supply technologies is presented in Table 3. As can be seen from these tables,

- The cost of off-grid options is generally higher than that of conventional energies;
- The cost of supply reduces as the size of plant increases. Electricity supplied from small-sized off-grid plants tends to cost much higher than the bigger sized plants of same technology.
- Some renewable technologies are either cost effective or reaching cost effectiveness.

However, the levelised cost approach, despite its wider use, is a one-dimensional indicator and fails to capture any dimension beyond costs. In addition, the external costs due to environmental effects and security of supply were not captured in the above study. This limitation needs to be kept in mind while using this comparator.

2.2.2 Weighted Score System

Lhendup [63] presented a weighted score system where a number of aspects (such as technical, regulatory features, environmental and social aspects) related to rural energy supply options are considered. A set of indicators is then identified for each aspect and a weight is given based on the importance of the indicator. Each option is tested against a set of indicators and a score is given depending on the performance of the option against the indicator. The product of the score and the weight for a particular indicator gives the weighted score. The process is repeated for all indicators and the sum of the weighted scores for any option gives its total score. Supply options were ranked based on their weighted scores (see Table 4). Lhendup [63] used a performance scale of 1 (low) to 5 (high) and a total weight of 100 for 18 indicators. The paper explains the justification for each ranking in an appendix and indicates that the methodology can be implemented in a simple spreadsheet model.

Table 2 Levelised cost for renewable energy technologies (2005, US cents/kWh)

Technology		Rated output kW	Levelised cost components				
			Capital	Fixed O&M	Variable O&M	Fuel	Average
Solar PV		0.05	45.59	3	13		61.59
		0.3	45.59	2.5	8		56.09
		25	42.93	1.5	7		51.43
		5,000	40.36	0.97	0.24		41.57
Wind		0.3	26.18	3.49	4.9		34.57
		100	13.55	2.08	4.08		19.71
		10,000	5.85	0.66	0.26		6.77
		100,000	5.08	0.53	0.22		5.83
PV-wind hybrid		0.3	31.4	3.48	6.9		41.78
		100	22.02	2.07	6.4		30.49
Solar thermal	With storage	30,000	10.68	1.82	0.45		12.95
	Without storage	30,000	13.65	3.01	0.75		17.41
Geothermal	Binary	200	12.57	2	1		15.57
	Binary	20,000	5.02	1.3	0.4		6.72
	Flash	50,000	3.07	0.9	0.3		4.27
Biomass gasifier		100	4.39	0.34	1.57	2.66	8.96
		20,000	3.09	0.25	1.18	2.5	7.02
Biomass steam		50,000	2.59	0.45	0.41	2.5	5.95
MSW/Landfill gas		5,000	4.95	0.11	0.43	1	6.49
Biogas		60	3.79	0.34	1.54	1.1	6.77
Pico/Micro-hydro		0.3	14.24	0	0.9		15.14
		1	12.19	0	0.54		12.73
		100	9.54	1.05	0.42		11.01
Mini-hydro		5,000	5.86	0.74	0.35		6.95
Large hydro		100,000	4.56	0.5	0.32		5.38
Pumped storage		150,000	34.08	0.32	0.33		34.73

Source ESMAP [33]

Table 3 Levelised cost of conventional/emerging technologies (2005, US cents/kWh)

Technology		Rated output kW	Levelised cost components				
			Capital	Fixed O&M	Variable O&M	Fuel	Average
Diesel/Gasoline generator		0.3	5.01		5	54.62	64.63
		1	3.83		3	44.38	51.21
		100	0.98	2	3	14.04	20.02
	Baseload	5,000	0.91	1	2.5	4.84	9.25
	Peakload	5,000	7.31	3	2.5	4.84	17.65
Microturbines		150	1.46	1	2.5	26.86	31.82
Fuel cell		200	5.6	0.1	4.5	16.28	26.48
		5,000	5.59	0.1	4.5	4.18	14.37
Combustion turbines	Natural gas	150,000	5.66	0.3	1	6.12	13.08
	Oil		5.66	0.3	1	15.81	22.77
Combined cycle	Natural gas	300,000	0.95	0.1	0.4	4.12	5.57
	Oil		0.95	0.1	0.4	10.65	12.1
Coal steam with FGD and SCR	Sub-critical	300,000	1.76	0.38	0.36	1.97	4.47
		500,000	1.67	0.38	0.36	1.92	4.33
	Supercritical	500,000	1.73	0.38	0.36	1.83	4.3
	Ultra-super critical	500,000	1.84	0.38	0.36	1.7	4.28
Coal IGCC (without FGD and SCR)		300,000	2.49	0.9	0.21	1.73	5.33
		500,000	2.29	0.9	0.21	1.73	5.13
Coal AFBC (without FGD and SCR)		300,000	1.75	0.5	0.34	1.52	4.11
		500,000	1.64	0.5	0.34	1.49	3.97
Oil steam		300,000	1.27	0.35	0.3	5.32	7.24

Source ESMAP [33]
Note FGD: Flue gas desulphurisation; IGCC—Integrated Gasification Combined Cycle
AFBC—Atmospheric Fluidised Bed Combustion
SCR—Selective Catalytic Reduction

Although this attempts to capture various dimensions in a simple way, there is some inherent subjectivity involved here in terms of weights attached and ranking given to each factor. However, if a participatory approach is used in deciding the weights and the ranks for a given locality or case, this method can turn out to be a useful tool.

2.2.3 Sustainability Indicators

Ilskog [51] presented a set of 39 indicators for assessing rural electrification projects. These indicators considered five sustainability dimensions—technical sustainability, economic sustainability, social/ethical sustainability, environmental sustainability and institutional sustainability. This is presented in Fig. 1.

Table 4 Provides the criteria suggested by Lhendup [63] and their weights

Criteria	Weight	Ranking	
		1	5
1 Technical features	60		
1.1 Energy density of the system	8	Low	Very high
1.2 Ability to meet the anticipated demand	8	Not at all	Fully
1.3 Energy payback ratio	8	Low	High
1.4 Lifespan of the system	7	Short	Very long
1.5 Quality of supply	6	Poor	Very good
1.6 Weather and climatic condition dependence	6	Fully dependent	Not at all
1.7 Availability of local skills and resources	5	Not available	Available
1.8 Incremental capacity of the system	5	Difficult	Easy
1.9 Dependence on fossil fuels	4	Fully dependent	Not at all
1.10 Other infrastructure development	3	Required	Not required
2 Government regulations	15		
2.1 Tax incentives	5	Not provided	Provided
2.2 Regulation on use of local resources	5	Regulated	No regulation
2.3 Opportunity for private participation	5	Low	High
3 Environmental and social aspects	25		
3.1 Public and political acceptance	6	Not acceptable	Acceptable
3.2 Land requirement and acquisition	6	High	Low
3.3 Hazard rating	5	High	Low
3.4 Environmental pollutants	5	High	Low
3.5 Interference with other utility infrastructure	3	High	Low
Total	100		

Source Lhendup [63]

Ilskog and Kjellstrom [52] presented an assessment of rural electrification cases using 31 of these indicators. Each indicator was scored on a scale of 1–7, with 7 representing the best performance. The total score was obtained by simple averaging of the scores, implying that all indicators received same weight. The scoring was based on interviews with 800 randomly selected stakeholders and the performance comparison of projects was done using these indicator scores. The paper considered that 'the technical sustainability is facilitated if the technical infrastructure locally available meets the requirements of the technology installed, if the technology used can provide the service needed and if favourable technical performance leads to low costs for the services' [52]. Table 5 compiles the indicators under five sustainability dimensions.

This method was clearly applied keeping the stakeholder participation in mind. Therefore, it is compliant with the participatory approach and can be used either at the local level or at a higher level of aggregation. It can also capture qualitative factors and offers the flexibility that additional factors can be added while any factor that is not relevant for a given application can be removed. This can be easily adapted to include concerns of energy security and governance. However, several factors will require additional estimation/calculation and therefore will require a systematic tool to implement this approach.

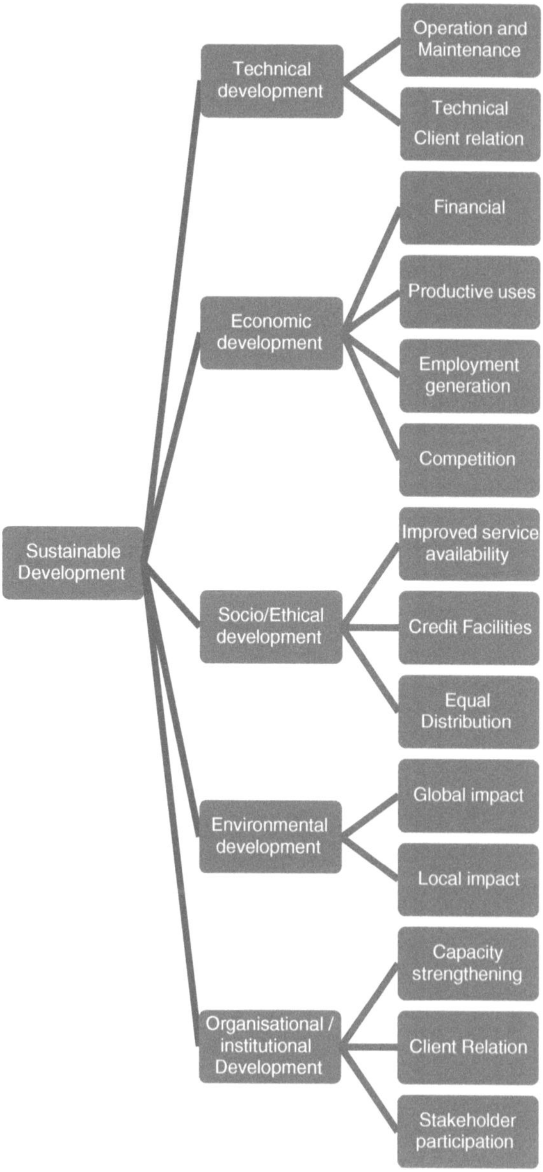

Fig. 1 Sustainability indicators process. *Source* Ilskog [51]

2.2.4 Optimisation Techniques

Optimisation has been widely used in energy analysis since the 1970s but applications for rural energy supply systems started in the 1980s. Parikh [76] presented a linear programming model to capture the interactions between energy and agriculture in rural areas of developing countries. The model considered 12

Table 5 Indicators used for rural electrification analysis

Technical sustainability	Economic sustainability	Social/Ethical sustainability	Environmental sustainability	Institutional sustainability
Operation and maintenance—conforms with national standards	Financial perspective—profitability (%)	Improved availability of social electricity services—number of street lights in the area, number per 1,000 population	Global impact—share of renewable energy in production, %	Capacity strengthening—share of staff and management with appropriate education, %
Technical losses, %	Financial perspective—O&M costs USD/kWh	Credit facilities—Micro-credit possibilities available for electricity services connection (yes/no)	Global impact—Emission of CO_2 from production, kg CO_2/kWh	Degree of local ownership, %
Technical client-relation issues—daily operation service (%)	Financial perspective—Costs for capital and installation, USD/kWh	Equal distribution—share of population with primary school education (%)	Local impact—share of electrified households where other energy sources for lighting has been replaced, %	Number of shareholders (number)
Technical client-relation issues—availability of services (%)	Financial perspective—Share of profit set aside for reinvestment, %	Share of population with access to electricity, %	Local impact—share of electrified households where other energy sources for cooking main meals has been replaced, %	Share of women in staff and management, %
	Financial perspective—Tariff lag, USD, kWh	Share of electricity consumers in high income category, %	Any serious local environment impact identified, yes/no	Number of years in business (number)
	Development of productive uses—share of electricity consumed by business, %	Subsidy offered for electricity services, USD, kWh		Client-relation—Share of non-technical losses (%)

(continued)

Table 5 (continued)

Technical sustainability	Economic sustainability	Social/Ethical sustainability	Environmental sustainability	Institutional sustainability
	Share of electricity used by households for income-generating activities, %			Level of satisfaction with energy services (%)
	Competition—number of electricity service organisations in the area			Stakeholder participation—Auditing of reports on an annual basis (yes/no)

Source Ilskog and Kjellstrom [52]

different energy sources and several conversion technologies. It captured the supply and demand for energy and agricultural outputs to find an optimal mix for a given rural condition. Ramakumar et al. [85] developed a linear programming model for integrated rural energy systems. The idea is simple here—the system will be designed to minimise the total cost subject to a number of constraints related to energy availability, energy demand, and other technical/system constraints. There were many different methods of optimisation used to analyse rural energy systems namely: Linear programming, Geometric programming, Integer programming, Dynamic programming, Stochastic programming, Quadratic programming, Separable programming, Multi-objective programming, Goal programming and Hybrid methods. The model presented by Ramakumar et al. [85] was suitable for stand-alone systems. Ashenayi and Ramakumar [5] and Ramkumar et al. [84] expanded these models to include a knowledge-based design tool and other factors such as power reliability (loss of power probability) and scenario design.

Sinha and Kandpal [98–100] designed an optimisation model for rural energy supply considering lighting, cooking and irrigation demands. These models were applied to an Indian context to analyse the cooking, lighting and irrigation needs of rural areas and this showed how such a technique can be effectively used as a decision tool. These models considered a number of alternative technologies with different cost characteristics and found the optimal combinations using linear programming.

Iniyan and Jagadeeshan [53] developed an optimal renewable energy model (OREM) which considered 38 different renewable energy options. The model minimised the cost/efficiency ratio and used resource availability, demand, reliability and social acceptance as constraints. Devdas [24–27] presented a linear programming model for analysing rural energy for local-level development. The program optimised the revenue of rural output subject to energy and non-energy constraints. The model was applied to the District of Kanyakumari of Tamil Nadu, India.

Parikh and Ramanathan [77] presented the INGRAM model to analyse the interactions between energy, agriculture and environment in rural India. This program maximised the net revenue of the rural energy system subject to constraints that included crop residue balance, animal feed balance, dung balance, fertiliser nutrients balance, as well as energy balance to ensure adequate supply to meet the demand. The model was calibrated for the year 1990–1991 and applied for the year 2000. Although this is not a rural energy supply model per se, it incorporated elements that are relevant for a sustainable rural energy system development, although it did not consider the use of modern renewable energies in the analysis.

Akella et al. [1] used an optimisation framework to analyse the optimum renewable energy use in a remote area in India. The paper considered solar photovoltaic systems (SPV), micro hydropower (MHP), biomass energy supply (BES) and wind energy supply (WES). The model provides the least-cost combination of different renewable energies that could be used to meet the need. However, the model considers the overall energy use in the village and does not consider the projects that would be used to supply such energies. The model used the following equations (Box 1) where Z is the total cost of providing energy, MHP, SPV, WES and BES represent four renewable resources indicated above. All sources are used to meet the demand subject to availability constraints. The problem was solved using Lindo and Homer. The HOMER analysis produced somewhat higher results because the software takes into account the cost of the local grid, cost of batteries and cost of conversion, while LINDO only considered the renewable system costs.

A very similar study was reported by Kanase-Patil et al. [55] where the case study was performed at a different location in India. This study considered micro-hydro, biomass, biogas, PV and wind. A number of alternative scenarios have been presented, and the least-cost supply option is determined through the optimisation process. The optimisation programme was run using Lingo and HOMER packages.

Box 1: Equations used by Akella et al. [1]

Minimise:	$Z = 1.50\ \text{MHP} + 15.27\ \text{SPV} + 3.50\ \text{WES} + 3.10\ \text{BES}$
Subject to:	$\text{MHP} + \text{SPV} + \text{WES} + \text{BES} = D$ kWh/year
	$\frac{\text{MHP}}{0.9} \le 12.8.166\ \text{kW h m}^{-2}\ \text{year}^{-1}$
	$\frac{\text{SPV}}{0.9} \le 22.363\ \text{kW h m}^{-2}\ \text{year}^{-1}$
	$\frac{\text{WES}}{0.80} \le 15{,}251\ \text{kW h m}^{-2}\ \text{year}^{-1}$
	$\frac{\text{BES}}{0.85} \le 641{,}385\ \text{kW h m}^{-2}\ \text{year}^{-1}$
	MHP, SPV, WES, BES ≥ 0

Source Akella et al. [1]

Howells et al. [47] presented a study of a hypothetical, non-electrified, low-income South African village using MARKAL/TIMES—an optimisation model used for deciding the least-cost option for meeting the energy needs of the village. In contrast to other models, this application considered the entire range of end-use energy demand—cooking, lighting, space heating, water heating, refrigeration and other. It also considered electricity, diesel, LPG, solar, wind, candle, paraffin, coal and wood.

The study suggests that such a detailed analysis is feasible. Such a framework can be used to capture multiple objectives—such as cost minimisation, minimisation of environmental effects, etc. It could capture the dynamic aspect of energy transition by considering a long-term perspective. But data availability is a constraint—especially data on appliance stocks, efficiencies, is not readily available. The model complexity increases as more technological options are included and the long-term dynamics is considered.

2.2.5 Multi-Criteria Decision-Making Method

The multi-criteria decision making (MCDM) is a decision support system that is used to capture multiple dimensions of a project or a policy, some of which may be conflicting with each other. There are three schools of thoughts in this area—the American School (or value measurement models), the European School (or outranking models), and the goal, aspiration reference level models [65]. The American school focuses on the Analytical Hierarchy Process and multi-attribute value/utility theory. This assumes that the decision-maker is aware of the preferences and can express and rank them unambiguously. On the other hand, the European school does not assume full knowledge of preferences by the stakeholders and is therefore less restrictive. The third category, on the other hand, tries to find alternative solutions that are closest to achieving a desired goal or aspiration level. These methods compare options relative to an ideal solution and the option closest to the ideal is chosen.

Each of these has been applied to energy and renewable energy issues—Greening and Bernow [42] point out the potential of MCDM in energy and environment studies, while Ref. [110] presents a recent review. Some other studies include Refs. [28, 65, 75, 78]. Some studies have also applied these techniques to decentralised energy systems. A brief review of these approaches and applications to rural energy issues is presented below. Loken [65] insisted that in choosing an appropriate MCDM approach, it is better to avoid 'black-box' models as they are poorly understood by the decision-makers. Transparent approaches generate better acceptability of results and outcomes as decision-makers trust the results.

AHP Method

The analytical hierarchy process is a powerful and flexible decision-making process developed by Saaty [89] that is used to solve complex problems involving interactions of different criteria across different levels. It is a multiple criteria decision-making technique that allows subjective as well as objective factors to be considered in decision-making process. The AHP allows active participation of decision-makers/stakeholders in reaching agreement, and gives managers a rational basis on which to make decisions.

The AHP is a theory of measurement for dealing with quantifiable and intangible criteria that has been applied to numerous areas, such as decision theory and conflict resolution [108]. AHP is a problem-solving framework and a systematic procedure for representing the elements of any problem [91]. AHP is based on the following three principles: Decomposition, comparative judgements, and synthesis of priorities.

Formulating the decision problem in the form of a hierarchical structure is the first step of AHP. In a typical hierarchy, the top level reflects the overall objective (focus) of the decision problem. The elements affecting the decision are represented in intermediate levels. The lowest level comprises the decision options.

Once a hierarchy is constructed, the decision-maker begins a prioritisation procedure to determine the relative importance of the elements in each level of the hierarchy. The elements in each level are compared as pairs with respect to their importance in making the decision under consideration. A verbal scale is used in AHP that enables the decision-maker to incorporate subjectivity, experience, and knowledge in an intuitive and natural way. For pair-wise comparison, a scale proposed by Satty [89] is commonly used (see Table 6). After comparison matrices are created, relative weights are derived for various elements. The relative weights of the elements of each level with respect to an element in the adjacent upper level are computed as the components of the normalised eigenvector associated with the largest eigenvalue of their comparison matrix. Composite weights are then determined by aggregating the weights through the hierarchy. This is done by following a path from the top of the hierarchy to each alternative at the lowest level, and multiplying the weights along each segment of the path. The outcome of this aggregation is a normalised vector of the overall weights of the options. The mathematical basis for determining the weights was established by Saaty [89].

Dyer and Forman [31] describe the advantages of AHP in a group setting as follows: (1) both tangibles and intangibles, individual values and shared values can be included in an AHP-based group decision process, (2) the discussion in a group can be focused on objectives rather than alternatives, (3) the discussion can be structured so that every factor relevant to the discussion is considered in turn and (4) in a structured analysis, the discussion continues until all relevant information from each individual member in a group has been considered and a consensus choice of the decision alternative is achieved. A detailed discussion on conducting AHP-based group decision-making sessions including suggestions for assembling the group, constructing the hierarchy, getting the group to agree, inequalities of power, concealed or distorted preferences, and implementing the results can be found in Saaty [90] and Golden et al. [39].

AHP is the most commonly used MCDM approach. Zangeneh et al. [116] used the AHP method to prioritise distributed energy options using a case study of Iran. The first level sets the prioritisation goal. At the second level, four factors are considered—technical, economic, environmental attributes and regional primary energy resources. At the third level, sub-criteria for each of the factors are introduced. For economic aspects, two factors are considered: Costs and market. For cost, two further factors are considered—investment cost and operating cost. For market, two other factors are considered—potential of making money through supply and potential of ancillary service supply.

For technical issues, three factors are considered—operational issues, structural issues and technical requirements. Six factors are considered under operational issues—power quality, forced outage rate, response speed, efficiency, start-up time, and capacity factor. Under structural issues four factors are considered—footprint, lifetime, modularity and installation lead time. Similarly, three factors are included in the technical requirement category—maintenance, domestic technical knowledge and interconnection equipment.

Table 6 AHP pair-wise comparison scale

Score	Description
1	Equally preferred
3	Weak preference
5	Strong preference
7	Very strong or demonstrated preference
9	Extreme importance
2, 4, 6, 8	Intermediate values

Under the environmental dimension, three factors have been considered: noise emission, pollution emission and aesthetics. While noise and aesthetics did not have any further factors, under pollution emission, five factors are considered—PM10, SO_2, CO_2, NO_x and CO. The case study considered PV, wind, fuel cell, micro-turbine, gas turbine and diesel engines as alternative technologies. The prioritisation was done based on a survey of expert views (51 participated but 37 results were retained). The analysis was performed using Expert Choice software. The paper presented the results of the prioritisation exercise with detailed tabulation of criteria values at each level.

Other studies using AHP for energy analysis include Wang et al. [109], Wang and Feng [111], Limmeechokchai and Chawana [64], and Kablan [54].

Multi-criteria Decision-Making for Renewable Energy Sources (MCDM-RES)

Polatidis and Haralambopolous [80] presented an integrated renewable energy planning and design framework that they applied to a Greek island in the Aegean Sea. The planning activity is considered to be a multi-dimensional activity that takes technical, economic, environmental and social aspects into consideration. The authors highlight the complexity of energy planning process due to the presence of structural aspects, multiplicity of actors and multi-dimensional sphere of interactions. This is captured in Fig. 2. In addition, they are mindful of the fact that technologies and decision-making processes have further temporal and spatial dimensions, and these aspects along with social dynamics have to be taken into account.

Given the presence of multiple objectives and multiple actors, the authors use a multi-criteria decision-making framework. The decision-making process involves the following eight steps:

- Problem identification and initial data collection.
- Institutional analysis and stakeholder identification.
- Creation of alternatives.
- Establishing evaluation criteria.
- Criteria evaluation and preference elicitation.
- Selection of the MCDA technique.
- Model application.
- Stakeholder analysis of the results and feedback.

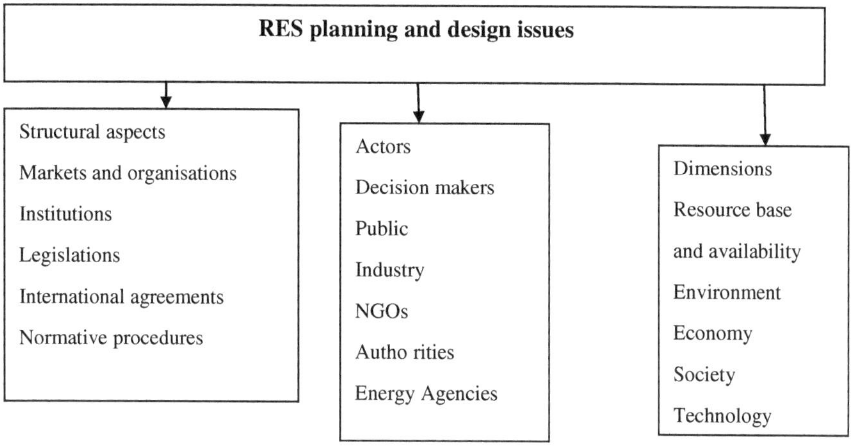

Fig. 2 Renewable energy planning issues. *Source* Polatidis and Haralambopolous [80]

At the last stage, where the results are shared with the stakeholders, if the level of acceptance is found to be low, the process will be repeated to generate a more acceptable outcome and the solution can then be implemented.

The application to the Greek case showed how the above steps were followed. Four alternative options were considered: (1) installing a 2 MW diesel generator, (2) a 4 MW hydro plant in conjunction with 4 wind turbines of 0.6 MW each, (3) a 4 MW hydro plant in conjunction with 8 wind turbines of 0.6 MW each, and (4) a 2 MW diesel generator along with a 4 MW hydro plant and 12 wind turbines of 0.6 MW. The evaluation criteria were initially developed through a literature review. This was then discussed with the stakeholders and finalised. These are presented in Table 7. The analysis was performed using PROMETHEE II package, a widely used program following the European tradition of MCDM analysis. The PROMETHEE suit of packages has been used in energy planning decision-making (see Table 8 for a list of such applications).

Many other studies have been reported in the literature in the field of rural energy or renewable energy supply using the MCDM approach. Such studies have covered a wide range of applications (see Table 9) and have used established software packages or have presented alternative tools. Cherni et al. [18] presented a multi-criteria analytical tool SURE for rural livelihood decision analysis which can be used to decide appropriate rural energy supply options that can be used for enhancing rural livelihood. The package was developed through the support of DfID (Department for International Development, UK) and incorporates technical and non-technical aspects such as financial, social, human and environmental dimensions. The paper reports the application of the model to a Colombian rural community.

Cherni et al. [18] indicate that although MCDM has been used in analysing rural energy issues, the technical aspect received a privileged treatment in the

Table 7 Evaluation criteria used in the Greek analysis

Energy resource criteria	Economic criteria	Environmental criteria	Social	Technological
Amount of imported oil avoided (toe/year)	Payback period (years)	Land used by the project (m^2)	Employment creation	Reliability and safety
Potential for reducing blackouts (% of peak load)	NPV of investments (€)	Compatibility with other activities	Public acceptance	
Amount of electricity produced (kWh/year)	Installation cost (€/kW)	Noise creation		
	Operational costs of electricity generation (€/kW h)	Visual impact		
	Net economic benefit for the community (€/year)	Construction of electricity networks and access roads (km)		
	Entrepreneurial risk (risk of project failure)	CO_2 reduction potential (tons of CO_2 avoided)		

Source Polatidis and Haralambopolous [80]

Table 8 List of applications of PROMETHEE in the area of energy

Application field	References
Comparing CSP (Concentrating solar power) technologies	[14]
Regional energy planning with a focus on renewable energies	[81, 103, 105]
Analysis of national energy scenarios in Greece with a focus on renewable energies	[29, 37]
Designing energy policy instruments	[30], (Madlener and Stagl 2005)
Evaluation of different heat supply options	[38]
Prioritisation of geothermal energy projects	[40, 41, 43]
Participatory analysis of national renewable energy scenarios in Austria	[60, 67]
Evaluation of biomass collection and transportation systems	[61]
Siting of hydropower stations	[68]
Comparing cooking energy alternatives	[78]
Evaluation of residential energy systems	[87]
Comparing energy technologies based on renewable, fossil or nuclear resources	[104]
Evaluation of energy research projects	[107]

Source Oberschmidt et al. [75]

Table 9 Application of MCDM in rural energy issues

Reference	Area of application	Tool/Method
Cherni et al. [18]	Rural energy supply for rural livelihood	SURE
Haurant et al. [44]	PV on farming land in an island	ELECTRE
Georgopoullou et al. [36]	Renewable energy planning in a Greek Island	ELECTRE III
Kablan [54]	Rural energy in Jordan	AHP
Pokharel and Chandrashekar [79]	Rural energy	Goal programming (STEP)
Karger and Hennings [56]	Sustainability of decentralised options	AHP

decision-making process. The issue of sustainable rural energy supply and rural livelihood issues received little attention. SURE they claim have overcome this challenge. Buchholz et al. [12] argue a similar participatory process for wider use of biomass energy in a sustainable energy context.

2.2.6 Systems Analysis Approach

For any analysis of a rural energy system, it is important to understand and capture the complex interrelationships that exist with the society, environment, technologies and governance aspects. One of the main issues related to any decision-making is the failure to incorporate appropriate feedback from various interactions. Also, the analysis has often been carried out at a highly aggregated level, which in turn removes the possibility of capturing important issues or aspects that influence the system performance [101]. Further, additional issues related to conventional modelling include the following [83]:

(a) Inter-temporal interactions and feedback are not modelled explicitly;
(b) The disequilibrium framework for modelling is missing.
(c) Time delays and other distortions of the energy system variables are not explicitly modelled.
(d) Non-linear responses to actions are not explicitly represented.

The systems approach has attempted to remedy this problem by understanding the information feedback structures in systems [20, 35] and by representing such feedbacks through casual loops and analysing them quantitatively. In the area of rural energy and rural electrification, various studies have been reported in the literature. For example, Alam et al. [4] reported a model for rural energy system of Bangladesh. The model considers crop production, biogas production, and rural forest and agro-based industries and analyses how output can be optimised to improve the quality of life. Similarly, Alam et al. [2] have reported an application of their previous model for farming in Bangladesh, while Alam et al. [3] reported the analysis of rural household energy use. The model incorporates feedback loops, non-linearity and time-lag features commonly found in real systems. Xiaohua et al. [115] presented a model to analyse the interaction between the rural energy system and the economy for a Chinese rural community. It considered a basic feedback structure of rural energy- economy and analysed the factors affecting rural energy use.

2.3 *Practice-Oriented Literature for Decentralised RE Supply Analysis*

The literature is also well developed for this category of studies but we shall review four studies—Cabraal et al. [13], CEEP [15], ESMAP [32] and World Bank [113] as they include some form of frameworks for analysing off-grid energy supply.

2.3.1 Cabraal et al. [13] Study

This study focused on the best practices for household rural electrification using solar PV systems. The study reported a number of case studies and provided a framework for economic/financial comparison of alternative options for household electrification. The cash flow analysis using economic and financial evaluation methods was used as the framework of analysis and was performed in Excel. An annex to the report enlists the following steps for such an analysis:

- Energy demand estimation: As electricity is used for lighting and appliances, the demand for these services has to be considered for decentralised electricity supply;

- Village selection criteria—identifying a set of conditions/parameters for village selection is the second step in the process. The parameters normally considered are number of households to be connected, distance from the grid, size and number of productive loads, and load growth prospects. Each of these factors affects the cost of supply and hence is an important consideration for supply decision.
- Alternative technology options—The third step is to identify alternative technologies that could be used to supply the energy service. The study considered kerosene/battery schemes, solar home schemes, isolated grid and central grid extension. The sizing of the alternative systems to satisfy the needs is also done at this stage.
- Least-cost comparison of options—This step provides the economic and financial cost comparison of alternative options using the cash flow method. The levelised cost of each option is calculated to determine the least-cost option. The break-even analysis is used to identify the range of economic viability of alternative options for villages of different size (household numbers) and household density.

The study also evaluated the experience of a number of case studies by considering the technical, financial and institutional performances using a set of criteria. For technical evaluation, the factors considered were: system size selection, system quality, installation repair quality, training/maintenance, availability of spare parts, and battery recycling. For financial evaluation, the performance was evaluated based on credit supply, financial sustainability, system pricing and tax/subsidy structures. For institutional performance, the institutional structure, marketing strategy, information dissemination and sustainability of institutional structure were considered. The study while focusing on practical applications, has not reported any systematic tool for the distributed off-grid systems.

2.3.2 Worksheet-Based Tool in CEEP [15]

CEEP [15] relied on a worksheet-based package, called rural renewable energy analysis and design tool (RREAD) for RE system analysis for decentralised operations. The model considered—economic, social, technical and environmental aspects. The programme analyses the energy availability, technical viability, economic feasibility and social—environmental value of a PV, wind and PV-Wind hybrid systems (see Fig. 3 for the model structure).

> The input module consists of six sets of data: Resource data, load data, system configuration, capital and operating costs, financial data and policy scenario information (e.g. the existence of tax credits, subsidies, and program initiatives to internalise social benefits/costs, etc.). The resource, load data and system configuration data are used to evaluate the system's overall energy performance, including energy output, resource-load matching capacity and service reliability. Cost, financial and policy data measure the economic, social and environmental values of using renewable energy systems [15].

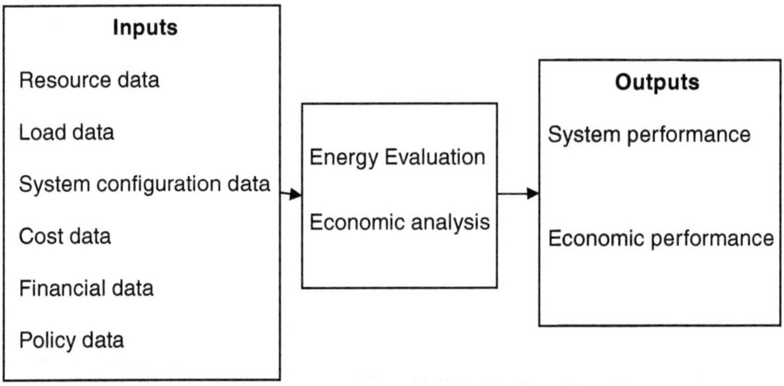

Fig. 3 Structure of RREAD. *Source* CEEP [15]

This method was used to determine the levelised costs of supply for different systems under different conditions in China. The study then conducted a socio-economic analysis using a local survey. The report also used a logistic regression model to analyse the survey results. This study considered all relevant dimensions in a systematic way using a practical tool but it was designed only for PV-based supply. This limited technological choice is a constraint for its generic use.

2.3.3 ESMAP [32] Study

The ESMAP [32] study provided the best practice guidelines for implementing decentralised energy systems for project managers. The decentralised option is an alternative approach to 'production of electricity and the undertaking and management of electrification project that may be grid-connected or not'.

It provides a step-by-step approach to project implementation which focuses on five steps.

- Step 1: Definition of institutional and regulatory environment—This stage will make sure the government commitment to the activity is enlisted and the institutional arrangement for decentralised energy services is established. Key activities thus include assessing government commitment, role and responsibilities of the agency responsible for decentralised energy supply, verification of legal provisions for off-grid supply and public sector involvement, identification of market barriers, assessment of technical capacity, assessment and identification of local financial organisations, identification of local institutions/NGO and their roles/responsibilities and assessment of private sector interest.
- Step 2: Market assessment and identification of project concept: This is the preparatory step for developing the project idea. Key steps include: Collection and review of existing information and/or initiation of a market analysis, collation of market information and completion of market analysis, analysis of competing products or services, identification of cost of service and disposable

income of consumers, and the identification of possible distribution paths. This step provides market information that can be used in the next stage.

- Step 3: Appropriate technology and product choice—This step selects the appropriate technology option to provide a reliable and cost-effective supply to meet the local needs. Key steps include: Identification of available technologies, energy demand estimation in terms of energy use, form, and quantity, determination of the most appropriate technology option, selection of a product delivery option, and product testing and specification preparation. At the end of this step, a tested product option is available for delivery.
- Step 4: Selection of a delivery mechanism—The study focuses mainly on two types of delivery mechanisms—cash and credit system or leasing through a dealer and delivery through an energy service company (ESCO). The key steps involved at this level are: Assessment of credit availability, review of distribution infrastructure, assessment of affordability of decentralised energy options, selection of a product and its delivery channel, and initiation of distribution channel establishment. This step provides a delivery mechanism for an identified technology option and initiates the process for establishing the delivery of the product.
- Step 5: Review and evaluation of financial options—This step evaluates whether the financial sector is geared to meet the financing needs of decentralised supply and identifies appropriate options to meet the local needs. Key steps here include: Identification of financial needs, evaluation of the rural banking system and the availability and cost of credit, mobilisation of the banking sector, identification of local partners, identification of financing options and programmes and determination of the terms of financing.

This study covers the entire set of activities related to a decentralised energy supply and is quite generic in terms of technology choice or country of use. It suggests a sequential framework of analysis with a detailed list of criteria that can be used in each step. Two areas are under-represented—environmental aspects and social dimension of the problem. Also, the potential for conflicts at each stage is not adequately captured.

2.3.4 World Bank [113] Study

World Bank [113] identifies the following critical factors for an off-grid project design:

(a) Comparing technology options—the first task is to determine the suitable technology option or options. The general guidance given is as follows:

(1) if the consumer size is small and dispersed, and if their main need is lighting, individual systems like SHS works; Other technologies for individual demand are pico-hydro systems where water resources are available while wind home systems are now being piloted.

(2) Where consumers are concentrated and can be economically inter-connected, a mini or micro grid is commonly used. Diesel, RET (renewable energy technologies) or hybrid options are used.

(b) Social safeguards and environmental considerations—Although off-grid projects are generally environmentally beneficial, some components such as use of batteries can have some environmental effects. These need to be carefully integrated with the national policies on waste recycling and hazardous wastes. Similarly, off-grid projects must adhere to national guidelines or regulations on watershed protection, land use and land acquisition.
(c) Productive and institutional applications—Off-grid communities could engage in many productive activities such as agricultural production and processing, fishing, animal rearing if quality energy is provided. Similarly, institutional or community-level uses such as schools, clinics or community centres could form another main use of off-grid electricity. The project design should take advantage of such 'systematic and pragmatic approaches' [34].
(d) Enhancing affordability—Generally, only 2–3 % of the consumers are found to afford cash purchases of off-grid solutions. The customer base can be expanded to 20–30 % with micro-credit. With leasing, the customer base can be increased to 40–50 %, while long-term fee-for-service options could increase the customer base further. Subsidies, consumer financing, low-cost technology options and support to businesses and commerce can also be considered to improve affordability.
(e) Business models for off-grid solutions—The supply of off-grid solutions could be provided using a number of players—private entities, individuals, community-based organisations, NGOs and state agencies. Appropriate incentives are required to attract players into this business. Various alternatives have been experimented—including dealerships, ESCO, leasing arrangements, medium term service agreements, community-based supply and hybrid forms. The local market condition plays an important role in deciding the appropriate business model.
(f) Regulating the off-grid supply—This is an important aspect of supply where the state has to play an important role to ensure that consumers are not overcharged and that they receive a quality supply. The regulatory framework for traditional grid supply may not be appropriate for such services and needs a special attention.
(g) International co-financing support—For providing access to energy using renewable energies, various co-financing options are available, including funding from the Global Environment Facility (GEF), Global Partnership for Output-based Aid (GPOBA), the climate investment funds and the Clean Development Mechanism (CDM). These sources could be tapped to reduce the financial burden of the national agencies.

The elements of a sustainable off-grid electrification project are presented in Fig. 4.

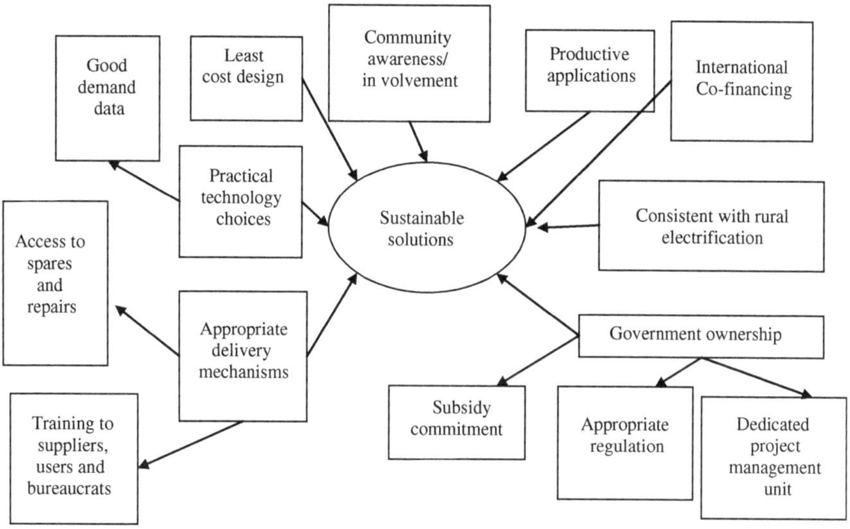

Fig. 4 Elements of a sustainable off-grid electrification project. *Source* World Bank [113]

The alternative analytical frameworks presented above provide ways of analysing the mini-grid system. However, each tool or analytical approach has its own strength and weakness. While the case-based approaches tend to provide practical guidance, they do not always rely on academically rigorous frameworks. On the other hand, academic studies are not always practically oriented, thereby reducing their potential for real-life applications. Thus, there is need for an integrated approach.

However, before such an integrated approach is considered, it is important to take a look at the practical experience at the project level to understand and identify essential factors that may be affecting their performance. With this objective in mind, we present insights from a field-level study in the next section.

3 Insights from the Field

This section goes into the project level evaluation and strives to understand the basic conditions that have largely shaped off-grid project viability. Based on secondary sources, the evaluation covers 74 cases of off-grid/decentralised project interventions from India[3] that include a range of technologies such as biomass,

[3] The information collection and analysis is carried out under a specific methodological framework. The initial activity was the scoping exercise of case profiles of off grid energy interventions by mining all the possible secondary information. This was followed by selected field visits and a series of expert interviews, both direct and telephonic. In order to gain further insights at a collective level, a participatory workshop was conducted and preliminary findings of this study were presented and discussed with different stakeholder representatives in the workshop.

Table 10 Details of off-grid sites visited

Project name	Technology	Size of the plant (kW)	Ownership	Funding
Rampura	Solar	8.7	Community	Non-govt.
Radhapura	Biomass	10	Community	Govt.
Tamkuha	Biomass	33	Private	Private
Amthagouda	Micro-hydro	20	Community	Non-govt.
Karlapeta	Micro-hydro	25	Community	Non-govt.
BERI	Biomass	1,000[a]	Government	Mixed

[a] Cumulative capacity

solar, micro-hydro, small wind and hybrid systems. There is considerable variation among the projects in terms of geographical locations, use of technology, size of the plants, tariff structures, ownership and management arrangements, among others.

The desktop research on the above off-grid cases was complemented by field visits. We conducted field visits to six project sites and investigated the fundamental conditions shaping the project operation and management. The details of the field sites and key operational attributes of the projects are presented in Table 10.

A statistical profile of the database of selected cases is presented in Table 11 with information related to the type of technology, size of the plant, implementing entity, source of finance and energy applications. A majority (53 %) of the projects are based on biomass gasification technology and mid-sized plants, ranging between 11 and 50 kW, constitute more than 50 % of the cases. There is variety in terms of project implementing entities (e.g. village energy committees (VEC), NGOs, state nodal agencies and in some cases private business entities) but in majority of cases either VEC or NGO is the implementing agency. Though private entrepreneurship is believed to possess a lot of potential in the rural energy business, the database has only 10 % of cases implemented by private entities. We categorise the source of funding into three major groups, i.e. government, non-government and mixed. Nearly half of the cases in the database have funding from non-government sources. On the basis of energy applications, the database has 40 % of the selected cases demonstrating clear income-generating links.

A key challenge in preparing the database for analytical work was to identify each project's current operational status. This was an intensive exercise and we have largely relied on expert information to categorize projects as operational, partially operational and non-operational. The database has 61 % of cases under the operational category, which allows us to take the analysis to the next level of identifying key determinants (see Table 12).

The operational viability of off-grid projects in the Indian context is largely determined by policy support, social acceptance in the form of community participation, linkages with income-generating opportunities, and technological appropriateness. To incorporate the policy dimension to our database, we took the

Table 11 Statistical profile of the selected off-grid projects

Technology	%	Plant size	%	Implementing agency	%
Biomass gasification system	52.7	Less than or equal to 10 kW	24.3	VEC and NGO	66.2
Micro-hydro	24.2	11–50 kW	51.4	State agency	23.0
SPV and others	23.1	Above 50 kW	24.3	Private company	10.8
Source of funding	**%**	**Policy link**	**%**	**Income-generating link**	**%**
Government	37.8	States with policy links	62.5	With links	39.2
Non-government	48.6	States without any policy link	37.5	No links	60.8
Mixed	13.5				

Source This study

Table 12 Distribution of selected off-grid projects in terms of operational status

	Case types	Operational (61 %)	Non-operational and partial operational (39 %)
Technology	Biomass	47	79
	Otherwise	53	21
Size	<10 kW	22	28
	>10 kW	78	72
Funding source	Govt.	24	59
	Non-govt. and Mixed	76	41
Community participation	Exists	24	45
	Not exist	76	55
Income-generating linkage	Exist	47	28
	Not exist	53	72
Policy linkage	Exist	67	41
	Not exist	33	59

geographical location of each off-grid case and mapped the project to state-specific decentralised/renewable specific policies (or elements of it).[4] It was found that in 62.5 % of cases, there exists some kind of policy support to promote and develop off-grid energy development in the concerned State. This percentage figure goes up to 67 % if we consider only those projects which are currently operational; more interestingly, nearly 60 % of the projects currently non- or partially operational seem to have suffered because of lack of policy support.

[4] Even though it is hard to find any specific policy operating at the state level to promote decentralized interventions, policy 'strings' can be traced in other provincial policies (mostly state-specific renewable energy policies) specifying the need and importance of decentralized energy development in the state.

The critical role of policy support in determining the success of a project was strongly emphasised in our interactions with project developers during the field visits. The consensus is that macro-economic policy enunciation must be strengthened with local mechanisms for continuous monitoring and accountability frameworks, specifically for government owned and operated projects. Further, the interactions with stakeholders reveal that regulatory uncertainty and vacuum can be stifling the growth of these projects and that the current paradigm of ad hoc and piecemeal dispensation to regulation produces sub-optimal outcomes. Policies at the local level are perceived as major determinants of project outcomes and often developmental policies integrative of off-grid elements form the policy foundation for many of the off-grid energy systems. The flip side of this is that in-congruencies in the local governance structures act as impediments to project development.

It is generally accepted that the success of off-grid/decentralised energy systems largely depends on the ability of such projects to generate local income-generating opportunities. Our case-based analysis supports this proposition and the field visits made it clear that often people prioritise livelihood opportunities over other energy needs. For instance, in Radhapura, energy requirement for irrigation purposes is given high preference over lighting needs. Even in projects run by private initiatives such as Husk Power Systems operating in Uttar Pradesh and Bihar, there is every effort to create additional sources of income through the project intervention.

Another crucial element which has been prominently highlighted in the literature is to community participation in project activities, which in some cases is expected to culminate with community ownership of the project. In the current paradigm of off-grid systems in India, largely dominated by government funded schemes, community participation has been assigned a significant role in project management. Such a premise, however, seems to be challenged in practice by the reality of capacity constraints among communities, local level conflicts, and 'elite capture'. In our analysis of the 74 case studies, only a fourth of the operational projects are characterised by community participation and the remaining are largely led by private enterprise. Interestingly, the majority (55 %) of the non- and partially operational cases in our database also does not have community participation; hence, it would be erroneous to make any straightforward conclusion regarding the role of communities. From our site visits, we found that in two cases—Radhapura and BERI—there is evidence to suggest that community participation has failed its premise. While in Radhapura societal conflicts have inhibited community participation, the BERI project operational in Karnataka has experienced a shift from community-led management to a professional management body. On the other hand, the success of private initiatives like HPS in Bihar suggests that there are viable alternatives to community participation in off-grid project design.

Unlike community participation, the database gives reason to link more definitively a project's source of funding with the project's long-term viability. For operational projects as well as the category of non- and partially operational projects, government funding has clearly not been fortuitous in the majority of cases. The reason may be found in the soft budget syndrome identified in public

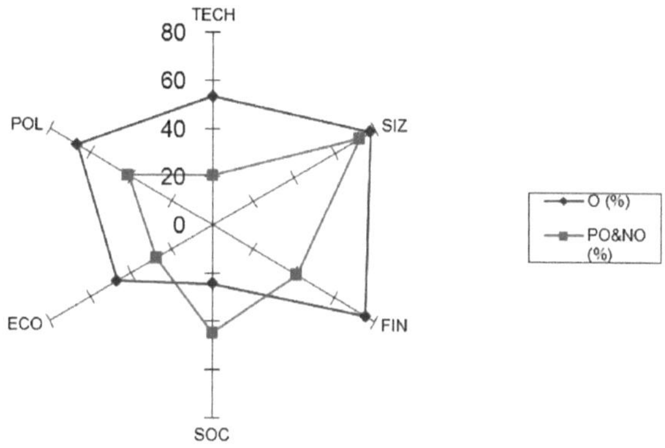

Fig. 5 Comparative analysis of performance of Indian off-grid projects. *Source* this study

finance literature. With guaranteed funding and lax monitoring, government funded projects are said to operate under soft budget constraint conditions, which inevitably leads to weak budgetary discipline by project developers and implementers. Moreover, a majority of public funded projects grossly fail on after sale service and poor maintenance structures.

It can be discerned from Fig. 5 that size of the plant impacts the success of the project. It appears from the analysis that there is a minimum defining size, which can have definite impact on the success of the project. Projects with a mean size more than 20 kW are found to be more viable indicating that very small-sized plants may not be able to fulfil the basic energy needs of a community, especially in a dynamic context when such needs grow and change over time.

Next step in the analysis is to examine the comparative position of both the groups, i.e. operational as well as non- and partially operational in a holistic manner. Figure 5 reveals some interesting insights. Operational projects are found to be dominantly spread across all the variables save community participation. This indicates that projects that are operational are often large sized, with primary source of funding from non-governmental agencies and having strong income-generating and policy links. Comparatively, the performance of non-operational and partially operational cases on the six key variables appears to be skewed with weak policy support, minimal income-generating links and modest community participation.

4 Towards an Integrated Approach for Off-Grid Analysis

The above analysis makes it clear that successful off-grid interventions are based on much more than techno-economic assessment. A meaningful business model would need to factor in the whole set of constituent elements such as choice of technology, scale of the project, type of policy support, role of community, income linkages and funding sources. The integration, moreover, would need to feature in a dynamic framework that is able to account for multiple interacting drivers at different scales of the social, economic and institutional context in which the intervention is planned.

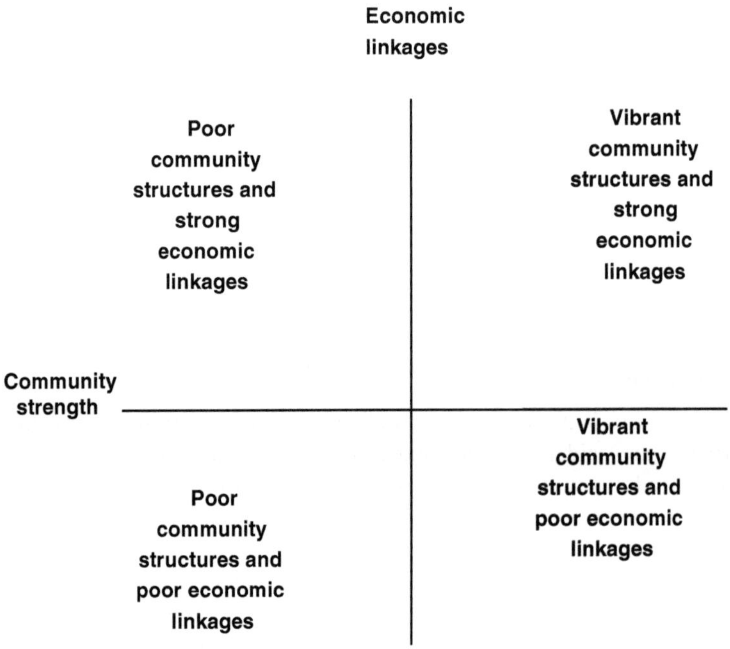

Fig. 6 Possible combinations of community structures and economic linkages

We suggest that the business case for an off-grid intervention should start by looking at the local context. A useful approach may be to distinguish between the set of determinants that are generally context dependent and those which are context neutral. Choice of technology, social acceptance and ownership, and economic linkages are essentially context specific outcomes in the project cycle of an off-grid intervention. At a more macro level and as exogenous influences, we have policy support and institutional financing.

Considering technology choice to be partially neutral and to some extent driven by external determinants, the choice of intervention would primarily be determined by two crucial locally embedded elements, i.e. strength and ability of local community structures and economic linkages. In terms of a conceptual framework, community structures can have two extreme forms, i.e. one deeply cohesive, well organised and having genuine interests to participate in the project operation and management, and the other largely disorganised, fractured, sabotage prone and passive to the project matters. Economic linkages can have similar characterisation with two extreme types: One with easy market access, vibrant local economy, and the other remote and opportunity constrained. Figure 6, with its four quadrants, not only captures the combinations of these extreme varieties of the two context dependent determinants, but also allows us to contextualise in terms of intermediate combinations with varying degrees of heterogeneity. Such contextualisation,

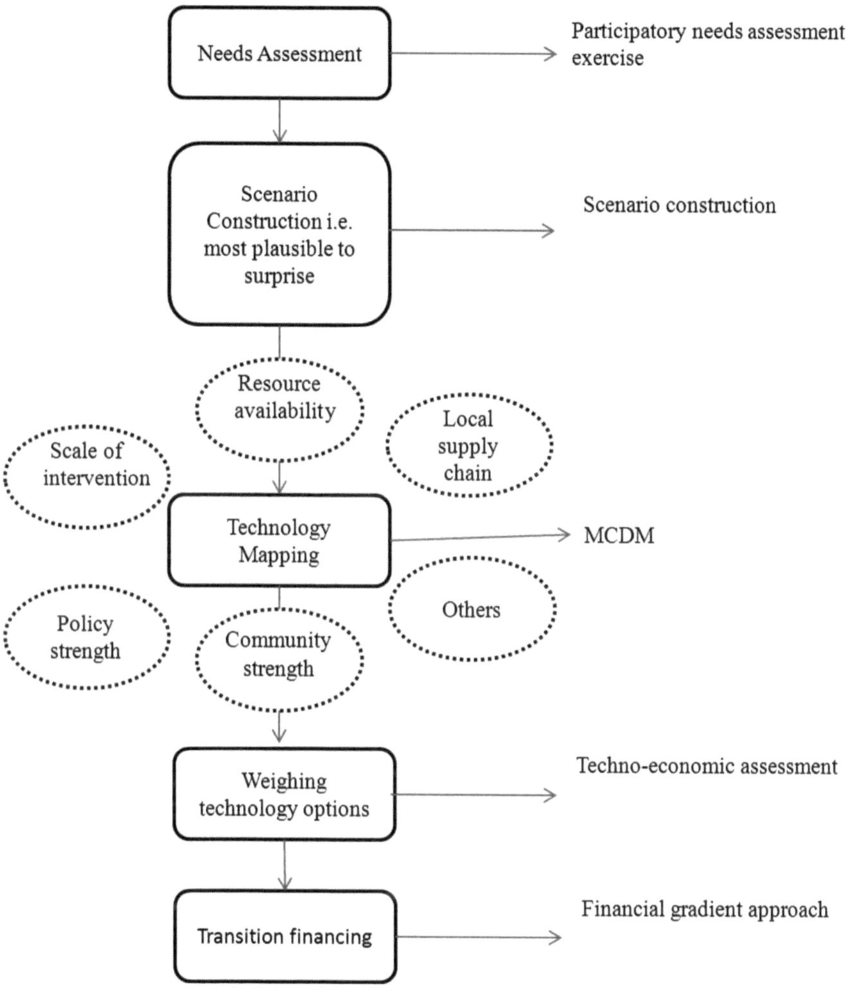

Fig. 7 An integrated approach for off-grid interventions

in turn, establishes the need for a non-uniform and context-relevant approach to decision-making in off-grid energy interventions.

Figure 7 presents a decision hierarchy or 'tree' for sustainable off-grid energy projects. It consists of six essential stages and, importantly, decisions arrived at each stage are fed into the next stage in the process. Given the framework of Fig. 7, one can possibly map the various contextual reference points to a continuum of off-grid energy services ranging from one extreme of that of a 'merit' good to the other extreme of fully marketable services. Once the nature of services is identified, the next step in the hierarchy is about the decision on the scale of project intervention. This can be facilitated through a baseline assessment of the specific context for its energy needs and productive potential; the dynamic character of the context may be introduced through the technique of scenario building and future scenarios may range from the most plausible to a 'surprise'.

Step three in the decision tree is the evaluation of appropriate technology, which would require the application of a multiple criteria technique. The criteria would include resource availability, local supply chain, community capabilities, policy strength, scale of the intervention, and other supporting factors. An appropriate multi-criteria decision tool considering the above identified factors can decide on the appropriateness and relevance of technology type/s with respect to the context and scale. Following the multi-criteria analysis, alternative technology types can be made subject to cost effectiveness analysis to arrive at a final decision on technology selection.

The final stage of decision-making involves issues related to finance and the choice of an appropriate financing mechanism. Here, we would need to refer back to the mapping done in stage one, in which the context defines the nature of off-grid energy services to be provided. It is hypothesized that the choice of finance would be determined by the characterisation of the off-grid energy service. If the context defines the services to be fully marketable, one would expect private investment to be forthcoming given the appropriate policy support. On the other hand, if the context is such that the off-grid energy services are best viewed as a 'merit' good, it would be justified to expect public investment in the project. In between these two extremes, off-grid energy services would require a mixed form of financing which may be linked to varying gradients of transition depending on the dynamism of the context.

5 Conclusions

The review of literature presented in this chapter shows that researchers and practitioners have used various analytical tools and frameworks of different levels of complexity to analyse decentralised mini-grid systems. However, the academic literature often focuses on abstract cases having limited real-life linkages. The practice-oriented literature, on the other hand, relies on financial calculations to a large extent without necessarily taking an integrated perspective.

To overcome such issues, this chapter has presented a field-level analysis that looked into the experience of 74 off-grid electricity projects from India using mini-grids. The desk-based analysis was supplemented by field visits to six project sites. The experience from these projects reveals that operational viability of off-grid projects is largely determined by policy support, social acceptance in the form of community participation, linkages with income-generating opportunities, and technological appropriateness. The key pointers coming out of our analysis are that: (a) macro-economic policy instruments must be aligned with local accountability mechanisms, specifically for government owned and operated projects; (b) at the project level, key choices related to technology selection and scale of intervention need to emerge from an understanding of the context; (c) the context itself is a variable depending on community structure and local economy; and (d) financing of a project may be linked to the nature of energy services defined by the context.

We have proposed an analytical process consisting of six stages to capture the multiple dimensions in the analysis. The framework is quite flexible as it does not prescribe specific tools or choices, and therefore it is suitable for application in a wide range of contexts and for various alternative technologies. Various applications of this integrated approach are also included in the subsequent chapters of the book. Although we are not reporting any tool development based on this framework, this remains an area for further work. Similarly, there is further scope for research to develop real-life case studies using such a framework.

References

1. Akella AK, Sharma MP, Saini RP (2007) Optimum utilisation of renewable energy sources in a remote area. Renew Sustain Energy Rev 11:894–908
2. Alam MS, Bala BK, Huq MZ (1997) Simulation of integrated rural energy system for farming in Bangladesh. Energy 22(6):591–599
3. Alam MS, Islam KK, Huq AMZ (1999) Simulation of rural household energy consumption in Bangladesh. Energy 24(8):743–752
4. Alam MS, Huq MZ, Bala BK (1990) An integrated rural energy model for a village in Bangladesh. Energy 15(2):131–139
5. Ashenayi K, Ramakumar R (1990) IRES—a program to design integrated renewable energy system. IEEE Trans Energy Convers 15(12):1143–1152
6. Banerjee R (2006) Comparison of options for distributed generation in India. Energy Policy 34:101–111
7. Bekele G, Palm B (2010) Feasibility study for a sustainable solar-wind-based hybrid energy system for application in Ethiopia. Appl Energy 87(2):487–495
8. Bernal-Agustin JL, Dufo-Lopez R (2006) Economical and environmental analysis of grid connected photovoltaic systems in Spain. Renewable Energy 31(8):1107–1128
9. Bernal-Agustin JL, Dufo-Lopez R (2009) Simulation and optimization of stand-alone hybrid renewable energy systems. Renew Sustain Energy Rev 13(8):2111–2118
10. Bernal-Agustin JL, Dufo-Lopez R (2009) Multi-objective design and control of hybrid system minimizing costs and unmet load. Electr Power Syst Res 79(1):170–180
11. Bhattacharyya SC (2012) Review of alternative methodologies for analysing off-grid electricity supply. Renew Sustain Energy Rev 16(1):677–694
12. Buchholz TS, Volk TA, Luzadis VA (2007) A participatory systems approach to modelling social, economic and ecological components of bioenergy. Energy Policy 35(12): 6084–6094
13. Cabraal A, Cosgrove-Davies M, Schaffer L (1996) Best practices for photovoltaic household electrification: lessons from experiences in selected countries. World Bank, Washington
14. Cavallaro F (2009) Multi-criteria decision aid to assess concentrated solar thermal technologies. Renewable Energy 34(7):1678–1685
15. CEEP (2001) Off-grid renewable energy options for rural electrification in Western China, center for energy and environment policy. University of Delaware, Delaware

16. CERI (2004) Levelised unit electricity cost comparison of alternate technologies for baseload generation in Ontario. Canadian Energy Research Institute, Calgary, Alberta, Canada
17. Chakraborti S, Chakraborti S (2002) Rural electrification programme with solar energy in remote region—a case study in an island. Energy Policy 30(1):33–42
18. Cherni JA, Dyner I, Henao F, Jaramillo P, Smith R, Font RO (2007) Energy supply for sustainable rural livelihoods: A multi-criteria decision support system. Energy Policy 35(3):1493–1504
19. Clemens E, Raijal K, Takada M (2010) Capacity development for scaling up decentralised energy access programmes: lesson from Nepal on its role, Costs and Financing. A practical action publishing. http://cleancookstoves.org/wp-content/uploads/2011/02/UNDP-Capacity-Development-Energy-Access-Full-Report.pdf
20. Coyle G (1997) Systems dynamics modelling: a practical approach. Chapman and Hall, New York
21. Dalton GJ, Lockington DA, Baldock TE (2008) Feasibility analysis of stand-alone renewable energy supply options for a large hotel. Renewable Energy 33(7):1475–1490
22. Dalton GJ, Lockington DA, Baldock TE (2009) Feasibility analysis of renewable energy supply options for a grid-connected large hotel. Renewable Energy 34(4):955–964
23. Demiroren A, Yilmaz U (2010) Analysis of change in electric energy cost with using renewable energy sources in Gokceada, Turkey: As island example. Renew Sustain Energy Rev 14(1):323–333
24. Devdas V (1997) Rural system model for micro-level development. Energy Sources Part A 19(6):593–611
25. Devdas V (2001) Planning for rural energy system, Part I. Renew Sustain Energy Rev 5(3):203–226
26. Devdas V (2001) Planning for rural energy system, Part II. Renew Sustain Energy Rev 5(3):227–270
27. Devdas V (2001) Planning for rural energy system, Part I. Renew Sustain Energy Rev 5(3):271–297
28. Diakoulaki D, Henggeler Antunes C, Gomes Martins A (2005) MCDA and energy planning. In: Figueira J, Greco S, Ehrgott M (eds) Multiple criteria decision analysis: state of the art surveys. Springer, New York, pp 859–897
29. Diakoulaki D, Karangelis F (2007) Multi-criteria decision analysis and cost-benefit analysis of alternative scenarios for the power generation sector in Greece. Renew Sustain Energy Rev 11(4):716–727
30. Doukas H, Patlitzianas KD, Psarras J (2006) Supporting sustainable electricity technologies in Greece using MCDM. Resour Policy 31(2):129–136
31. Dyer RF, Forman EH (1992) Group decision support with the analytic hierarchy process. Decis Support Syst 8:99–124
32. ESMAP (2001) Best practice manual: promoting decentralized electrification investment. World Bank, Washington
33. ESMAP (2007) Technical and economic assessment of off-grid, mini-grid and grid electrification technologies, ESMAP technical paper 121/07. World Bank, Washington
34. ESMAP (2008) Maximising the productive uses of electricity to increase the impact of rural electrification programmes, ESMAP report 332/08. World Bank, Washington
35. Forrester JW (1961) Industrial dynamics. Pegasus Communications, Waltham
36. Georgopoullu E, Lalas D, Papagiannakis L (1997) A multicriteria decision aid approach for energy planning problems: The case of renewable energy option. Eur JOper Res 103(1):38–54
37. Georgopoulou E, Sarafidis Y, Diakoulaki D (1998) Design and implementation of a group DSS for sustaining renewable energies exploitation. Eur J Oper Res 109(2):483–500
38. Ghafghazi S, Sowlati T, Sokhansanj S, Melin S (2009) A multicriteria approach to evaluate district heating system options. Appl Energy 87(4):1134–1140

39. Golden BL, Wasli EA, Harker PT (1989) The analytic hierarchy process: applications and studies. Springer, New York
40. Goumas M, Lygerou V (2000) An extension of the PROMETHEE method for decision making in fuzzy environment. Ranking of alternative energy exploitation projects. Eur J Oper Res 123:606–613
41. Goumas MG, Lygerou VA, Papayannakis LE (1999) Computational methods for planning and evaluating geothermal energy projects. Energy Policy 27(3):147–154
42. Greening LA, Bernow S (2004) Design of coordinated energy and environmental policies. Use of multicriteria decision making. Energy Policy 32(6):721–735
43. Haralambopoulos DA, Polatidis H (2003) Renewable energy projects: structuring a multi-criteria group decision-making framework. Renewable Energy 28(6):961–973
44. Haurant P, Oberti P, Muselli M (2011) Multicriteria selection aiding related to photovoltaic plants on farming fields on Corsica Island: a real case study using the ELECTRE outranking framework. Energy Policy 39(2):676–688
45. Himri Y, Stambouli AB, Draoui B, Himri S (2008) Techno-economical study of hybrid power system for a remote village in Algeria. Energy 33(7):1128–1136
46. Hiremath RB, Kumar B, Balachandra P, Ravindranath NH, Raghunandan BN (2009) Decentralised renewable energy: scope, relevance and applications in the Indian context. Energy Sustain Dev 13(1):3–9
47. Howells M, Alfstad T, Cross N, Jeftha L, Goldstein G (2002) Rural energy modeling, working paper 11, program on energy and sustainable development, Stanford University, California
48. IEA (1998) Projected costs of generating electricity, 1998 update. International Energy Agency, Paris
49. IEA (2010) Projected costs of generating electricity. International Energy Agency, Paris
50. Kaygusuz K (2011) Energy services and energy poverty for sustainable rural development. Renew Sustain Energy Rev 15(2):936–947
51. Ilskog E (2008) Indicators for assessment of rural electrification—an approach for comparison apples and pears. Energy Policy 36:2665–2673
52. Ilskog E, Kjellstrom B (2008) And they lived sustainable ever after? Assessment of rural electrification cases by means of indicators. Energy Policy 36:2674–2684
53. Iniyan S, Jagadeesan TR (1997) On the development of reliability based optimal renewable energy model for the sustainable energy scene in India. Int J Ambient Energy 18(3):153–164
54. Kablan M (1997) Prioritization of decentralised electricity options available for rural areas in Jordan. Energy Conserv Manag 38(14):1515–1521
55. Kanase-Patil AB, Saini RP, Sharma MP (2009) Integrated renewable energy systems for off-grid rural electrification of remote area. Renew Energy 35(6):1342–1349
56. Karger CR, Hennings W (2009) Sustainability evaluation of decentralised electricity generation. Renew Sustain Energy Rev 13(3):583–593
57. Kaundinya DP, Balachandra P, Ravindranath NH (2009) Grid-connected versus stand-along energy systems for decentralized power—a review of literature. Renew Sustain Energy Rev 13:2041–2050
58. Khan M, Iqbal M (2005) Pre-feasibility study of stand-alone hybrid energy systems for applications in Newfoundland. Renewable Energy 30(6):835–854
59. Kolhe M, Kolhe S, Joshi JC (2002) Economic viability of stand-alone solar photovoltaic system in comparison with diesel powered system in India. Energy Econ 24:155–165
60. Kowalski K, Stagl S, Madlener R, Omann I (2009) Sustainable energy futures: methodological challenges in combining scenarios and participatory multi-criteria analysis. Eur J Oper Res 197(3):1063–1074
61. Kumar A, Sokhansanj S, Flynn PC (2006) Development of a multicriteria assessment model for ranking biomass feedstock collection and transportation systems. Appl Biochem Biotechnol 129(1–3):71–87

62. Lau KY, Yousof MFM, Arshad SNM, Anwari M, Yatim AHM (2010) Performance analysis of hybrid photovoltaic/diesel energy system under Malaysian conditions. Energy 35(8):3245–3255
63. Lhendup T (2008) Rural electrification in Bhutan and a methodology for evaluation of distributed generation system as an alternative option for rural electrification. Energy Sustain Dev 12(3):13–24
64. Limmeechokchai B, Chawana S (2007) Sustainable energy development strategies in the rural Thailand: The case of the improved cooking stoves and the small biogas digester. Renew Sustain Energy Rev 11(5):818–837
65. Løken E (2007) Use of multicriteria decision analysis methods for energy planning problems. Renew Sustain Energy Rev 11(7):1584–1595
66. Madlener R, Stagl S. (2005) Sustainability-guided promotion of renewable electricity generation. Energy Economics 53(2):147–167
67. Madlener R, Kowalski K, Stagl S (2007) New ways for the integrated appraisal of national energy scenarios: the case of renewable energy use in Austria. Energy Policy 35(12):6060–6074
68. Mladineo N, Margeta J, Brans JP, Mareschal B (1987) Multicriteria ranking of alternative locations for small scale hydro plants. Eur J Oper Res 31(2):215–222
69. Nandi S, Ghosh HR (2010) Prospect of wind-PV-battery hybrid system as an alternative to grid extension in Bangladesh. Energy 35(7):3040–3047
70. Nfah EM, Ngundam JM, Vandenbergh M, Schmid J (2008) Simulation of off-grid generation options for remote villages in Cameroon. Renew Energy 33(5):1064–1072
71. Nouni MR, Mullick SC, Kandpal TC (2008) Providing electricity access to remote areas in India: an approach towards identifying potential areas for decentralized electricity supply. Renew Sustain Energy Rev 12:1187–1220
72. Nouni MR, Mullick SC, Kandpal TC (2006) Photovoltaic projects for decentralized power supply in India: a financial evaluation. Energy Policy 34:3727–3738
73. Nouni MR, Mullick SC, Kandpal TC (2006) Biomass gasifier projects for decentralized power supply in India: a financial evaluation. Energy Policy 35:373–385
74. Nouni MR, Mullick SC, Kandpal TC (2006) Techno-economics of small wind electric generator projects for decentralized power supply in India. Energy Policy 35:2491–2506
75. Oberschmidt J, Geldermann J, Ludwig J, Schmehl M (2010) Modified PROMETHE approach to assessing energy technologies. Int J Energy Sect Manage 4(2):183–212
76. Parikh J (1985) Modeling energy and agricultural interactions 1: a rural energy systems model. Energy 10(7):793–804
77. Parikh JK, Ramanathan R (1999) Linkages among energy, agriculture and environment in rural India. Energy Econ 21(6):561–585
78. Pohekar SD, Ramachandran M (2004) Application of multi-criteria decision making to sustainable energy planning—a review. Renew Sustain Energy Rev 8(4):365–381
79. Pokharel S, Chandrashekar M (1998) A multi-objective approach to rural energy policy analysis. Energy 23(4):325–336
80. Polatidis H, Haralambopolous DA (2008) Renewable energy sources planning and design: a multi-criteria approach, chapter 11. In: van den Bergh JCJM, Bruinsma FR (2008) Managing the transition to renewable energy: theory and practice from local, regional and macro perspectives, Edward Elgar, Cheltenham
81. Polatidis H, Haralambopoulos DA (2007) Renewable energy systems: a societal and technological platform. Renewable Energy 32(2):329–341
82. Prodromidis GN, Coutelieris FA (2011) A comparative feasibility study of stand-alone and grid-connected RES-based systems in several Greek Islands. Renew Energy 36(7):1957–1963
83. Qudrat-Ullah H (2007) Modelling for policy assessment in the electricity sector of Pakistan. Int J Energy Sect Manage 1(3):240–256
84. Ramakumar R, Abouzhar I, Krishnana K, Ashenayi K (1994) Design scenarios for integrated renewable energy systems. IEEE Trans Energy Conserv 10(4):736–746

85. Ramakumar R, Shetty PS, Ashenayi K (1986) A linear programming approach to the design of integrated renewable energy systems for developing countries. IEEE Trans Energy Convers 194: 18–24
86. Reddy AKN, Sumithra GD, Balachandra P, D'sa A (1990) Comparative costs of electricity conservation, centralized and decentralized electricity generation. Econ Polit Wkly 25(2):1201–1216
87. Ren H, Gao W, Zhou W, Nakagami K (2009) Multi-criteria evaluation for the optimal adoption of distributed residential energy systems in Japan. Energy Policy 37(12):5484–5493
88. Royal Academy of Engineering (2004) Cost of generating electricity. The Royal Academy of Engineering, London
89. Saaty TL (1980) The analytic hierarchy process. McGraw Hill, New York
90. Saaty TL (1982) Decision making for leaders. Lifetime Learning, New York
91. Saaty TL (1983) Priority setting in complex problems. IEEE Trans Eng Manage EM-30:140–155
92. Sen R, Bhattacharyya SC (2014) Off-grid electricity generation with renewable energy technologies in India: an application of HOMER. Renew Energy 62:388–398
93. Setiawan AA, Zhao Y, Nayar CM (2009) Design, economic analysis and environmental considerations of mini-grid hybrid power system with reverse osmosis desalination plant for remote areas. Renew Energy 34(2):374–383
94. Shaahid SM, Elhadidy MA (2007) Technical and economic assessment of grid-independent hybrid photovoltaic-diesel-battery power systems for commercial loads in desert environments. Renew Sustain Energy Rev 11(8):1794–1810
95. Shaahid SM, Elhadidy MA (2008) Economic analysis of hybrid photovoltaic-diesel-battery power systems for residential loads in hot regions—a step to clean future. Renew Sustain Energy Rev 12(2):488–503
96. Shaahid SM, El-Amin I (2009) Techno-economic evaluation of off-grid hybrid photovoltaic-diesel-battery power systems for rural electrification in Saudi Arabia—a way forward for sustainable development. Renew Sustain Energy Rev 13(3):625–633
97. Sinha CS, Kandpal TC (1991) Optimal mix of technologies for rural India: the irrigation sector. Int J Energy Res 15:331–346
98. Sinha CS, Kandpal TC (1991) Optimal mix of technologies for rural India: the lighting sector. Int J Energy Res 15:653–665
99. Sinha CS, Kandpal TC (1991) Optimal mix of technologies for rural India: the cooking sector. Int J Energy Res 15:85–100
100. Sinha CS, Kandpal TC (1991) Decentralised versus grid electricity for rural India. Energy Policy 19(5):441–448
101. Smil V (1993) Global ecology: environmental change and social flexibility. Routledge, London
102. Corporation SunPower (2008) The drivers of the levelised cost of electricity for utility-scale photovoltatics. SunPower Corporation, USA
103. Terrados J, Almonacid G, Pqrez-Higueras P (2009) Proposal for a combined methodology for renewable energy planning. Application to a Spanish region. Renew Sustain Energy Rev 13(8):2022–2030
104. Topcu YI, Ulengin F (2004) Energy for the future: an integrated decision aid for the case of Turkey. Energy 29(1):137–154
105. Tsoutsos T, Drandaki M, Frantzeskaki N, Iosifidis E, Kiosses I (2009) Sustainable energy planning by using multi-criteria analysis application in the Island of Crete. Energy Policy 37(5):1587–1600
106. Turkey BE, Telli AY (2011) Economics analysis of standalone and grid-connected hybrid energy systems. Renew Energy 36(7):1931–1943
107. Tzeng G-H, Shiau T-A, Lin C-Y (1992) Application of multicriteria decision making to the evaluation of new energy system development in Taiwan. Energy 17(10):983–992

108. Vargas LG (1990) An overview of the analytic hierarchy process and its applications. Europe J Oper Res 48(1):2–8
109. Wang B, Kocaoglu DF, Diam TU, Yang J (2010) A decision model for energy resource selection in China. Energy Policy 38(1):7130–7141
110. Wang JJ, Jing YY, Zhang CF, Zhao JH (2009) Review on multi-criteria decision analysis aid in sustainable energy decision-making. Renew Sustain Energy Rev 13(9):2263–2278
111. Wang X, Feng Z (2002) Sustainable development of rural energy and its appraising system in China. Renew Sustain Energy Rev 6(4):395–404
112. Weis TM, Ilinca A (2008) The utility of energy storage to improve the economics of wind-diesel power plants in Canada. Renew Energy 33(7):1544–1557
113. World Bank (2008) Designing sustainable Off-Grid rural electrification projects: principles and practices. The World Bank, Washington
114. World Bank (2011) India: biomass for sustainable development: lessons for decentralised energy delivery. A report from world bank India country management unit. World Bank, South Asia Region
115. Xiaohua W, Yunrong H, Xiaqing D, Yuedong Z (2006) Analysis and simulation on rural energy-economy on Shouyang county in China. Renew Sustain Energy Rev 10(2):139–151
116. Zangeneh A, Jadid S, Rahimi-Kian A (2009) A hierarchical decision-making model for the prioritisation of distributed generation technologies: a case study for Iran. Energy Policy 37(12):5752–5763

Demand Management for Off-Grid Electricity Networks

P. J. Boait

Abstract For isolated mini-grids powered mainly or wholly by renewable energy-based generation, balancing electricity supply and demand is a considerable challenge. This chapter reviews the scope and practical methods for active demand management as part of the solution to this problem. It highlights the wide range of possible methods and their applicability and limitations for different mini-grid technologies and operating environments. The need for suitable tools to design and plan demand management schemes is identified.

1 Introduction

One of the most prominent benefits from innovation and cost reduction in renewable energy-based generation technologies has been the ability to provide electricity to communities, typically in rural or island locations that do not have access to a national or regional electricity distribution grid. However, providing an affordable and reliable service without the efficiencies of scale available to a large network is a considerable engineering challenge. Any system for generation and distribution of electricity must continuously balance electricity supply and demand. It can do this in only three ways whether the system is a national grid or a small-scale off-grid network. The options are:

- Varying the output of the generators to match demand. This is practical where there is a prime mover such as a biogas engine driving the generator but less so for a photovoltaic panel or wind turbine.

P. J. Boait (✉)
Institute of Energy and Sustainable Development, De Montfort University, Leicester, UK
e-mail: p.boait@dmu.ac.uk

S. C. Bhattacharyya and D. Palit (eds.), *Mini-Grids for Rural Electrification of Developing Countries*, Green Energy and Technology,
DOI: 10.1007/978-3-319-04816-1_6,

- Drawing on a store of electricity such as a battery bank. This works well but battery capacity is expensive and the charge state must be carefully managed to ensure a reasonable working life.
- Adjusting demand to match available supply, for example by dropping low priority loads or by scheduling and limiting demand rather than allowing it to arise unconstrained.

In practice, the first two options are similarly dependent on storage—in the first case it is the gas holder or wood pile that is providing the energy store allowing flexibility in the generator output. It is the third option that is the subject of this chapter. It is particularly necessary where generator output is partly or wholly dependent on the sun or wind. Then the ability to dynamically manage demand can allow the supply of electricity to be sustained to a larger number of consumers, or provided more reliably, than by battery or generator back-up alone.

The theoretical constraints, feasibility and merits of a range of demand management techniques for small 'off grid' electricity supply networks are discussed in the next section. We conclude that there is a need for simulation techniques to assist mini-grid design and planning. Then practical experience of demand management techniques is reviewed and possible lines of further development discussed, with particular emphasis on the different issues faced by power-limited and storage-limited systems.

2 Demand Management Methods

The supply of electricity to a consumer only has three dimensions which can be varied to adjust their demand—voltage, current and time. Voltage can be quickly dismissed as a useful variable in this context because a safe electricity supply system has to maintain voltage within narrow (usually statutory) limits. Even voltage reduction is not usable as a demand management method because its effect on consuming appliances is unpredictable—typically they will either stop working at some unknown threshold, or act as constant power devices and increase the amount of current drawn. As examples, fluorescent lights fall into the former category and 'black plug' DC power converters the latter. Time (in the sense of scheduling in some way when demand occurs) and current are therefore the effective variables that can be subject to management. However, voltage can be useful as a signal—a drop in voltage may be used to detect a generator overload condition and trigger a demand management action.

Time dependent solutions are well established in national scale systems, such as the UK's Economy 7 scheme which offers a reduced tariff for a 7-h period overnight. The history of this scheme illustrates a basic problem with price signals intended to influence the timing of demand—when first introduced there was an excessive peak in demand at the start of the cheap period. To suppress this peak a system known as Radio Teleswitch [11] had to be introduced. This sends a signal hidden in a BBC radio broadcast that controls and spreads the switch-on time of thermal storage appliances and water heating. Subsequent trials and research around the world [7, 12] have shown that a price signal on its own is unsatisfactory as a demand management measure because of the tendency to induce peaks at times with the most attractive combination of price and convenience.

The simple solution to this problem is for the system operator to physically control the timing of demand, as is achieved both by Radio Teleswitch and by mini-grid operators such as Husk Power Systems [6] who use a timed meter to limit demand from their poorest consumers to an evening time window. However, such control is unattractive to consumers who, unless they have no choice, prefer to be able to make their own decisions. A way to make time-dependent demand management more acceptable is to separate out loads according to the impact of supply interruption. Appliances performing heating or cooling can often accept an interruption to supply or scheduling of supply with no degradation of the service provided to the user, particularly if there is thermal mass (such as building fabric or a hot water tank) associated with the heating or cooling system that ensures temperatures will only change slowly. Similarly where an electricity supply is provided for charging the consumer's own batteries, such as in mobile phones or laptops, the supply is potentially interruptible. If at the same time, a reliable supply is maintained to lighting and entertainment loads then consumer needs and expectations can be satisfied alongside a substantial level of demand management.

Turning to current flow as a demand management parameter, all electricity distribution systems limit current flow (and consequently power) to each consumer using fuses or trips as a safety measure. Tripping at a contracted power level is often used as a crude demand management method for off-grid systems, for example by the rice husk system operators and on the Scottish island of Eigg where consumers are penalised with a 'fine' of £25 to reset their trips if demand exceeds 5 kW for domestic consumers and 10 kW for commercial use [3]. The other simple way for the supplier to regulate current is to provide all the consuming appliances—in effect provide energy services rather than electric power. There are advantages to this in that the supplier is incentivised to choose efficient devices such as LED lighting, but for a mini-grid to meet the aspirations of its customers and stimulate electricity-using enterprise the option for the consumer to choose their own equipment must be available.

A critical limitation on operators of small electricity distribution systems arises from the mathematical properties of an aggregation of consuming appliances whose demand is variable or intermittent. The central limit theorem states that the means of n random samples drawn from any distribution with mean m and standard deviation σ will have an approximately normal distribution with a mean equal to m and a standard deviation equal to $\sigma/\sqrt{n}$.

This implies that as the number of such appliances n served by a grid increases, the variability (standard deviation) of their total electricity consumption will decrease by a factor of $1/\sqrt{n}$. Figure 1 shows the radical effect of this in practice. It shows the results of multiple simulations of a number of refrigerators or freezers running with a 20 % compressor duty cycle over a year, with randomisation of

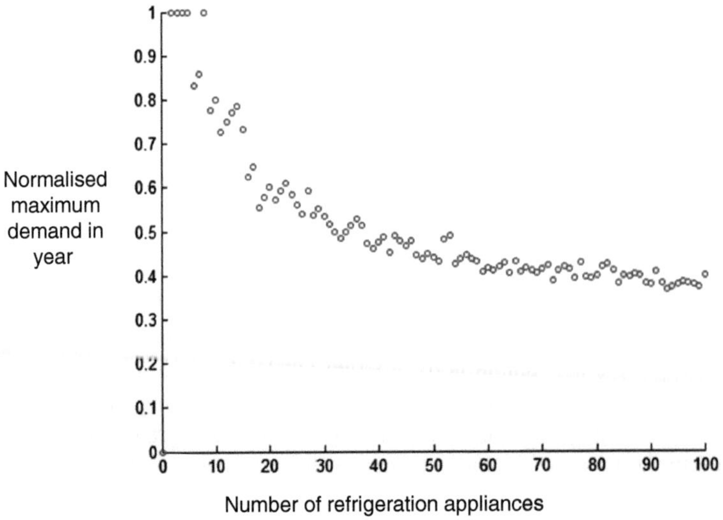

Fig. 1 Effect of number of consuming appliances on observed peak demand as a proportion of maximum possible demand (*source* this study)

their relative operation as would occur in practice. Each point shows the maximum load presented at any time during a simulated year of operation by that number of refrigerators, expressed as a fraction of the total load that would occur if all their compressors operated simultaneously. The much higher proportionate load presented for numbers below 20 is clear. What this means in practice is that the current tripping limits for each consumer on a small network have to be set much tighter relative to average consumption than for a network operating on a regional or national scale. Consequently, given fixed limits for each consumer, there is less opportunity for capacity that is not being used by one consumer to be picked up by another.

For systems where the critical constraint is battery capacity and management of the state of charge, the demand management problem is slightly different. This typically applies to PV or wind powered systems. In this case, there is usually less difficulty in meeting short-term demand peaks such as the coincident freezers discussed earlier because it is not expensive to size the inverters supplying the distribution network generously. The problem is ensuring the battery does not discharge too deeply on days when the collection of renewable energy is low. This translates into placing a daily limit on the total energy that each consumer can draw. The stringency required for this limit is dependent on the size of the consumer population in exactly the same way as current limiting for a system constrained by generator capacity.

Thus the operator of any mini-grid system, whether generator or battery limited, has to make a careful and difficult judgement on allocation of power and/or energy limits which balances the risk of brownouts or outages against the need to make efficient use of his capital investment. The ratio between the total possible connected load and the load that a system component is expected to operate at or can support is referred to as a diversity factor. There are many resources that an engineer working with a large-scale grid system can draw on to decide this, but for mini-grids each will present a unique case. Simulation tools which include the kind of calculation shown in Fig. 1 could assist in making this judgement particularly prior to initial deployment of a system.

3 Flexible Demand Management for Mini-Grids

The preceding discussion leads to the conclusion that for off-grid networks at the scale of large or multiple villages that aim to serve a variety of commercial as well as domestic consumers there is a need for flexible demand management methods that can mitigate the inherent mathematical constraints of a small network. In principle, this could be achieved by continuously adjusting the allocation of available system capacity to the consumer population in a way that best reconciles the interests of the operator and the community.

With power limits varying during the day, it is clear that occasions when consumers exceed their allowed load and cause a trip will happen much more often than with fixed power limits. It will therefore be more accommodating to consumers if supply is restored automatically when demand falls below a lower threshold. There are safety issues in this form of 'soft tripping' mechanism, clearly a high current fault must be detected and the supply must remain tripped until reset manually. Automatic restoration should only apply if the correct limit is modestly exceeded. If combined with a simple 'traffic light' or 'speedometer' display that indicated when demand was approaching the tripping point this should give consumers the ability to manage their own demand in response to time varying limits.

Figure 2 shows a 'smart' adjustable tripping profile that is possible with an electronic circuit breaker. Essentially, it has three current thresholds, which cause a trip to occur in successively shorter timescales as the current rises as a multiple of the nominal trip level. The width of the vertical ribbon reflects the variation in actual trip points due to manufacturing tolerances—this would have to be taken into account in the allocation of capacity to consumers. The first of these thresholds (at the top of the diagram) where the trip operates close to its rating, would be suitable for automatic reconnection if exceeded, while the thresholds at 5 times and 10 times the rating (and the intermediate values) would be subject to manual resetting that would only take effect if the current drawn was below the nominal trip level.

For battery-constrained systems, it makes more sense to widen current tripping limits to be consistent with the peak power capacity of the system as determined by the inverters or batteries and include a daily energy limit above which tripping occurs. The energy limit should be capable of adjustment depending on the availability of sun or wind as appropriate and to allow flexible re-allocation of limits between consumers.

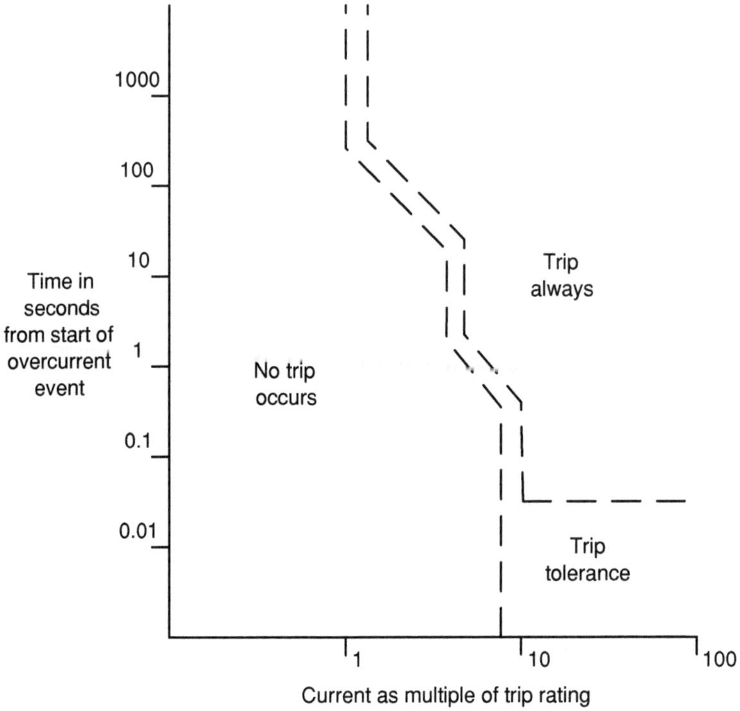

Fig. 2 Typical electronic trip characteristics (*source* this study)

4 Signalling Methods

To make current or energy limits dynamic, some form of signalling is required which communicates the need to vary tripping constraints at the point of use. Since user action may be needed to respond to the change in constraints, typically some form of visual signalling to the consumer is also needed. There are two main approaches to this:

(a) Detection of the need for change from one of the parameters of the electricity supply itself—voltage or frequency are the practical options.
(b) A central apparatus able to analyse, or be informed of, the system state transmits control signals to the sub systems at the points of consumption.

Use of voltage or frequency trips to initiate load shedding is one of the traditional methods of demand management under stress for large-scale grids. These methods are capable of refinement for a mini-grid but they do have limitations. In the case of voltage signalling, the problem is that the voltage variation seen by a given consumer depends on their distance from the power source and the load presented by their neighbours as well as the actual voltage presented to the network by the generation system. This makes the setting of voltage trip thresholds quite difficult, and it is only possible on a system capable of operating with a wide voltage range so that overload conditions can be reliably discriminated from low voltages arising from location at the end of a feeder. If voltage deviation is limited by regulation to a few percent either side of a nominal level then this method is not capable of delivering efficient demand management on a mini-grid.

Frequency tripping avoids the location sensitivity that arises with voltage as a signal and can be implemented with low cost electronics embedded in appliances particularly refrigeration [8]. However, there are risks particularly for small networks of emergent unstable conditions where frequency drops due to overload, appliances automatically disconnect, the frequency rises again allowing reconnection, and the cycle repeats. Kremers et al. [5] shows how this can arise dependent on parameters such as the penetration of frequency-sensitive devices. Similar instability can occur with voltage sensing. A simple method of mitigating instability risks is to require a randomly assigned timeout between recovery of frequency or voltage into their normal range and reconnection of an appliance. This is employed by the Gridshare device described in the following sections.

Centralised demand management avoids these instability risks by ensuring that permitted demand can be supplied with a known reliability depending on the risk/return balance chosen by the operator when calculating the diversity factors to use. The disadvantage is that a telecommunications medium is required to collect data on the network state at the centre and distribute decisions. Where the business model requires a data communications path to allow service to be enabled at the consumer's connection point when pre-payment is made, then there is a natural synergy with centralised management. Mini-grid concepts following this approach include SharedSolar [13] and Powerhive [9].

5 Examples of Mini-Grid Demand Management

In response to the constraints and opportunities discussed above, a number of research teams and vendors have developed products and systems that aim to deliver 'smart' demand management for mini-grids, although for reasons of commercial confidentiality not all are explicit about the techniques used. One which is well documented and illustrates a solution for a generator-limited system is the GridShare devised by Quetchenbach et al. [10]. The GridShare concept relies on detection of voltage drop at the consumer's connection point to indicate when a generator is close to being overloaded. Its first deployment is on a mini-grid in Bhutan serving 90 households with a 40 kW micro hydro generator. The critical appliances on this grid are rice cookers each requiring about 600 W. Before the introduction of GridShare, simultaneous operation of these cookers in many households prior to mealtimes caused brownouts (defined in this case as a voltage drop below 190 V) with the effect of prolonging cooking times unacceptably and degrading the operation of other appliances.

The GridShare device provides a 'traffic light' display to users, indicating normal operation of the grid with a green light and potential overload with a red light. When the red light is ON, if users switch on a rice cooker or similar appliance a trip disconnects the supply. If this load is disconnected, the supply is automatically restored after a random delay to enable lighting and similar low level loads to operate. If users have a rice cooker in operation when the grid status changes from green to red, their cooker load is allowed to continue for 1 h or until it is turned off. Thereafter the constraint to light loads applies. This mechanism and its potential benefits were thoroughly explained to all households. As a result, the duration of brownouts dropped from an average of 45 min per day to 8 min, and there was a general improvement in satisfaction with the service. Interestingly, at the aggregate level, there was little change in the timing of demand, as maintaining a higher voltage enabled rice cooking to proceed more quickly in each household resulting in more effective sharing of the generator resource.

The use of flexible daily energy limits for energy-limited systems has been successfully implemented by Graillot et al. [4] in the form of the Circutor Energy Dispenser [2]. This device is installed in place of a conventional meter and circuit breaker at the supply connection point. Each consumer has a contract captured on a contactless smart card which specifies the number n of kWh per day that may be drawn. This energy budget is entered into an account within the dispenser device using the smart card, but not as a single daily amount. The budget is translated into an equivalent power flow of n/24 kW which is credited continuously to the user's energy account. Any usage of electricity is debited from this account. If the available credit within the meter is exhausted, the supply is disconnected. If the user then reduces their load below the rate at which credit is being accumulated the supply can be restored. If a user consumes less than the daily rate, they can accumulate up to 3 days credit and use that to meet exceptional loads.

A signal from the central management system can increase the credit rate when the system batteries are fully charged and there is ample power available from the PV generator. A display informs the user allowing them to use this additional energy at no extra cost. Conversely if poor weather has caused the batteries to discharge below a threshold a restricted credit rate is signalled. Credit can also be transferred between meters using the smart card, which is the basis of the payment system. This allows community events in shared facilities to be donated an electricity budget from the meters of the consumers taking part.

This technology has been installed in remote PV and diesel hybrid mini-grids in Spain, Morocco, Senegal and the Galapagos islands. The 'Urja Bandu' (Energy Friend) deployed by CAT [1] in India and Australia has similar characteristics to the Circutor but without the ability to accept remote adjustment of the energy budget.

6 Conclusions

This review of possible demand management techniques, and actual mini-grid implementations of those techniques, has shown that practically every possible option has been tested somewhere in the world with generally useful results but it is difficult to be confident that a particular solution is optimum. Probably the only clear-cut design guidance that can be given is that dynamic current limits such as those provided by an electronic trip are appropriate for generator-limited systems while an energy budgeting scheme is appropriate for battery-limited systems.

The severe implications of the central limit theorem on the statistical properties of demand for small networks imply that simulation should have a role in mini-grid design allowing demand management options to be compared and optimised energy or power budgets constructed for potential consumers. The development of suitable easy-to-use design and planning tools alongside technology for dynamic management of power and energy limits will help operators and consumers obtain maximum value from the investment in each mini-grid.

References

1. CAT Projects (2011) Bushlight India Project. http://www.catprojects.com.au/bushlight-india-hardware#.UsAaLmfuPcs. Accessed 29 Dec 2013
2. Circutor (2013) Electricity Dispenser B II. http://pdf.directindustry.com/pdf/circutor/microgrids-self-generation-renewable-energies/11644-242213-_11.html. Accessed 23 Jan 2014
3. Eigg Electric (2013) Island going green. http://islandsgoinggreen.org/about/eigg-electric/. Accessed 23 Jan 2014
4. Graillot A, Briganti M, Solano-Peralta M, Vallvé X (2012) 15 years of field experience with the "daily energy allowance" concept as the basis for load control and guide for social behaviour in rural micro grids. In: Proceedings of 6th European PV-Hybrid and Mini-Grid conference
5. Kremers E, González de Durana JM, Barambones O (2013) Emergent synchronisation properties of a refrigerator demand side management system. Appl Energy 101:709–717
6. Krithika PR, Palit D (2013) Participatory Business models for off-grid electrification, Chapter 8. In: Bhattacharyya S (ed) Rural electrification through decentralised off-grid systems in developing countries. Springer, London
7. Mohensian-Rad A, Leon-Garcia A (2010) Optimal residential load control with price prediction in real-time electricity pricing environments. IEEE Trans Smart Grid 1(2):120–132
8. Open Energi (2013) Open dynamic demand. http://www.openenergi.com/our-technology/dynamic-demand/. Accessed 23 Jan 2014
9. Powerhive (2013) Cloud-based management software. http://powerhive.com/technology/. Accessed 23 Jan 2014
10. Quetchenbach TG, Harper MJ, Robinson J, Hervin, KK, Chase, NA, Dorji C, Jacobson AE (2013) The GridShare solution: a smart grid approach to improve service provision on a renewable energy mini-grid in Bhutan. Environ Res Lett 8:014018 (11 pp)

11. Radio Teleswitch (2013) Radio teleswitch services. http://79.171.36.154/rts/index.asp. Accessed 23 Jan 2014
12. Rastegar M, Fotuhi-Firuzabad M, Aminifar F (2012) Load commitment in a smart home. Appl Energy 96:45–54
13. SharedSolar (2013) SharedSolar solution. http://sharedsolar.org/solution/. Accessed 23 Jan 2014

Business Issues for Mini-Grid-Based Electrification in Developing Countries

Subhes C. Bhattacharyya

Abstract The off-grid electrification business in general and mini-grids in particular face investment as well as commercial challenges arising from, among others, locational disadvantage, limited customer base, poor paying capacity and ambiguous business environment. This chapter provides an overview of the business environment and discusses the financial and economic viability of projects. It then presents examples of alternative business models used for delivering such mini-grids. This chapter suggests that project design (i.e. sizing, technology choice and standardised delivery model) and appropriate utilisation of plant play a crucial role. Careful funding and risk mitigation finally improves the long-term viability of the project.

1 Introduction

Providing electricity services to rural areas through decentralised mini-grid systems is not an easy business, as it not only involves specialised activities of electricity production, distribution and supply (discussed in "Technical Aspects of Mini-Grids for Rural Electrification"), but also requires cautious investment decisions, careful business planning and watchful project management. This is so because the usual challenges of investment and business risks are compounded by additional barriers such as locational disadvantage (rural areas in remote parts of a region), limited consumer base, weak paying capacity, poor business environment and skill shortages. Moreover, the delivery business depends on the nature of the service provider, whereby the ownership and organisational arrangement of the business significantly influence the intervention. Despite the

S. C. Bhattacharyya (✉)
Institute of Energy and Sustainable Development, De Montfort University, Leicester LE1 9BH, UK
e-mail: subhesb@dmu.ac.uk; subhes_bhattacharyya@yahoo.com

S. C. Bhattacharyya and D. Palit (eds.), *Mini-Grids for Rural Electrification of Developing Countries*, Green Energy and Technology,
DOI: 10.1007/978-3-319-04816-1_7,

demanding business conditions, there are innovative business activities taking place in mini-utility-based rural electricity supply.

This chapter provides an overview of the business challenges and suggests the ways of addressing them. It also considers alternative business delivery options and provides examples of alternative ways of organising such activities.

The chapter is organised as follows. The second section presents the business environment for mini-grids. The following section identifies the main challenges and suggests options to address them. The fourth section discusses alternative business models and their characteristics. Finally, suggestions for improvement and innovation are provided in the concluding section.

Different players are involved in the rural electricity supply, with different motives, perspectives and therefore catering to different market segments. Their outlook requires them to consider different issues and challenges. For example, a local community interested in electrifying their area through a one-time intervention will have a different set of considerations than a profit-oriented social organisation interested in rural electricity supply as a business proposition. Given the wide range of possibilities, it is not possible to outline the entire range of concerns in this chapter. However, important differences in the considerations will be noted in the chapter.

2 Business Environment for Rural Mini-Grids

The potential for developing any local mini-grids for electrification exists in areas where the grid has not reached so far and is unlikely to reach in the near future. Depending on the status of grid coverage of a country, such non-electrified areas can be found in urban/peri-urban areas[1] as well as in rural areas. Generally, the rural areas are likely to have greater opportunities for mini-grids. However, depending on the level of electrification of a country, such opportunities may exist in a variety of locations. Even in countries with a good level of rural electrification, remote areas with difficult geographical conditions (hilly terrains, forest areas) as well as hamlets which are not recognised as census villages will offer an opportunity for the mini-grids. In other countries where electrification has not progressed much, opportunities may be more widely available.

Traditionally, the business environment is analysed at different levels considering the micro (or local) and the macro-environment. Accordingly, the local business environment for a mini-grid will depend on its specific location.

[1] According to IEA (2013), more than 9 % of urban population in developing countries lack access to electricity. Although 95 % of the urban Asia has access to electricity, the rate falls to only 65 % for urban Africa.

(a) Urban/peri-urban areas or locations close to urban centres are likely to offer advantages in terms of consumer mix, consumer base, demand profile, ability to pay, local infrastructure (communication link, transport facility), access to energy resources and availability of human capital. A larger consumer base and the likelihood of existence of commercial activities would be attractive to any potential investor. Some form of informal electricity supply may also exist in such cases, which may provide a ready market for a mini-grid business. Availability of resources and physical infrastructure influences the project positively, ensuring low-cost delivery of project outputs. Similarly, a better consumer mix with a larger share of middle class consumers offer better project viability through a relatively lower cost sensitiveness. However, the proximity to urban centres also carries a higher risk of grid extension, which when materialises can jeopardise the cost recovery of the investment within the project lifetime. This uncertainty acts as a major threat for any business development (see below for further discussion on this issue). Given the widely varying rates of electrification across countries, existence of peri-urban areas and smaller towns without electricity cannot be overruled.
(b) In rural locations on the other hand, unfavourable local conditions in terms of poor local infrastructure, limited consumer base, adverse consumer mix, limited paying capacity and limited availability of human capital are likely to prevail. Given the heterogeneity of rural areas within a country and across countries, each case needs to be considered separately. However, a few general points can be noted from the electrification perspective. Business attractiveness will be better if any location offers a special advantage, which can arise due to its resource richness (a stream of water, a hot water source, windy area, a mining base or availability of any specific biomass resource), a strategic location or some other form of advantage that makes it stand out.[2] Similarly, rural areas with a significant concentration of population and having a favourable income mix would also stand a better chance of electrification. These preferred areas are likely to be electrified earlier than the rest. The dispersed settlements, on the other hand, can itself make local electricity distribution a challenging task, as carrying small amount of electricity at low voltages over long distances is not a cost-effective proposition. The remoteness of the settlements as well as their relatively small size (in population and consumer terms) acts as a hindrance. The problem is further compounded if the population has a poor paying capacity that may arise, among others, from dependence on subsistence activities, limited livelihood opportunities and involvement in non-monetary transactions. However, the challenges found in rural areas offer opportunities for mini-grid-based electrification, as such areas may remain outside electricity map otherwise.

[2] In other words, if the location offers an opportunity for differential rent in Ricardian sense, it is likely to generate greater interest from investors.

The local business environment will be quite familiar to a local community or a local business person or an NGO interested in mini-grid-based electrification. The local aspirations for electricity use for local development will inform technology selection and system design (as explained in "Smart Design of Stand-Alone Solar PV System for Off Grid Electrification Projects" for solar PV) in this case but the main issue is how the local aspirational design can be transformed into a local supply overcoming the constraints that prevent realisation of such investments. Thus the focus of one-off type of initiatives will revolve around the project design and implementation issues. On the other hand, local conditions will be critical for the viability of a project being considered by any outside investor and accordingly, favourable locational attributes will positively influence investment decision. Following Porter's five forces analysis [18] in such a case, the threat of substitutes or alternatives to mini-grid supply (such as from solar home systems and fossil fuels), the bargaining power of buyers (i.e. the potential for locking-in buyers to the new supplier or buyers' ability to influence the price), the potential for competition from other suppliers (e.g. existing diesel generator-based suppliers), the potential threat of entry to the business to increase competition and the supplier power to squeeze profitability will be important considerations.

At the macro-level, a PESTLE[3] analysis is relevant. The policy environment does not often provide much support to the mini-grid-based electrification in many countries.[4] The off-grid electrification policy may be weak or non-existent and the policy awareness may be missing altogether. The political weakness of a country in terms of unstable political environment, unfavourable investment climate and weak governance structure can deter investors.

The legal and regulatory weaknesses can also hinder off-grid electrification. As discussed in Ref. [7], any non-electrified area may already be included in the service area of the existing electricity utility and unless the off-grid service area is carved out legally, the mini-grid deployment as a miniature version of the utility model may not be legally tenable. As any new local grid can amount to a potential duplication of service within a single area of service, new entry often requires consent of the incumbent, which may decide to extend the central grid instead. This threat of grid extension always acts as a deterrent for new investment in mini-grids as there is the risk of stranded assets in such an eventuality. This issue thus feeds into the analysis of micro-environment as well. Although this issue is important for a local-level initiative as well as for the investor-owned initiative, the threat of alternative supply from the grid is a destabilising factor for any private investor. Some form of assurance against possible grid extension will facilitate investment in mini-grids.

[3] This stands for political, economic, social, technological, legal and environmental.

[4] The policy environment varies significantly across countries and some countries are promoting off-grid solution through policy support. For example, India's recent thrust on solar energy and consequent declaration of state-specific solar policies are laying emphasis on off-grid energy development in respective states. This is unlike in the past, where state policies on renewable energy were primarily focusing on grid-based renewable energy development.

Unclear or ambiguous regulatory arrangements can also exist in respect of licence or permissions for doing business, tariff systems, connection to the grid, safety requirements and quality of supply. Such ambiguities can arise due to a number of reasons.

(a) For example, the governing act may not distinguish between urban and rural areas of supply, and therefore does not recognise the problem itself—a problem that seems to be receiving some attention now. As the energy access issue has gained wider recognition in the past decade and the electricity industry in many countries has been reformed to improve performance, cognizance has been taken of the electricity access issue. For example, many countries in Africa have created a separate Rural Electrification Agency to enhance rural electrification.
(b) The nascent nature of the off-grid business and perhaps the sentiment of an inferior solution has not led to the adoption of a specific approach for off-grid solutions, including the mini-grids.
(c) Some ambiguities may be deliberately kept considering the available regulatory capacity and to reduce the regulatory burden on a nascent business. It is sometimes argued that regulatory inaction can promote innovative outcomes, that may not emerge in the presence of an overzealous and proactive regulatory system.

Whatever the reason may be, incomplete or unclear rules make business riskier. The legal legitimacy of such an activity can be weak, thereby discouraging the investors. As mentioned earlier, the clarity about the coverage of a mini-grid service area is a moot issue (see Box 1).

Box 1: Ambiguities in the Indian Electricity Act regarding off-grid coverage

The Electricity Act 2003 governs the activities of the Indian electricity industry. Section 12 of the Act makes transmission, distribution and trading of electricity a licenced activity, unless an appropriate regulatory commission exempts, for a specified period and subject to specified conditions, from this requirement through notification under Section 13. The exemption from the licence requirement can apply to local authorities, user associations, co-operatives, non-governmental organisations and franchisees. Section 14 further provides that no licence is required to generate and distribute electricity in a rural area notified by the State Government to be fit for this purpose, subject to the safety requirement specified in Section 53.

The Rural Electrification Policy of Government of India [10] clarifies that all rural areas defined by the 73rd amendment of Constitution of India will come under the purview of exemption under Section 14 of the Electricity Act. But the universal service obligation of the distribution licensee continues inspite of the above policy. This implies that the conflict over the service area is not resolved and the distribution licensee can extend the grid in any mini-grid area as and when it wishes to do so. Moreover, while the distribution licensee benefits from government subvention and cross-subsidy potential, an operator exempted under Section 14 of the Act does not enjoy similar benefits. This creates an uneven playing field for two types of operators, a formal licensee and an exempted supplier serving the same area.

Similarly, clarity on the legal and the eligibility requirements of the entity undertaking the activity can also be important. A case in point is the developments in India in this respect. Among the variety of players found to be operating in the mini-grid business at the moment, the legal status of some is not clear. This is particularly true of community-based systems where the community manages generation and distribution activities but it does not have any formal standing as a corporate body. Although the Electricity Act 2003 allows such informal setups to undertake the activity, it is difficult to monitor their activities and ensure compliance with safety requirements, consumer issues and reporting requirements. Moreover, it is difficult for such informal groups to raise funds and access financial resources from the formal sector, thereby creating hurdles for their integration with the mainstream and making their long-term viability difficult. On the other hand, interventions by the Renewable Energy Development Agencies in specific states (such as West Bengal and Chhattisgarh) and the private sector players (such as Husk Power) as the owner of the project and implementation agency are corporate[5] bodies having a clear legal entity and an organisational structure, and hence may be easy to keep track of and to integrate.

The economic factor cannot be ignored either. Given the importance of tariff of any electricity service for the business viability, any uncertainty and risk in the tariff setting process can adversely affect the business interest. On one hand, mini-grid supply will always be compared with the grid supply and therefore there will be a logical tendency to benchmark the tariff with the grid counterpart. But the dissimilarities between the two cannot be overlooked. The central grid allows integration of a large consumer base, thereby providing an opportunity to average out the cost and tariff. It also benefits from cross-subsidies and subventions. Cross-subsidies are practically impossible for a small system and operating subsidies may not flow to mini-grids if they are not regulated. Moreover, the cost of supply is generally higher for most mini-grids. Thus ensuring a tariff parity with the central grid system is very challenging without financial support. If the price is left uncontrolled, the supplier, when profit motivated, is likely to charge a much higher rate. Given the bargaining power rests with the supplier in such cases, they are

[5] These are legal entities created in accordance with prevailing rules and may be companies, associations, charities, and the like.

likely to impose their views on the consumers. At the same time, very different tariffs compared to the peers are likely to create consumer resentments and thereby causing harm to the business in the long term. A local community initiative, on the other hand, is unlikely to be profit-driven and therefore a negotiated tariff may take members' paying capacity into consideration but achieving financial viability, even just operational viability, in such cases may be difficult to achieve. Moreover, raising finance for capital investment, which is a specialised activity, can be a major challenge. As the funding will be a prerequisite for any project implementation, a streamlined approach to funding support can be useful.

Although our focus in this chapter is on business issues, it is important to highlight here that the economic consideration from the demand-side or users' side is also crucially important for any successful intervention. We highlight two issues:

(a) First, integrating the intervention with the rural economic environment by creating backward and forward economic linkages assumes importance. If the mini-grid business creates demand for inputs in the local economy that provides income generating opportunities and if electrification opens up opportunities for employment generation through productive use of electricity, the business environment improves. However, the opportunities for creating economic linkages depend on the technology choice. A diesel-generator or a solar PV-based mini-grid offers limited backward linkage opportunities while it is more pronounced in biomass-based interventions. For example, the Husk Power Systems in India is generating local-scale employment through their fuel demand for rice husk. Forward linkage on the other hand depends on the project design, technology choice and the willingness of the project developer to establish such opportunities by creating additional market linkages and supply chains. However, such activities of rural development go beyond the normal scope of business of a traditional electricity supplier and hence may limit their interest.
(b) Second, poor access to finance by the end users is a limiting factor in most rural areas to expand electricity demand. Electricity demand, like any other forms of energy, is a derived demand which has to be used in conjunction with an appliance. Any demand creation is thus dependent on the level of penetration of electricity using appliances in rural areas, whether for household uses or productive uses. However, as durable goods such appliances require substantial initial investments, which the rural population may not be able to afford as a one-time payment. As formal banks and financial institutions do not reach many rural areas, micro-finance organisations have emerged to offer micro-credit options as a solution. A stake holder survey on private capital flows to energy access by Monroy and Hernandez [16, 17] finds micro-finance as an important factor for project viability. Although various financing arrangements have come up (such as financing provided hand-in-hand with a technical solution such as a solar home system, micro-finance provided directly by energy companies, and the like),[6] no specific support has yet

[6] See [4] for a detailed review of financial options for energy access.

emerged for mini-grid-related businesses. Consumers here would need a general credit line to buy products of their choice but the absence of a regular income flow, lack of credit record and their inability to meet the guarantee requirements hinder their access to credits. Also, the interest for micro-credits lies between 2 and 3 % per month, making the cost of credit very high for the poor.

The technical dimension of a mini-grid-based electricity supply has been covered in "Technical Aspects of Mini-Grids for Rural Electrification" and "Smart Design of Stand-Alone Solar PV System for Off Grid Electrification Projects" of this book. From the business perspective, the scale of operation will be an important issue. Typically, a micro-utility business, being a small activity that locks the supplier into a long-term business, is unlikely to attract large investors, unless there exists a scaling-up or replication potential. Replication requires one or more standard models that can be repeated in different locations, thereby generating some sort of scale economy. A cluster of villages in a rural area may provide an opportunity for (as discussed in "Viability of Husk-Based Mini-Grids in South Asia") for this. However, standardised operation may also tie the business to a specific technology (say solar PV for lighting or biomass gasifiers) thereby creating a certain path dependency. Scaling-up on the other hand requires a larger area of service and integration of other commercial activities that require significant amounts of energy. It is not clear whether increasing the size of the system is beneficial for decentralised technology options, as each technology has it own feature. Solar, being very modular, might be easily scalable to cover just lighting applications as well as larger demands, whereas a gasifier may be more appropriate beyond a certain threshold because of its inherent minimum sizing requirement. Moreover, developing commercial activities can be a different project in itself. Both of which may not materialise in the short term and the investor may not have appetite for such additional non-core activities. For a local community-led initiative, the project size and technology choice can also be very crucial for project implementation. Instead of aiming for a large complex project, it may be appropriate to develop the electrification project in stages, starting with a small project to meet the basic needs and enlarging the scope subsequently to cater to other needs. This allows for confidence building through hands on experience with an initial activity, which if successful may open up opportunities for further expansion.

Finally, the importance of social and environmental dimensions in such interventions cannot be overemphasised. The support and involvement of the local community is essential but the complex social power structure and the possibility of divisions in the community due to political, religious, ethnic, economic or other factors may need careful consideration. The participation of women in the decision making is also another important consideration and precaution against any form of exclusion is important as well. The social implications of lack of energy access and the environmental benefits of electrification have been widely discussed in the literature. Put simply, the use of local and renewable resources of energy reduces environmental impacts and offers health and climate-related benefits. Their use can also open doors for some financial support and hence can be beneficial for the project realisation.

It is important to realise that the mini-grid business is one of the many options available for investment to an investor and therefore, the attractiveness of the investment in terms of potential benefits is important. The business environment analysis is the first step in this direction and unless the first impression is positive, it is unlikely that investments will flow.

3 Viability and Cost-Benefits of Mini-Grids

Depending on the size and physical characteristics of the rural areas (income, distribution of households by income, access to resources and infrastructure, availability of human resources, and energy potential), and considering the prevailing macro-environment, the mini-grid electrification potential of villages can be classified as potentially viable, marginally viable and non-viable cases from an investor's perspective. However, from the society's perspective, the benefits of electrification through mini-grids may be higher than the costs of electrification. This section briefly presents the financial and economic evaluation of mini-grids.

3.1 Financial Viability

A financial analysis of any investment project determines whether the cash inflows from the project are sufficient to meet the cash outflows on a regular basis and whether the investor is likely to receive an adequate return on the investment. Thus the operational profitability and attractiveness of the investment compared to other investment opportunities are two important considerations from the investor's perspective.

The financial analysis is based on accounting cash flows. The cash flow during the project lifecycle is considered in determining the financial viability of any investment. The analysis generally takes into account the company tax rules (such as tax payable on profit, tax credit and deferment options, and carrying of financial losses forward), any allowances given for specific types of projects (e.g. accelerated depreciation, tax holidays and the like) and any grant/support available from any source.

The revenue or benefit from the project arises from the sale of electricity but additional revenue opportunities from any sale of byproducts and co-benefits (carbon credits for example) are also included. The expenditure side on the other hand includes all capital-related costs (e.g. return on equity capital, interest on loan, return of capital either through loan payments or depreciation charges), operating costs (employment costs, fuel costs, operation and maintenance costs) and taxes. Care needs to be taken to avoid double counting of any costs or benefits. As it captures the investor's perspective, the weighted average cost of capital for the investor is generally used as the discounting factor. The analysis can either be done in current prices (i.e. taking inflation into consideration) or in constant prices (where inflation is removed from the analysis).

As the objective is to ensure that the costs are at least equal to the benefits, often the levelised cost of electricity is determined from such an analysis which if charged throughout the project life would generate adequate revenue to meet the costs. The discounted net present value of the cash flows must be equal to the product of discounted price and electricity sold. This is an important indicator used in project analysis.

The financial analysis of any mini-grid project is greatly influenced by the following factors.

(1) Capacity utilisation of the plant: The electricity produced by the power-generating plant is directly related to its rate of utilisation. Often this is decided by the amount of resource availability and the duration of service. In many cases, mini-grids are used only during the evening hours and hence their rate of utilisation remains limited to 20–30 %. In case of biomass or small-hydropower projects, only 50 % of the rated capacity is put to use, thereby reducing the capacity utilisation. As the cost of the installation has to be recovered over a limited level of use, the unit cost becomes higher. On the other hand, where the mini-grid relies on a continuous flow of energy and the service is run day and night, it achieves a higher rate of utilisation and hence the unit cost reduces. Integration of productive demand or an anchor load can help improve the utilisation rate and reduce the cost of supply. The limited time operation of the mini-grids contributes to the high cost of electricity from such services.

(2) Capital cost of the project—Irrespective of technology choice for electricity generation, a local mini-grid system is a capital intensive intervention in terms of unit capital cost. First, it may be difficult to get appropriate land in the village for setting up the system, and the land acquisition can be a costly process. Second, it is not uncommon to find the capital requirement for the power-generating capacity to range between $1,000 and $3,000 per kW of capacity. Diesel generator sets are the least capital intensive option while hybrid options cost much more and mini-hydro options cover a wide range of costs over the entire range [12]. Third, the distribution network can easily cost $5,000–8,000 per km and the final connection to the consumer premises requires additional investment. The cost of distribution increases when the generating stations are located inappropriately due to non-availability of land. While some technologies allow modular sizing, for others discrete standard size of the generator is the norm. The remote location further adds to the cost due to higher cost of transportation of equipment, damage risks during transportation and higher labour costs due to non-availability of skilled

personnel locally. The investment is often likely to be quite significant compared to the local economic activities in the rural area concerned but by any investment project standard, it will be a small project.

(3) Capital structure of the project—The overall financial viability of any project critically depends on the source of funding and the cost of securing them. Funding in the form of debt and equity finance forms the capital structure in most cases but in the case of rural electricity projects, capital grants and donations can also be important. Any capital grant reduces the cost of investment as there is no return on and of the capital to be charged to the project. Capital grant to the extent of 90 % of the capital cost has been provided in many Indian projects to keep the costs down. But our case studies in the second part of the book show that one-time capital grant may not be sufficient to ensure project viability unless adequate revenue is generated to pay for regular replacement of assets (such as batteries).

Debt funding has traditionally supported grid-based electrification in many countries and the creation of rural electricity infrastructure has often benefited from cheaper loans made available from government agencies or international funding agencies. The rate at which lenders are willing to lend money depends, among others, on the benchmark interest rate prevailing in any economy, the share of borrowed funding in the overall capital structure, availability of appropriate assets as security, additional costs related to the transaction and creditworthiness of the borrower. A wide range of rates can exist at any given time depending on the source, type of borrowing entity and the nature of the project. Even a borrower may be able to secure loans from different sources, thereby allowing averaging potential for the overall cost of borrowing. The risk of non-recovery and the high operational cost of doing business in rural areas often make lending costly. High interest rates of 15–30 % per year are not uncommon in this business [12]. In addition, the loan term can be quite short—5 to 10 years of loan term is quite common [9, 12]. The high interest rate and shorter term of repayment adversely affect the cost of electricity supply because of the front-loading of costs. The mismatch between project debt terms and project life remains an issue.

The owners' contribution (or equity share) supports the balance of the funds required. While in the utility mode of operation, consumers are not expected to contribute to the capital cost of the supply infrastructure, in the case of local mini-grids or off-grid access system the local participation is an important consideration to develop some sense of local ownership. These contributions may come in-kind (e.g. labour), in physical assets (e.g. local land) or in the form of connection fee. It is important to account for such contributions and give credit as appropriate. The private owner would also expect a return on its investment, which is commensurate with the risks taken by the investor. Accordingly, the wedge between the interest rate and the return on equity can be substantial.

Access to funding for rural energy projects is an issue in itself and the challenge remains significant for local community-led projects, as well as private investor-led projects. The size of most rural-scale projects is small compared to the normal business transaction thresholds of traditional financial institutions (such as investment banks), making their direct involvement quite unlikely. While intermediaries or special purpose organisations may exist, they often deal with formal organisations. The informal nature of the community organisations, absence of past credit record for such organisations, limited own-funding ability and limited security for guarantee purposes make borrowing difficult and costly. As most lenders require at least 20–30 % of the cost as down payment, mobilising such funds locally can become a Herculean task at the village level. This adversely affects the financial viability of a project. The private investor on the other hand may have easier access to the capital market because of its previous track record and its ability to use its balance sheet for securing loans.

Often reference is made to the carbon finance as a potential source of funding for energy access. However, it is less attractive for small individual projects as the procedure is complicated and the transaction cost is high. In India, both Husk Power Systems and DESI Power have registered projects under the Clean Development Mechanism. However, at the prevailing carbon price of about €2.5 in 2014, there is little income generation potential from the sale of certified emission reductions.

(4) Operating cost—The cost of system operation and maintenance is another factor that can affect the project viability in certain cases. While such costs can be quite low for hydropower projects or solar PV-based supply, operating cost, particularly the fuel-related costs can be important in fossil fuel (diesel)-operated generation or biomass-based technologies. The cost of fossil fuel, being an internationally traded good, often depends on the international market conditions and the national fuel tax policies. Cost of transport and storage can be high for remote locations. The employee-related costs can also be important in labour intensive type of activities and if outsiders have to be employed to carry out the work in rural areas.

(5) Tariff or electricity charges—Being the most important revenue-related factor, the price charged for electricity supply is crucial for the financial analysis. The tariff may be regulated by any competent authority or the government or is set through bilateral negotiation or imposed by the supplier unilaterally. Whatever may be the case, the revenue from tariff ultimately decides whether or not the business recovers its costs, leading to a viable business. Where the supply is predominantly catering to domestic consumers with limited demand, a flat-rate is a simpler option but it does not provide any incentive to limit consumption and may encourage users to consume more. A metred supply either through a prepaid card or a regular meter charges consumers according to their electricity use. Consumers with a limited demand will face a relatively small monthly cost for their supply while availability of a meter can allow the supplier to charge different rates for different levels of consumption (block tariff). But in a residential-only system, the scope for cross-subsidy in a mini-grid is quite limited.

On the other hand, when the supply caters to a mixed set of consumers, different rates may be charged, thereby offering the possibility of cross-subsidy to a limited extent. But a high differential tariff, particularly when this makes alternative options (e.g. diesel-based electricity) viable, can be detrimental to project viability. Thus there is practically a ceiling price beyond which a differential tariff cannot be extended. This is an important aspect in tariff design. When a large anchor load is catered to, the business has the advantage of a reliable flow of income but being a dominant buyer, the anchor load can dictate the terms, thereby reducing the possibility of a monopoly tariff. Simultaneously, the dependence on a large anchor load enhances business risks, as the threat of exit of such a buyer can have a devastating effect on the business.

For the viability of a business, revenue generated from the tariff has to cover the cost of the business. However, when the alternative source of energy, such as kerosene, is highly subsidised, or the price of grid-based electricity is maintained artificially low, the financial viability of a mini-grid is difficult to achieve. Either an operating subsidy will be required or the consumers would have to pay a higher price if supply is desired. On the other hand, where the alternative fuel is not subsidised (i.e. kerosene is not subsidised or dry cell batteries are costly to buy as is found in some African countries), the price comparison does not get distorted and the mini-grid stands a better chance of success.

A financially viable mini-grid business essentially minimises the costs while maximising the revenue. However, the scope for such adjustments may not be great in rural contexts, making the business viability challenging.

3.2 *Economic Cost-Benefits of Mini-Grid-Based Electrification*

As opposed to an investor's perspective, an economic analysis of investment projects considers the society's perspective and investigates whether it is worth investing in the specific project. An economic analysis can take the financial analysis as the starting point and make necessary adjustments to reflect the economic perspectives on costs and benefits (as opposed to a financial perspective). In general, adjustments are required in identifying the cost or benefit elements and their appropriate economic valuation. An economic analysis considers the additional costs imposed by the project and additional benefits derived from the project. Accordingly, (1) sunk costs or cost already incurred, or (2) taxes or subsidies that represent transfer benefits or costs are not considered as real economic costs or benefits. Similarly, the source and structure of funding is not important for the economic analysis—only the economic investment cost at the start of the project is sufficient. On the other hand, external costs or benefits of a project are relevant for the economic analysis. The valuation of costs and benefits differs from their financial values. In economic analysis, the opportunity cost is appropriate and if the market shows signs of distortions, then shadow prices are used to correct for imperfections. The economic analysis uses willingness-to-pay and willingness-to-accept compensation rather than the price actually paid for a good or service.[7]

In the case of mini-grid projects, the above considerations will affect the investment-related costs, operating costs and the benefits from the project. If a diesel-powered system is used, it is likely that some taxes and subsidies on the price will distort the market price and depending on the country's importing or exporting status, the local selling price may be much different from the international border price. However, given that diesel is a traded good, it is appropriate to use the international benchmark price adjusted for local transportation costs to deliver the fuel to the project site. Moreover, if the foreign exchange market is not fully competitive, the conversion of international price of diesel to the local currency using the regulated exchange rate will not capture the true cost. The appropriate shadow price for the exchange rate has to be used. Thus the adjusted price will be very different from the prevailing market price—whether it is lower than the financial price or not depends on the specific case.

Moreover, the environmental damage caused by burning diesel in terms of local pollution and carbon emissions are not adequately reflected in the fuel price, thereby creating an external cost. In the financial analysis, this would not appear as there is no money transaction actually taking place from the project. However, as the society suffers from this damage, it is appropriate to include the external cost in the economic analysis.

Similarly, electrification may lead to a number of benefits in terms of avoided costs of other energy use, improved economic productivity and hence higher income potential, reduced healthcare costs due to improved indoor conditions and reduction in drudgeries, opportunities for education (and hence human capital development), improved security and the like. However, it is not always easy to estimate economic benefits: (1) well-defined markets may not exist for such benefits, and putting a price or value may be difficult; (2) quantification of such benefits may be difficult, (3) they may occur in the future and there may be significant uncertainty about the nature and extent of such effects. In addition, when willingness-to-pay or willingness-to-accept is used as the basis for valuation, there can be systematic biases in the responses of the survey participants, which can create further difficulties in the valuation process.

Therefore, care needs to be exercised in carrying out an economic analysis of mini-grid investment for rural electrification to avoid double counting errors, inappropriate identification of cost or benefit items and inappropriate valuation of costs and benefits. Studies generally tend to suggest that electrification can often be justified on economic grounds (see [8] for some relevant studies) but studies on mini-grids are limited at this point of time. However, unless a project becomes commercially viable, private investors are unlikely to be interested in such ventures.

[7] For further information, please refer to [3]. See [12] for some examples of economic analysis related to mini-grids.

4 Business Models for Mini-Grid Electrification

Despite challenging business conditions, the electricity supply through mini-grids has been taken up in various parts of the world. In the second part of this book, we have included examples from India and Nepal. Here the general trend of such activities is briefly presented.

Mini-grid or local grid has a long history. In fact, in the early days of electricity industry development, electricity was only available through local grids in urban areas or areas where resources were available. Local grids have also been used for rural electrification in China. In the first phase of China's electrification, strong emphasis was laid on small-scale hydropower development in rural areas to promote self-reliance and the electricity so produced was distributed locally through low voltage distribution systems. These systems were developed through the policy of "self-construction, self-management and self-consumption" where rural residents and collectives funded and managed the systems with some support from the state [6]. In Brazil, diesel generator-based mini-grid systems have been widely used in the Amazon region where more than 1,000 diesel generator sets are in use [5]. In fact, diesel generator-based local mini-grid is most common around the world [1].

Business models for rural electrification in general and for mini-grid-based electrification can be classified using different criteria [2, 11]. The ownership-based classification—that is whether it is privately owned (or commercial), quasi commercial (using corporate social responsibility or private–public partnership approach) and publicly owned (or non commercial)—is very commonly used [13]. However, here a typology based on the size of the system is used to capture the differences in the service level.

(a) *Micro-lighting utility model* (3 W *service*). These are very small systems essentially for providing the basic lighting service in households, where the plant size is typically less than 1 kW but slightly bigger systems of 5–6 kW are also used. Typically, solar PV-based electricity generation is used and the distribution to households is done through DC or AC grids depending on the size of the system. Generally, DC systems are used for up to 1 kW plant capacity while AC distribution is used for bigger plants. The system typically services a small village or a hamlet with 30–50 households. Examples of both public and private ownerships can be found for this model:

    (1) Mera Gao Power (MGP) is a private entity providing lighting services to the poor in Uttar Pradesh in India. It follows a standardised system design where solar PV systems are being used to generate electricity, low voltage DC grids are being used for distribution and consumers benefit from two light points and a mobile charge point at their premises for a fee. For a village of 100 households, MGP requires an investment of Rs. 248,000 (or invests around $4,200 at an exchange rate of Rs 60 for one dollar).

The tariff per month for two light points is Rs 70 per household,[8] generating an income of Rs 84,000 per year. It is reported that about Rs 75,000 is generated annually to pay for the investment costs, yielding a payback period of just about 3 years [15]. MGP has successfully electrified more than 17,000 households following its standardised, utility-like approach and has managed to keep the costs down for its supply. Other such examples of private micro-lighting utility models from India include Naturetech Infrastructure (operating in Uttar Pradesh, India) and Husk Power Systems (in Bihar).

(2) On the other hand, the solar PV mini-grid in Chhattisgarh State represents the public sector model of this category. While the private sector is not relying on operational subsidies, the state led model in Chhattisgarh is ensuring tariff parity with the grid-based supply by availing the state subsidy. This case is discussed in more detail in "Energizing Rural India Using Distributed Generation: The Case of Solar Mini-Grids in Chhattisgarh State, India" and is not elaborated here.

By just catering to the basic lighting needs through a reliable limited period supply, the initial cost and the operating costs are being controlled. The private players are highlighting the quality and reliability of supply despite its limited period of supply and charging a tariff that the residents would be willing to pay or be able to bear. Once the service gets acceptance, further services can be added and the quantity restrictions can also be removed for those who are willing to pay more. This prepares the market and also keeps the risk of stranded assets to a minimum if the grid is extended in the meantime. But using the conventional metric of tariff per kWh, the cost of supply becomes quite high for a private supply, which may be an issue in the long-term. Moreover, MGP received a capital grant from USAID to start its operation in 50 villages, which has helped it to overcome initial funding hurdle. They have also secured equity funding from Insitor Management, an impact investing firm.[9] However, it is not clear whether it will be able to raise funds for its future operations based on its initial micro-utility business despite reporting operational profit. As its operations are recent, it is not yet known whether it is able to retain sufficient funds for future battery replacements.

(b) *Lighting-plus systems*. Similar to the micro-lighting utility, these mini-grid systems generally provide a limited level of service to a normal customer but can also provide customised supply to consumers requiring a higher level of supply or support some commercial load (hence, "light-plus" service category). While a micro-lighting utility uses LED bulbs to economise on lighting load, these small mini-grid systems tend to rely on CFL bulbs for lighting. They can also cater to small household demands such as a television or a fan

[8] However, collection is done weekly to avoid non-payment.

[9] OASYS South Asia project is also supporting Mera Gao as part of the demonstration activity to electrify nearly 3,000 households.

load. For those willing to pay for a higher level of consumption, additional loads (such as for a refrigerator or other household appliances) can also be accommodated. The Husk Power System (HPS) in India with operations mainly in Bihar offers such a service in the private sector, where normally 20–30 kW[10] capacity plants based on rice husk based are being used to provide a basic supply of 30 W per household. The HPS case is analysed in "Viability of Husk-Based Mini-Grids in South Asia" and is not repeated here. A public sector example of this category comes from the electrification of the Sunderban Islands in West Bengal by the state renewable energy agency, WBREDA. A number of island villages have been electrified using either solar PV systems or hybrid systems (solar PV and wind or biomass combinations) of varying capacity (generally less than 30 kW but some with higher sizes [20]). Each household here gets a connection for a 70 W or a 120 W load. In addition, community systems and commercial loads are also catered to. The lighting service provided here is comparable to lighting obtainable from an energy saving lighting system in a central grid-based supply and some additional needs (like fan, TV, radio and mobile charging) can also be met. Moreover, commercial and productive activities can also be supported to some extent.

While HPS follows a tariff system that ensures cost recovery, the public sector-led supply in the Sunderbans is aimed at recovering operating and maintenance costs where the capital cost is subsidised by the government. The cost minimising innovations of HPS along with its commercial and social skills have so far ensured a financially viable operation. On the other hand, the socially-oriented service approach in the Sunderbans leads to a dilemma between financial viability through cost recovery and the social objective of electrification by keeping the charges affordable [20]. Without capital subsidy and perhaps some operating subsidy, such a supply is unlikely to materialise whereas replication of the private investor-led model faces the funding challenge as the initial capital is difficult to secure from commercial sources.

(c) *Service with an anchor-load.* A variation to the lighting-plus model has also emerged where a major consumer (e.g. an industrial load, a telecommunication service load, an agricultural estate such as a tea/coffee farm or something similar) provides the base load and supports the system development. The residential lighting and other basic demand is also catered to in the vicinity of the major load.[11] A further extension of the service to the local commercial activities can also be considered where such a potential exists.

The symbiotic relationship with an anchor load (individual or a group load) can offer a win-win case for both the parties. The anchor load customers may already be running their businesses using alternative fuels and may be willing

[10] HPS has also installed larger capacity plants in certain locations.

[11] If the anchor load is only serviced, it becomes a stand-alone solution and does not really address the electricity access issue.

to pay the price for an alternative but reliable service that reduces their efforts on a non-core activity (power supply). The alternative supply may also have positive benefits in terms of carbon footprint reduction for the anchor load. On the other hand, the electricity supplier can enter into a power selling agreement with the major user, thereby providing security to the income stream. Such an agreement can also be used as a guarantee for securing loan funds from the financial institutions, thereby enhancing the bankability of the project. Moreover, the base load is likely to allow a better utilisation of the plant capacity, thereby reducing the average cost of supply. However, the supplier faces a dominant purchaser, who is likely to influence the sale price, thereby reducing the profit margin. Also, the project viability remains locked-into the business viability of the anchor load and any changes in the fortune of the anchor business can adversely affect the mini-grid business. Moreover, where the power comes from an intermittent source, the system is likely to require a stand-by facility to meet the reliability of supply required by the purchaser, which may add to the capital cost.

DESI Power in India (discussed in "Viability of Husk-Based Mini-Grids in South Asia") and OMC Power (India) have followed this approach. Moreover, DESI Power has also participated in the SPEED[12] programme supported by the Rockefeller Foundation to test the concept of telecom tower as an anchor load. This test demonstrated that based on the prevailing support system available in the Indian off-grid electrification sector and considering the business tested, it is possible to earn a return on equity of over 20 % [19]. OMC Power in India is also following the same approach and providing electricity to telecommunication towers and communities. They have also created a micro-power business in a box where the local entrepreneur rents out appliances to the community. Using telecom as an anchor load has also been piloted in other countries: for example Grameenphone has teamed up with the University of Oslo to test this in Bangladesh and Safaricom in Kenya has also pursued this as part of its corporate social responsibility agenda [14].

As highlighted in IFC [13], the success in any of these mini-grid models rests on a few critical factors: (1) enlisting support of sufficient consumers to ensure a high level of system utilisation; (2) using a low-cost source of energy to meet the demand and (3) choosing a right delivery model. Appropriate size of the business plays an important role: a basic lighting system may be appropriate where limited commercial activities exist at present but a light-plus or an anchor load approach can work better for areas with higher commercial activities. Given the limited paying capacity of the consumers, ensuring affordability of supply is crucial. The supply can be privately owned, community-owned or publicly owned. As highlighted earlier, the mini-grid business involves all the activities of an electric utility at a smaller scale, starting from electricity generation to distribution and supply at the local level. Therefore, whatever organisation undertakes it, the

[12] Smart Power for Environmentally-sound Economic Development.

activity needs to be supported by adequate competences and skills so that the supply can be provided reliably. At the same time, access to finance for long-term debts and mitigation of business risks by creating a favourable environment is important.

5 Conclusions

Enhancing electricity access in the developing world through mini-grids is a challenging task. This represents a step change in electricity provision compared to the individual solutions and involves looking after the entire supply chain of a mini-utility system. Clearly, such a business is more demanding and accordingly, the commercial aspects require careful attention.

It is not easy to ensure financial viability of a business in a difficult rural environment. It is important to minimise the costs and achieve as much as income as is possible. This requires attention to the following five areas: (1) proper system design so that the system cost can be kept to a minimum to meet the needs and avoid excess capacity; (2) appropriate choice of technology to reduce capital costs and operating expenses; (3) careful structuring of funding so that the repayment burden can be spread over the life of the project and cost of funding is minimised; (4) appropriate charges for the service that is affordable but generates adequate funds for running the business and (5) appropriate management systems to deliver the supply, collect revenues and manage the system operation.

Despite the challenges, various innovative attempts have been made where the above factors have been applied successfully in practise. These examples include both public sector-led initiatives and private initiatives and cover micro-lighting utilities, lighting-plus services and services with an anchor load. The financial model remains different in each case but where the private sector is providing the supply, the financial viability has been achieved whereas the public sector has followed a socially-oriented approach without necessarily focusing on full cost recovery. As the business environment is difficult and it is not the fault of the rural people to live there, support for infrastructure development can be justified on various grounds. With such helping hands, it is possible to enhance access to electricity in rural areas through mini-grids.

If mini-grids have to emerge as an important solution in the fight against electricity access, a rapid expansion of the business is required globally. This requires moving from the pilot and demonstration phase to the replication and scaling-up phase. This opens up new challenges for the business in terms of financing, organisational growth and appropriate supply chain management and consumer satisfaction. Further research is required in this area and more global effort is required for funding, capacity building and technology management.

References

1. APEC (1998) Analysis of renewable energy retrofit options to existing diesel mini-grids. Energy Working Group, Asia-Pacific Economic Cooperation, New York
2. ARE (2011) Hybrid mini-grids for rural electrification: lessons learned. Association for Rural Electrification, Brussels
3. Bhattacharyya S (2011) Energy economics: concepts, issues, markets and governance. Springer, London
4. Bhattacharyya S (2013) Rural electrification experience from South-East Asia and South America. In: BBhattacharyya S (ed) Rural electrification through decentralised off-grid systems in developing countries. Springer, London, pp 157–184
5. Bhattacharyya S (2013) To regulate or not to regulate off-grid electricity access in developing countries. Energy Policy 63:494–503
6. Bhattacharyya SC, Ohiare S (2012) The chinese electricity access model for rural electrification: approach, experience and lessons for others. Energy Policy 49:676–687
7. Bhattacharyya SC (2013) Financing energy access and off-grid electrification: a review of status, options and challenges. Renew Sustain Energy Rev 20:462–472
8. Cook P (2013) Rural electrification and rural development. In: Bhattacharyya S (ed) Rural electrification through decentralised off-grid systems in developing countries. Springer, London, pp 13–38
9. ESMAP (2001) Best practice manual: promoting decentralised electrification investment. World Bank, Washington
10. GOI (2006) Rural electrification policy. Government of India, New Delhi
11. GVEP (2011) The history of mini-grid development in developing countries. http://www.gvepinternational.org/sites/default/files/policy_briefing_-_mini-grid_final.pdf. Accessed 19 Feb 2014
12. IED (2013) Identifying the gap and building the evidence base on low carbon mini-grids, support study on Green Mini-grid development. Innovation, Energy Developpement, Francheville
13. IFC (2012) From gap to opportunity: business models for scaling up energy access. International Finance Corporation, Washington
14. Mary R, Ward C (2011) Harnessing the full power of mobile for off-grid energy. GSMA, London
15. MNRE (2012) Empowering rural India the RE way: inspiring success stories. Ministry of New and Renewable Energies, Government of India, New Delhi
16. Monroy CR, Hernandez ASS (2005) Main issues concerning the financing and sustainability of electrification projects in rural areas: international survey results. Energy Sustain Dev 9(2):17–25
17. Monroy CR, Hernandez ASS (2008) Strengthening financial innovation in energy supply projects for rural exploitations in developing countries. Renew Sustain Energy Rev 12(7):1928–1943
18. Porter M (2008) The five competitive forces that shape strategy. Harvard Business Review pp 2–18
19. Rockefeller Foundation (2011) SPEED: an agenda for action. The Rockefeller Foundation, New York
20. Ulsrud K, Winther T, Palit D, Rohracher H, Sandgren J (2011). The solar transitions research on solar mini-grids in India: learning from local cases of innovative socio-technical systems. Energy Sustain. Dev. 15(3):293–303

Part II
Case Studies

Approach for Designing Solar Photovoltaic-Based Mini-Grid Projects: A Case Study from India

K. Rahul Sharma, Debajit Palit, Parimita Mohanty and Mukesh Gujar

Abstract Having the largest rural population in the world, India confronts a huge challenge for rural electrification, especially for electrifying remote, forested and tribal habitations. Solar Photovoltaic-based mini-grids have emerged as a viable option for the provision of electricity in such remote rural locations, where grid extension is either not techno-economically feasible or electricity supply is intermittent. Very often such projects are purely technology-driven and several attempts at delivering electricity services to such remote locations have not succeeded, owing to the lack of adequate attention given to important socio-economic factors such as promotion of livelihoods or the creation of strong local institutions that can own, operate and manage the project over its lifetime. This chapter aims to present an interdisciplinary framework for the development of mini-grid projects in remote rural locations, developed from field experience of actual implementation of projects by TERI. Using this framework as a guide, TERI has commissioned solar photovoltaic-based mini-grids in a cluster of five villages in the state of Odisha. The detailed design methodology, including modifications to standardised practices in order to customise and improve the performance of these solar mini-grids is presented in this chapter as a case study. It is expected that the

K. R. Sharma (✉) · P. Mohanty · M. Gujar
Centre for Distributed Generation, The Energy and Resources Institute, Lodhi Road, New Delhi 110 003, India
e-mail: k.sharma@teri.res.in

P. Mohanty
e-mail: parimita@teri.res.in

M. Gujar
e-mail: mukesh.gujar@teri.res.in

D. Palit
Social Transformation Division, The Energy and Resources Institute, Lodhi Road, New Delhi 110 003, India
e-mail: debajitp@teri.res.in

S. C. Bhattacharyya and D. Palit (eds.), *Mini-Grids for Rural Electrification of Developing Countries*, Green Energy and Technology,
DOI: 10.1007/978-3-319-04816-1_8,

process followed and the resulting design will serve as a useful guide for renewable energy practitioners and researchers working in remote rural locations for provisioning of electricity services.

1 Introduction

Achieving universal electricity access has become a key target or policy objective for India as well for all other developing countries, especially after the launch of sustainable energy for all initiative [34]. This has been recognised both from the point of view of equitable access to resources, and the use of renewable energies to create the necessary access to modern energy services, in an environmentally benign manner. There are numerous studies closely examining the vital role of electrification for facilitating sustainable development especially in rural areas. They all point out how both human and economic developments rely on the access to reliable, affordable and socially acceptable energy services [1, 11, 18, 33], and lay special emphasis on decentralised modes of electricity delivery in developing countries [10, 13, 14, 19, 32].

For example, electricity contributes in a variety of ways to improving living conditions, from powering rural health clinics, to providing entertainment and communication services. Electricity allows improvements in farm and non-farm productivity through electric motors, the mechanisation of production to enhance scale, and power for water pumps, rice mills, agricultural or industrial production, among other benefits [5, 7, 9, 13, 17, 24, 26, 32, 33, 35].

A large number of projects and programmes, led by both Government agencies and private and non-governmental organisations have therefore taken up the responsibility of enhancing electricity access in remote rural areas through decentralised power plants based on locally available renewable energy. Very often, however, the emphasis is placed on delivery of the technology only, with lesser attention given to a demand led and customised approach to design, among a range of other factors impacting business models and institutional robustness, which add to the long-term sustainability of the project [16]. Lack of field expertise, absence of partnerships on the ground, standard approaches to implementation with an emphasis on speed of execution and the lack of long-term financing to maintain the system have frequently ended in frustration and disappointment on part of users and executing agencies.

Hence, it is not only important to focus on the whole range of technical, regulatory, socio-economic and financing issues that are linked to the implementation of rural energy access projects, but also prepare for the entire life-cycle of activities and costs that are associated with such projects. In this chapter, we present a standardised framework for the execution of off-grid rural electrification projects, based on the experience gained by TERI from designing and implementing such project over the last many years. Using this framework and by customising it to suit a particular context, we present and discuss the methodology used for the design of an off-grid electricity project in the Dhenkanal district of Odisha.

The case study of the project sites in Dhenkanal district of Odisha, being presented here, is being implemented as part of a multi-consortium action research project, titled 'Decentralised off-grid electricity generation in developing countries: Business models for off-grid electricity supply', known as 'OASYS South Asia Project'. The project aims to find appropriate local solutions, which are techno-economically viable, institutionally feasible, socio-politically acceptable and environmentally sound, for sustainable electricity supply to off-grid areas. An important component of the project is to develop an off-grid delivery model framework and implementation of a demonstration project covering un-electrified villages in order to test the framework. The methodology followed for this design process is presented in the chapter.

Section 2 outlines a step-wise standardised approach for the execution of off-grid village electrification projects. Using this approach from Sects. 2, 3 and 4 present the processes followed for the execution of the case study from Dhenkanal district of Odisha, focusing on the pre-installation, design and installation phases of the project. Since the project in Dhenkanal district has been commissioned during writing of this chapter, the post-installation experiences from the project has not been covered in this chapter.

2 Approaches for Execution of Village Scale Off-Grid Electrification Projects

A standardised approach for off-grid rural electrification projects is presented briefly in this section. The range of tasks/activities mentioned is not exhaustive, but gives an indication of the type of inputs or the investments that are needed in order to make an electrification project sustainable in the long run. The framework presented here is an adapted version of the approach for standardisation of off-grid electrification projects developed in 'Approach for standardisation of off-grid electrification projects' [16] through an extensive study of projects and programmes and inputs from TERI researcher's own experience in the design and implementation of similar projects.

The various phases of the project can be categorised as below, each stage involving different actors with varying degrees of contribution.

A. Project development and Pre-Installation
B. Design, Procurement, Installation and Commissioning
C. Post-commissioning and sustaining the project.

Each of the stages mentioned above can further be divided into specific activities as detailed out in Annexure A. It should be noted here that some of these

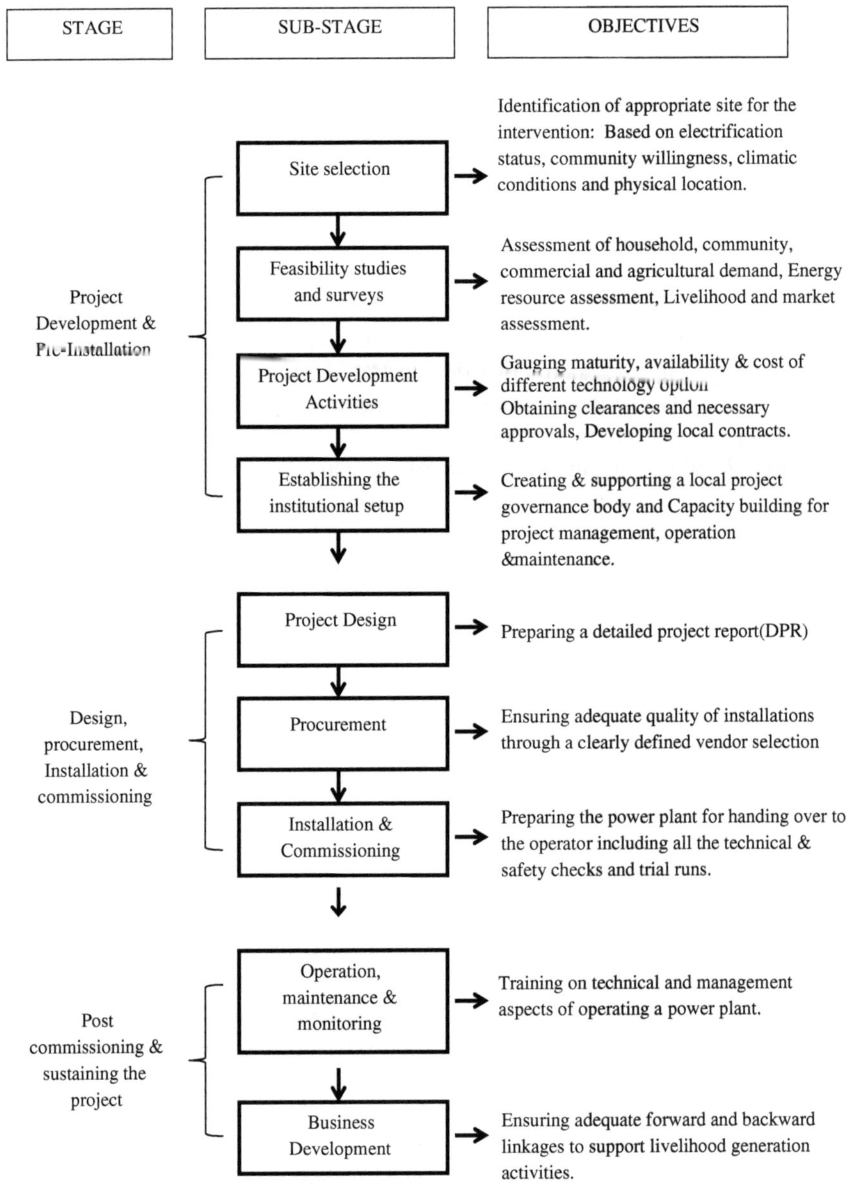

Fig. 1 Project execution methodology for rural electrification projects. *Source* Adapted from [16]

activities may not be strictly confined to the broad stage under which they are categorised and may run through the entire lifetime of the project. Figure 1 presents this process in the form of a flow chart for easy reference.

In the subsequent sections, the step-wise methodology described above is applied to the identified project sites in Dhenkanal district of Odisha. The installation and commissioning of the power plants in these sites has been completed in February 2014, and hence the focus of this chapter will be on the project development and pre-installation, and design, procurement, installation and commissioning activities described above. As mentioned in the previous section, the post-commissioning activities and impacts of the project are beyond the scope of this chapter and will be assessed over the next one year and the research outputs will be subsequently published.

It should be noted here that approach described above has been customised to suit the particular site conditions, ground and market realities and other externalities such as weather and project timelines. In this chapter, we attempt to present not just the basic steps followed for design, but also lay emphasis on and highlight the specific requirements of the sites and the manner in which they have been addressed, from a practitioner's viewpoint. We have attempted to illustrate the challenges faced and some possible solutions adopted during the actual implementation of the project, which are based on frameworks such as the one mentioned above. For readers looking for a more detailed methodology for technology design, we have cited other suitable sources of information through the chapter.

3 Project Development and Pre-Installation and Design

3.1 Site Selection

A critical component of the project, i.e. a structured framework for the selection of the sites has been developed as part of an analytical framework for off-grid electrification projects under the OASYS project [23]. Using this framework, a three-level process has been followed for the selection of sites, starting at the selection of the State and narrowing down to specific clusters of villages. While the focus of the analytical framework is largely on the local contexts and its characteristics, choice of the states was contingent upon multiple determinants operating at the sub-national scales.

Odisha was selected as the State for implementation based on the high number of un-electrified villages (approximately 10,000 un-electrified villages in the year 2010) and for being an experimental ground for many donor agencies, leading to greater opportunities for electricity development synergies on the field. Secondly, from the policy and institutional support perspective, Odisha was one of the earliest states in India to adopt power sector reforms and active participation was assured by various state agencies working in the electricity sector such as renewable energy development agency, electricity distribution companies (DISCOMS) and electricity regulators, which are important factors in determining the extent to which projects can get support and achieve scale.

Mapping Level 1: There are four DISCOMS operating in Odisha. In the first level of mapping, districts were divided into four regions based on the area of operation of these DISCOMs. Since the mandate for providing rural electrification lies with the DISCOMs, a mapping based on area of DISCOM operation was preferred over an administrative mapping based on district boundaries. This would aid in identifying DISCOMS, with a strong inclination to work on off-grid interventions and new business models, and enable future linkages with the DISCOM for possible scaling up of such projects. Two project sites (cluster of villages) from each DISCOM's area of operation were selected, based on criteria such as proximity to district/block headquarters, community interest, resource availability and potential for livelihood generation activities.

Mapping Level 2: At the second level of mapping, the eight identified village clusters were assessed based on the criteria mentioned in Box 1 and the list was narrowed down to three village clusters using the process of elimination.

Box 1 Level II criteria for mapping of villages

- Community related parameter:
  - Vibrancy
  - Participation
- Demand assessment:
  - Multi-resource availability
  - Robust supply chain of identified resources
  - Supply chain for spare parts
  - Availability of local skill-sets
- Financial gradients
  - Extent of equity contribution
  - Ability to pay for electricity services
  - Ability and willingness of local financial institutions such as rural banks, cooperative banks, etc.
- Status of grid connectivity
  - Strength of local institutions and possibility of private participation

Regarding the parameters in Box 1, it is worth mentioning that as all villages have to be electrified either by grid extension or using available renewable energy resources to provide universal access to all, the site selection criteria for the project was not technology dependent, as in many government funded projects, but technology was considered as neutral. The selection focused more on two locally embedded elements, which are more critical for success, i.e. (a) strength and ability of local community structures and (b) possible economic linkages. In case the local community structure is weak and immediate economic linkage is not possible, both these elements can be developed in a village cluster over a period of time. Depending on the availability of energy resources, the appropriate technology can be designed accordingly.

Mapping Level 3: The three village clusters were subjected to a third level of mapping. While these sites ranked similarly on most aspects discussed in the second level of mapping, key differences such as proximity to the district headquarter (for ease of monitoring a pilot project), extent of remoteness and spread of households (one site had very dispersed houses, making a mini-grid unfeasible) and extent of electrification (one site had reached stage of partial electrification towards the end of the mapping process) lead to the selection of a cluster of five villages and hamlets in the Dhenkanal district of Odisha, described in greater detail in the following sections.

3.2 Description of the Project Area

The village cluster, namely Rajanga village (and its Hamlet), Kanaka village, Chadoi village and Baguli village, having a total population of 555 inhabitants, are in the Dhenkanal district of Odisha. The district is identified as one of the backward districts in India by the Ministry of Panchayati Raj, Government of India [20].

The selected villages are all un-electrified villages and also not considered under the national rural electrification scheme, Rajiv Gandhi Grameen Vidyutikaran Yojana (RGGVY), steered by the Ministry of Power. These villages also have the least chance of getting access from central grid-based electricity in the coming decade as they lie inside a reserve forest (Kandhara Reserve Forest). As per forest regulations in India, taking electricity lines through the reserve forest is not permitted [22]. However, electricity grid may be drawn in the inhabited areas within the designated village. The demographic details of the villages are shown in Table 1.The map in Fig. 2 shows the location of the project district in the state of Odisha.

3.3 Institutional Arrangements

Owing to the remoteness of the sites under consideration, the first focus of the intervention was establishing a clearly defined institutional setup for managing and sustaining the project in the long term. An often neglected factor, but one which adds to the long-term sustainability of rural electricity systems, is the strength of the institution which is intended to manage the system over its lifetime. The local institution may or may not include the main project implementing agency, but it should include the local actors. There are a number of local actors in rural areas, leading to a variety of ways in which such institutions may be established. Some of

Table 1 Summary of information from selected sites in Dhenkanal district

Name of village	Rajanga	Kanaka	Baguli	Chaddoi	Rajanga hamlet (PuranaSahi)
Latitude/Longitude	N 20.56 E 85.27	N 20.54 E 85.26	N 20.55 E 85.29	N 20.54 E 85.27	N 20.57 E 85.27
Total households	34	43	35	12	12
Total population	178	189	142	46	Included in Rajanga

Source Authors' compilation

Fig. 2 Project location in Dhenkanal district of Odisha. *Source* TERI

these actors include members from the local community, energy entrepreneurs selected from within or outside the village, local government representatives, utility company representatives, NGO and CSO representatives and independent private operators of distributed generation power plant [3, 15].

3.3.1 The Village Energy Committee

In the case of the Dhenkanal project, it was observed that being a tribal community living inside the forest, the people in the villages under consideration were isolated from the local village level governing body (called the Panchayat) with its headquarter in the nearest large village, about 10 km from the tribal village cluster. Additionally, owing to the location inside a forest and without sufficient cash income for the villagers, it was found that private operators of distributed energy systems were not willing to install and operate systems in the location. However, a few NGOs (such as Wildlife Society of Odisha) had some presence at these sites owing to their development related initiatives.

It was thus decided that a village level group, comprising of representatives from these villages, will be constituted to hasten decision making and better co-ordination of project implementation. Since the community itself has no experience in managing electricity projects, representation from external agencies such as the local NGOs (active in agriculture development programmes and training and another one on forestry and wildlife, which have a strong presence in the villages owing to their location), was included in the group. The committee is called a Village Energy Committee (VEC) and a constitution defining its regulatory role over the operations has been formulated. Some of the responsibilities include:

- Identify and donate land for the construction of the community centre and power plant;
- Overseeing construction and installation activities and wherever possible, encouraging the community to contribute in the form of labour;
- Identify the operators of the systems who will be trained by TERI and the technology supplier;
- Be the point of contact for external agencies responsible for monitoring and maintenance of the system;
- Collect revenue from the project beneficiaries as per set tariff and maintain proper record of collections (see Box 2);
- Be responsible for payment of remuneration to the power plant operator(s), and petty maintenance of the solar mini-grid system and associated equipment, or other relevant developmental initiatives in the village.
- Grievance redressal of consumers and take suitable action in cases where the operator is not providing service, charging higher revenue than agreed upon, excluding certain users and so on.

In order to adequately orient the VEC to the finer points of operational management of an electricity business, monthly VEC meetings have been held throughout the process of design and commissioning and training was given through exposure visits to existing solar power plants, implemented by TERI in neighbouring districts. Training also included the basics of record keeping and banking, required for the long-term sustainability of the project. While working with VECs, this extended process of capacity building is often missing in the case of government driven projects leading to mismanagement of projects by the VEC, upon exit of the implementing agency [28].

While we have created the institution of VEC for this project, the suitability of the institutional setup must be gauged based on the specific features of the sites under consideration. Each institutional arrangement comes with its own pros and cons and associated costs. A VEC is in that sense, a low cost arrangement, but may not be the best suited institution for tackling technical challenges and some of the institutional challenges. Also, since the committee is local, efforts need to be made to ensure adequate representation and minimise the risk of takeover of the entire system by powerful members (either socially or financially) from among of the community. Literature indicates that such VEC have seen limited success in past projects [6]. However, taking into consideration site-specific conditions in the Dhenkanal case, the VEC model emerged as the most feasible option. Therefore, an important focus of the project has been on the development of a clear exit strategy, with potential for linkages with the local DISCOM, i.e., central electricity supply undertaking, to facilitate transfer of the power plant to a technically competent authority in the future.

Box 2 Tariff setting in the village

While the tariff collection process is yet to be initiated as the project has recently been commissioned, conversations regarding the recommended tariff have been ongoing with the VEC since project inception. This tariff includes a INR 500 (~USD 10) towards connection fee per household and INR 150 (~USD 3) as monthly tariff for two light points and one mobile phone charging point per household. The project aims to collect enough monthly revenue to cover operator salaries and to set aside a fixed sum of money in a savings bank account, to meet battery replacement costs at the end of the fifth year. A margin on this minimum amount is also factored into create some additional revenue for project expansion.

3.4 Demand, Resource and Socio-Economic Assessment

During and after the establishment of the VEC, the project team focused on an detailed assessment of resources, energy demand and livelihoods.[1] Due to the economic situation of the rural community and especially in regions like the

[1] A comprehensive assessment exercise requires several visits to the sites and adds to the cost of project execution. This may not be feasible for a private operator whose primary focus is the installation of the electricity production system and distribution/sale of electricity. However since this pilot project was a action-research based activity with a focus on integration of energy and livelihoods options which can assist in growth of local economy, the detailed survey was essential for selection of suitable livelihood activities.

selected cluster in Dhenkanal, there is a preference for fuel that can be acquired with little or no money transactions. This makes it difficult for modern sources of energy to compete with the traditional sources [16]. To enhance the abilities of rural poor to pay for electricity services from renewable energy power plants, it has been suggested that the power plant itself must contribute to the increase in overall income of the community, through the inclusion of productive applications and greater community contribution and participation [2, 4, 31, 36]. In other words choosing to promote off-grid electricity by strengthening the economic position of its future buyers means that the issue of rural energy supply cannot be tackled on its own, but only as part of a broader scheme for general development.

The livelihood options explored fell into two broad categories, as explained below:

(a) New livelihood options using locally available resources: these may require the purchase of certain appliances (such as grinders) and providing the local community the necessary training required to operate the system, process the raw material and package the finished product. It is important to note here that locally available resources and familiar raw materials are always preferred to the option of sourcing raw materials from elsewhere or introducing new agricultural or other practices. For example, if the local farmers are currently selling raw chilies to the local market at low rates, a potential opportunity lies in drying, grinding and packaging of these chilies and their direct sale to city markets. This limits the involvement of middle-men and the farmers earn higher revenues on their produce.
(b) Improving the efficiency of production of existing livelihood activities: It is very common to have small scale businesses operating in rural areas, with linkages to the market either established by NGOs or government marketing departments. Owing to lack of electricity, the producers may currently be using hand tools or may be travelling some distance to get their raw materials processed. In such a case, the provision of electricity and processing equipment can significantly hasten the production process, add scale, as well as reduce costs of transport and other costs associated with operations occurring at small scales.

A number of methods can be employed to gather data on resource availability, demand and other factors important for the design of mini-grids. For the purpose of this project, data on all three aspects, energy resources, energy demand and socio-economic status was required in order to make a comprehensive assessment of the development opportunities that an energy intervention can bring to the selected sites.

Resource assessment is a subject in itself and more details on the same can be found in technical books, hence an overview is presented in this chapter. "Technical Aspects of Mini-Grids for Rural Electrification" and "Smart Design of Stand-Alone Solar PV System for Off Grid Electrification Projects" of this book also provide some details of renewable energy resource assessment methodology. A solar resource assessment was done using average monthly values of horizontal solar radiation obtained from the online database [25] and data released by the India Meteorological Department (IMD).[2] During the village scoping, the potential for biogas and biomass energy was also evaluated. The absence of a sufficient number of livestock and restriction on use of biomass within a reserve forest, however, eliminated the option of using the bioenergy sources. As regards micro hydropower, there was no stream in the village cluster with sufficient head to generate electricity from water. In addition to this, it was also felt that solar PV would result in a lower requirement of maintenance and thus lesser dependence on external resources, as compared to other technologies with electric generators and other mechanical moving parts, an essential factor to consider owing to the remoteness of the community.

The rest of this sub-section will focus on the methodology used to collect data to assess demand, socio-economic conditions and livelihood activities that could be promoted with use of energy. Before delving into the assessment methodology, a short note on the linkages between energy and livelihoods is provided in the following section.

3.4.1 Assessment Framework

In the assessment process, the project team focused on the categories of data mentioned in Fig. 3. This list is indicative and must be adapted to suit the specific sites under consideration. While the main data categories and sub-categories are presented here, the significance of each of the mentioned categories is presented in Annexure B.

3.4.2 Summary of Information Collected for the Sites Under Consideration

The baseline survey conducted by TERI showed that the inhabitants of these villages are primarily dependent on subsistence agriculture, goat rearing and collection of forest produce for their income. Some are employed as labour by the Forest Department or MGNREGA[3] (Mahatma Gandhi National Rural Employment Guarantee Act).

While paddy is the main crop grown, horsegram, arhar (split red gramme), ginger, turmeric and other vegetables such as cucumber, ladies finger and eggplant are also grown. Villagers also collect Non-Timber Forest Produce (NTFP) from nearby forests which include mahua (Madhuca longifolia: seeds are used to extract oil), karanj (source of Pongamia oil), honey, saal leaf (Shorea robusta: the leaves

---

[2] India Meteorological Department, Pune. [15 Jan 2014] http://www.imdpune.gov.in/.

[3] A programme of the Govt. of India which guarantees a minimum of 100 days of paid work in the rural areas.

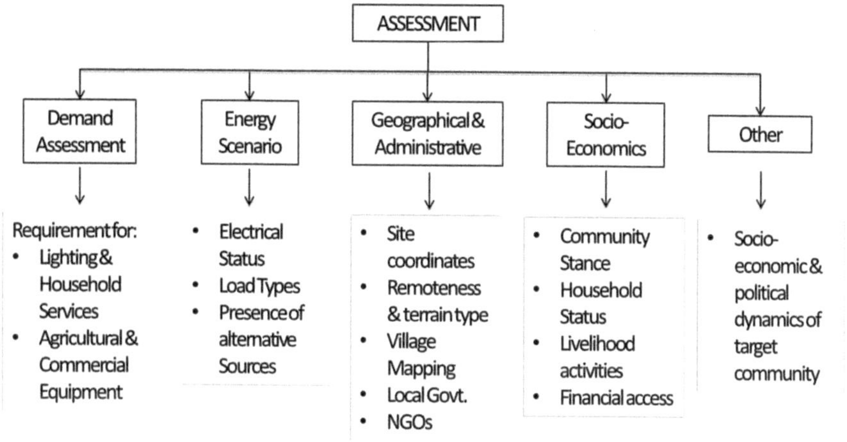

Fig. 3 Key components of the assessment framework

are dried and stitched together into plates), tamarind and amla (Indian Gooseberry). Most households also rear poultry and livestock, but in most cases the livestock belong to the relatively richer farmers from other areas who lease them out to the locals for grazing.

All the villages being inhabited by the tribal community and inside a reserve forest, limited development assistance has been channeled into these villages. Due to the lack of opportunities, in recent years the younger generation has started migrating to other states to look for better sources of income. Owing to the remoteness of the villages, the farmers are primarily dependent on middle-men, who come to the village to buy their produce and then sell the produce to towns and cities at higher prices. In such a process, the prices are usually dictated by the middle-men as farmers have little information on current market prices. In addition to this, with no provision for weighing or measuring the produce accurately, a barter system exists for many products, which is often disadvantageous to the local producers.

In addition to the lack of access to electricity and resulting limitations to growth, the area is also highly malaria prone and children also often suffer from water-borne diseases. The impacts of such childhood diseases usually lead to damaged immunity and growth over the life of the individual. Hence, there is a large scope for improving health services in this area.

The summary presents the broad findings of the survey. For the purpose of design of an intervention, the next section will focus on the methodology followed for the selection of certain livelihoods from the larger list of options identified during the field survey.

Table 2 Selection of appliances load for the village cluster in Dhenkanal district

Appliances	Capacity	Purpose
Grinder	1 HP at Rajanga	For grinding turmeric and chilli powder
Electronic weighing scale and sealing Machine	Weighing Scale—10–20 W at Rajanga Sealing Machine—150 W at Rajanga	For accurately weighing and packaging of ground turmeric and chili powder. These have wider application since users can use the machine for other products as well and bring about standardisation through accurate weighing, delinking them from the middle-men
Saal leaf plate pressing machine	0.5 HP at Rajanga	The community is currently engaged in the stitching of basic Sal leaf plates, called Khalis. By pressing two Khalis together in the machine, a firm plate can be moulded which has a much higher value in the market
Water pumps for agriculture	2 HP each at Rajanga and Kanaka	Agricultural is possible only during the monsoon and the farm lands lie unused during the dry season. Hence, to improve agricultural yield, especially of high value crops such as eggplant, water pumps have been provided to initiate activities during the non-monsoon period

Source Authors' compilation, 2013

3.4.3 Assessing Household, Community, Commercial and Agricultural Loads

Household load. for each household, based on the most prominent needs and the size of rooms, provision for two LEDs lights of 3 W each has been made in the design. In addition to this, it is noted that owing to low income levels, the household do not have any other high wattage appliances and therefore a household socket with a 10 W limit has been factored into the design. These household loads will also serve as the most consistent stream of revenue for the power plant as each household will pay a monthly fee for the use of such services.

Commercial load. based on a cost-benefit analysis of the potential livelihood options, the activities listed in Table 2 were selected for the design of the power plant. Some activities such as oil extraction and honey processing were excluded owing to their seasonality, low volumes and the involvement of small sections of the community in the collection of raw material from the forest.

Community load: In addition, provision for the following community services has also been included while designing the power plant:

- Street lights in all villages—the villages fall in the elephant corridor and so the street lights are expected to ensure safety at night;

- TV-DVD at the community centres (for educational and training purposes)
- Community centre lighting and fans (to create a resource/community centre for people to work and also for meetings, discussions, trainings, etc.)
- Water Purifier (50 LPH)—one each at Rajanga, Kanaka and Baguli for provision of potable drinking water and address some of the health related issues

4 Designing the Power Plant

This section focuses on the process by which data collected from the field has been used in the design of solar photovoltaic power plants. The design process presented here looks at the broad considerations that go into making certain decisions about how the power plant must be sized, and distributed, while presenting an overview of the technical details.

4.1 Solar Radiation Data

Accurate solar radiation data from ground weather stations is usually not available for remote locations. Therefore, in order to get an estimate of how much solar radiation is available, project designers usually use data from the nearest available ground weather station, or use a reliable online resource such as the NASA database, which requires only the coordinates (i.e. latitude and longitude) of the location as input. For the location described here, data from the NASA-RETScreen database has been used, which indicates that the daily average global horizontal solar radiation is close to 4.82 kWh/m^2/day (Table 3). For calculations, a more conservative figure of 4.5 kWh/m^2/day has been assumed. Although more accurate data from ground weather stations may be required for commercial scale solar power projects, factoring in the cost of establishing ground stations, readily available data may be utilised for village scale systems.

4.2 Optimising Loads

Once the range of possible loads and the estimated time during which the load will operate has been identified, the data needs to be mapped onto a load curve as depicted in Fig. 4.

Figure 4 depicts the energy required for domestic, commercial and agricultural loads throughout the year and also the maximum amount of energy that can be derived from a power plant to cater to these loads. This graph represents a scenario before the load has been optimised. We observe that the energy requirement is practically uniform throughout the year for the domestic segment (which will be

Table 3 Radiation data for the village cluster in Dhenkanal district

Month	January	February	March	April	May	June
kW h/m^2/day	4.53	5.26	5.82	6.34	6.18	4.50

Month	July	August	September	October	November	December
kW h/m^2/day	3.88	3.80	4.15	4.57	4.43	4.38

Source NASA-RETSCreen online database for coordinates 20.55 N, 85.28 E (i.e., for Rajanga village)

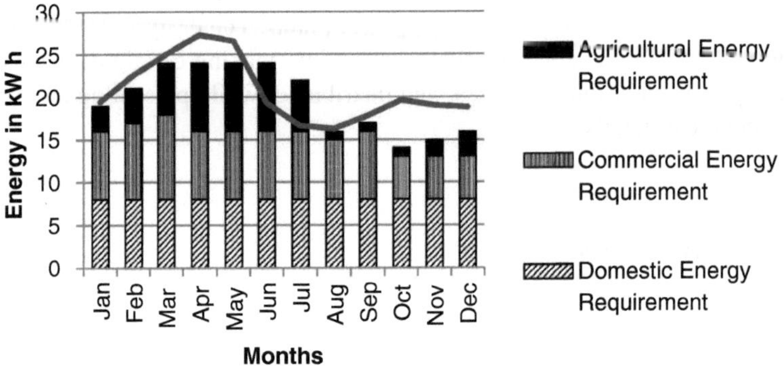

Fig. 4 Load curve before optimization. *Source* Authors' compilation

the case for lighting loads but will vary in case higher wattage equipment are being used in the homes), while there is variation in the agricultural and commercial segments. In the agriculture segment, water pump usage begins to drop during the rainy season (August to October), rises from November onwards and is high during the summer months (April to June). Commercial loads will vary depending on a larger number of factors, including seasons, the availability of raw materials for processing, the demand for finished goods, labour availability and so on.

For two months of the year, June and July, the demand is greater than supply and therefore all the demand would not be catered to. And in other months (October–December and April–May), the plant is not being utilised completely. Wherever possible, project designers must re-schedule loads to avoid these two situations. Although perfect matching of demand and supply may not be achieved, efforts to ensure that the demand is not greater than the supply are essential. In this case, such a shift has been indicated in Fig. 5, where the commercial loads have been either reduced or increased. Working with commercial loads is often simpler since domestic and agriculture demands are generally fixed (domestic loads may be constant through the year and agricultural loads may only be used during dry seasons). Such adjustments can be done by either moving a commercial activity from one month to another, finding alternative low energy intensive activities or finding more efficient machinery to carry out the same activity as before. This

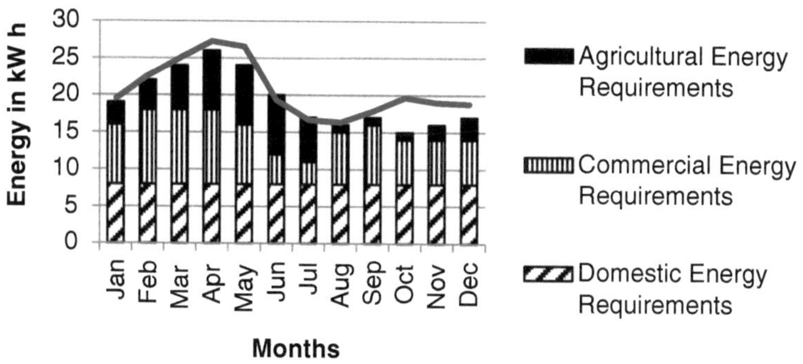

Fig. 5 Load curve after optimization of loads. *Source* Authors' compilation

helped in optimising the capacity of the power plant and at the same time provisioning energy for all identified loads, without the need to add extra capacity for the agricultural load (which may require energy only for few months).

4.3 Technical Design

It should be noted here that a detailed technical design (one which includes details on PV array/string sizing, wire selection, distribution losses and so on) is beyond the scope of this chapter and not presented here. However, the process described here is essential to preliminary project design and evaluation and can be used by donors, government agencies and project developers/implementers whose core area of expertise is rural development/energy access/implementation and not technical design. Readers can, however, refer to "Smart Design of Stand-Alone Solar PV System for Off Grid Electrification Projects" for details on off-grid solar power plant design. The following section briefly outlines the different components of a solar photovoltaic system and some design improvements that can significantly enhance the efficiency of a village scale power supply system.

4.3.1 Solar PV Plant Components and Assumptions for Design

Photovoltaic modules

There are a number of factors that impact module selection, apart from the efficiency and type of module technology. While deciding the minimum performance criteria for modules for the Dhenkanal sites, one important factor was the number of cells (wattage of each module). Larger modules with a higher number of cells are preferred as they usually have higher module efficiency and the relative length of wiring between modules is reduced, therefore reducing the associated losses in the system.

Also, the degradation of power over the module lifetime should be within acceptable limits. A good quality module should not degrade more than 10 % over the first 10 years of its life and by maximum 20 % over 25 years. Finally, the choice between these modules will also depend on their availability and cost in the country of installation.[4]

Battery Bank

Despite their long existence and widespread usage, lead-acid batteries have one of the lowest energy-to-weight and energy-to-volume ratios. In essence, this means they are too large in size and heavy for the energy they provide. However, their prime advantage is that they are cheaper, robust and well-tested for rural use. As an alternative, Lithium Ion batteries are more expensive, but have high charge density with longer useful life as compared lead-acid batteries and are significantly lighter in weight (refer "Smart Design of Stand-Alone Solar PV System for Off Grid Electrification Projects" for details).

In the Dhenkanal project, lead-acid tubular plate batteries have been utilised. Although sealed lead-acid batteries require little maintenance, flooded lead-acid batteries were selected, owing to their longer deep-cycle discharge life, high discharge rate capability and better performance at partial states of charge as compared to sealed batteries. In such batteries, the need for refilling the batteries with distilled water exists, which is often unavailable in rural areas. Hence, a solar still is being installed for the production of on-site distilled water. Also, the equipment being under warranty for 5 years, the technology supplier is expected to take care of the maintenance of the batteries.

Power Conditioning Unit (Charge Controller and Inverter)

The charge controller selected for off-grid application must have reverse polarity protection, overvoltage protection, deep-discharge protection and temperature compensation feature. Deep-discharge protection helps in protecting the battery from deep-discharge, and temperature compensation feature sets the charging point as per the operating temperature hence reducing the charging stress on batteries during charging.

Solar modules generate DC power which is either stored in the battery or can be directly fed to the load. However since most appliances available in the market operate on AC power, a conversion from DC to AC power is required, which is achieved through the inverter. Pure sine wave inverters are the most preferred option as they enhance the performance of appliances used and while an inverter's basic function is to convert one type of power to another; smarter inverters also play an important role in enhancing the life and efficiency of the system as a whole. Such inverters have an inbuilt battery management system, and a power–frequency control system which automatically adjusts the PV power from the PV inverter as

---

[4] Project implementers should choose solar PV modules that conform to the latest edition of IEC (International Electrotechnical Commission) Standards, such as IEC61215 for Crystalline Silicon Terrestrial PV Modules or IEC 61646 for Thin Film Terrestrial PV Modules.

per the desired load. Also these types of inverters are grid-interactive and therefore when the grid does reach an off-grid site, such inverters can feed extra power into the grid and in addition, the grid can be used for battery charging as well.

Smarter inverters also include a feature of load shedding wherein certain loads can be disconnected through a load shedding contactor in case the battery is drained beyond a recommended and set limit. The load shedding limit can also be set as per system requirements. While very high efficiency inverters are now available in many countries with efficiencies in the range of 97–98 %, in the ideal case for rural electrification project, an inverter should be selected such that its weighted average efficiency is over 90 %[5] (please refer "Smart Design of Stand-Alone Solar PV System for Off Grid Electrification Projects" for more detail on inverters).

4.4 Power Plant Design

This section highlights some of the key assumptions and design decisions taken by the authors for the design of the AC and DC grids. While the design methodology is adopted from existing IEEE standards, a step-wise design process is beyond the scope of this chapter [12]. The attempt here is to illustrate the ways in which standardised design processes may be altered to suit specific field conditions, user requirements, project budgets and timelines.

Cluster-based approach for decentralisation. The guidelines for decentralised distributed generation (DDG) released by the Ministry of Power, Government of India recommends, that to the extent possible, selection of the villages/hamlets is to be carried out in a cluster to take advantage of the clustering effect and the merit of setting up a local distribution grid covering all these villages/hamlets with a central power plant as against setting up of individual village/hamlet level systems [21]. Clustering could, however, be seen in different ways; technical clustering wherein a single power plant is installed and distribution network is drawn to supply power to different villages and institutional clustering wherein a single operating entity, such as the VEC with representatives from all villages, is established but discrete power plants are installed in each village.

While clustering has its own advantage of economies of scale and scope, it is imperative to assess the benefits—both from technical design and investment perspective as well as operational and maintenance costs and also ease of managing the project institutionally. Further, the feasibility of adopting cluster approach may also depend on the technology selected and the results may not be same for a solar PV project design and other renewable energy technologies such as biomass gasifier or mini hydropower-based projects.

Therefore a comparative study of the technical design and investment required in two scenarios was conducted, one with a central power plant at Rajanga village

[5] Minimum 95 % efficiency (at full load 0.8PF) and 80 % for partial load (at 50 and 75 %).

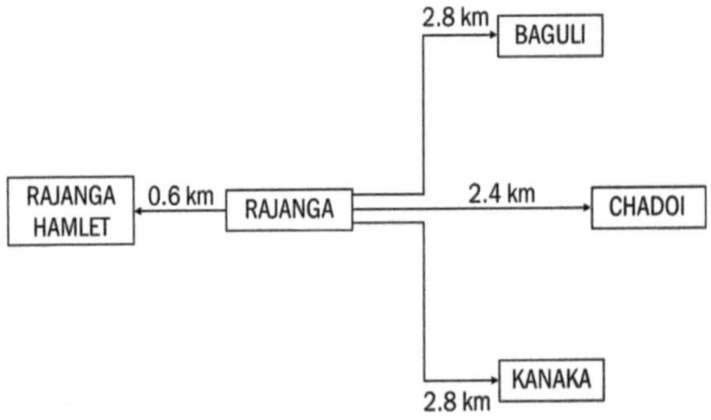

Fig. 6 Distances between villages in the project cluster in Dhenkanal district

Table 4 Costs of implementing five individual power plants at each site versus implementing a single power plant at Rajanga

Particulars	Costs (in USD)	
	Five distributed systems	Centralised system at Rajanga
Solar power plant	51,000	47,100
Distribution line	28,200	84,500
Household connections	4,400	4,400
Construction requirements	32,000	24,500
Annualised O&M cost over a 5 year period	3,900	1,600
Total	119,500	162,100

Source [28, 29]

with distribution lines to all other villages and a second scenario in which five individual power plants at each village (AC & DC) are to be designed [29]. Figure 6 shows the distances between villages in this cluster.

For this comparative study, the technical designs of the two scenarios were first developed and then using actual cost estimates quoted by the vendors, the costs of executing the two scenarios were calculated. Some important factors considered were: cost of salary of operator(s) (five operators for the case with five systems and a single operator for the centralised system), capital costs for solar power plants, civil construction and distribution network, and annualised operation and maintenance costs for a period of 5 years post-installation. While for a centralised system, the cost of civil construction, the operation and maintenance costs and solar power plant was found to be lower owing to the centralised installation and operation, this benefit was completely offset by the high cost of putting up the inter-village distribution network to all five sites (Table 4).

Table 4 clearly shows that there is high difference in cost (around 36 %) owing to almost 200 % increase in the cost of laying the inter villages transmission line cost from a centralised solar power plant. This is, thus, not a feasible option for such small power plants and low demand and so a decentralised approach is preferred from technical point of view vis-à-vis a clustered approach for electricity supply. However, there might be some operational benefits to having a centralised power plant, mainly accruing from ease of operation and maintenance of a single system. But with a good system design that ensures infrequent maintenance, even managing multiple power plants may not be a difficult proposition. Owing to such technical and economic limitations, a strong case can be made in favour of technical design of distributed systems. However for ease of management of the system, a single institutional entity can be formed covering the group of villages.

While the analysis carried out for solar PV technology clearly demonstrates the efficacy of having decentralised power plants with a possible institutional clustering for better management, the same may not be said for biomass gasifier technology. While solar PV technology is a truly modular technology with sizing from Wp to a scaled-up capacity of MWp, there are design capacity constraints existing for other renewable energy technologies. For example, the minimum commercial capacity available for a biomass gasifier in India is 10 kWe and so clustering of load for best utilisation of rated capacity may be useful for the gasification technology [27]. Similar is the case of small wind aero generators and mini-hydropower. Also, while this analysis was done for cluster of villages within a forested area and a small population, in case of area with higher population and density, a larger power plant may be more feasible.

Choosing between AC and DC systems. The choice between an AC or a DC system is based on the type of application being served, the scope for system expansion, requirements for future grid connectivity and budgetary constraints among other factors such as spread of households and distribution losses. DC micro grids for lighting and mobile charging (both of which can be operated easily on DC supply, especially with the availability of LED bulbs) have become a popular option in India when no other service is to be provided and the number of households is limited to a small cluster [30]. Since the inverter is excluded, this results in a cost saving while providing the same service quality to the users.

For two sites in the case presented here, Rajanga Hamlet and Chadoi, it was found that the number of households was limited to 15 (12 households when the survey was conducted). In addition, these two sites are not very close to the other larger sites where AC power plants were being constructed. Extending the AC grid from the larger sites to these smaller sites proved to be expensive proposition, as explained later in this chapter. Hence, considering all the above points, a DC micro grid was designed to cater to lighting and mobile phone charging need at these sites.

In Rajanga (the main village excluding the hamlet), Kanaka and Baguli where the number of households was larger and the scope for commercial and agricultural applications existed, an AC system was designed. It is to be noted here that the number of light points and quality of the final service (lighting) for households served by AC and DC power plants has been kept similar (i.e. the type of lamp and the lumen output of the lamps have been selected as same for households) so as to avoid any issues of deprivation. Thus, the metric chosen here for residential services is energy services (final output), and not energy (kWh) or how the energy is delivered, to ensure equitable services. This is important because users are often not concerned with what the source of power is,[6] but rather with the quality of the service they are receiving. Users getting services from the same project will compare the quality of service being received by them with others, which could potentially lead to conflict and hence it is important to ensure that the end-user is not impacted by the choice of system being designed.

Provision for demand growth. In the case of the project cluster, although the actual requirement is 3 W per light and 3 W per mobile phone plug point (in total 9 W), the design value per household is taken as 30 W. This has been done to address future increase in demand (some household may install fans) owing to enhanced income from the livelihood generation activities being promoted. Secondly, LED bulbs are expensive and not readily available in rural markets. Although at the beginning of this project, these LED bulbs are arranged by TERI as part of the project, there is no guarantee that in case the bulbs get damaged, the users will replace them with similar LED bulbs, owing to cost and availability constraints. The alternative bulbs available and used in the area are 7–10 W CFLs and users may choose to install these instead. In addition, some new productive or agricultural load may also be required in future or the number of households in the village may also increase.[7] Hence, to ensure that there is no overloading of the system in the future, some spare capacity has been built into the design.

However, any attempt to include additional capacity for demand growth must also consider the increase in project cost owing to this higher capacity. Instead of 30 W, as mentioned above, if 10 W per household had been considered as the household load, the project cost would have reduced by about 20 %. However, factoring in the reasons for increase in per household capacity mentioned above, leads to a definite need for a higher system size in the near future. For a system designed to cater to 10 W per household, this would require a significant investment in the future, not just in terms of increased solar module capacity, but also new inverters, an increase in battery bank size and possibly new cables to carry the higher current required by more households and higher loads.[8]

Factoring in days of autonomy of the solar power plant. The capacity of the solar modules in kW needs to factor in both day and night loads (i.e. after producing enough electricity to cater to the day loads, the solar panels must be of enough capacity to charge the battery during the day to cater to the night loads). In the design process used here, the solar panels have been sized to completely charge the battery during the day, to cater to loads being switched on the same night.

---

[6] However, in one of the DC micro grid sites, the operator of the system was initially concerned about the quality of service from the DC grid, as the electrical poles used for the AC grid were larger in size (as per standard AC distribution requirements). His perception was that the quality of power was related to the size of the pole. The issue had to be addressed carefully and through repeated engagement with the person and the community so as to avoid future conflicts.

[7] While the total number of households during the demand survey was around 130, it increased to 135 households during the actual installation of power plants, within a period of 6 months.

[8] Being an action-research project, the project team also want to understand the load growth phenomenon in the remote villages.

For the extra 'night' of autonomy, the solar panels have been sized to charge the battery over a period of 5 days, therefore lowering the total PV capacity required. This has been arrived at from noting the number of continuous sunny days in the region for most part of the year (i.e. the possibility of having a day with no sunshine at all, occurs on average once every 5 days for most part of the year), giving the plant sufficient time to charge the batteries for the extra night of autonomy.

Practical consideration for design of the DC micro grid. A DC micro grid is a visibly simple looking system working at low voltages. Hence, the chance of users sometimes interfering with the system does exist. For example, a motivated user may directly connect a high wattage device to the battery (such as another battery) for charging. To minimise such user-interference, the battery and the charge controller have been placed inside a secure wooden box, equipped with a number-lock. Additionally, a timer circuit has been included to automatically switch on and off the system so as to avoid any human interference on the run time of the systems.

Smart system designs for AC mini-grids. Since nighttime loads include household lighting, it is crucial that the energy being fed to the batteries is not compromised on account of over-usage of day time loads. Additionally, the system should be designed such that it remains partly functional even if the battery is almost fully discharged. This can be ensured in two ways:

(a) Over-design and load management—at the time of design, a little extra capacity of solar modules can be built in so that if the day load increases, it can be served directly during the daytime without draining the battery.

Through load management, the loads are segregated into essential and non-essential loads. The non-essential part of the load is connected to the system via a load shedding contactor which in turn is connected to the battery terminals. During cloudy or rainy days, when the battery is not fully charged, the non-essential loads can be automatically disconnected through the load shedding contactor as described in the section on smart inverters.

(b) Alternative solar PV system topologies: Alternative topologies not only improve the battery health but also improve system efficiency, reliability and flexibility for future expansion. The issues with standard configurations and one such improved configuration is presented below:

If we observe the dominant design for solar photovoltaic systems in India today (Fig. 7), a series configuration is usually adopted, which means the solar PV array is used to charge the battery first through the charge controller and then the DC power is inverted to AC power. In such design, the battery has to be healthy in order to provide the required power. The author's experience and other observations from the field show that with such system design, if the battery is in poor condition (low voltage owing to deep-discharge), it is likely that the entire system is also non-functional even on sunny days. Additionally, with such a design, since the entire power is provided through the battery the overall efficiency of such system is also not optimal (considering maximum watt hour efficiency of battery) because of battery losses.

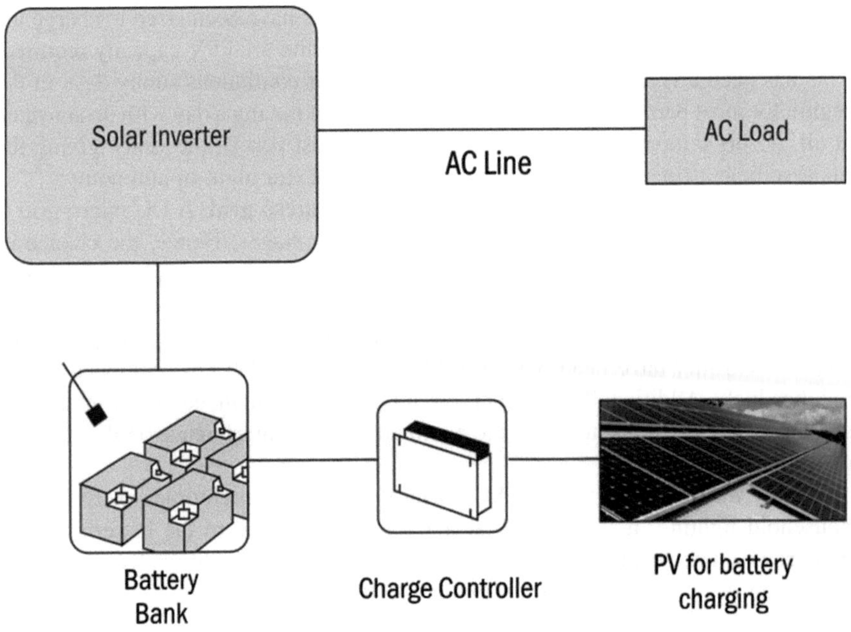

Fig. 7 Series configuration of solar photovoltaic system. *Source* TERI

Experiences clearly show that clean energy systems in rural and remote locations need to be designed taking into account that access to frequent service, repair and maintenance is difficult, that is, the system should be reliable. Also, since the capital cost of clean energy systems is high, it is imperative that the efficiency of such systems is optimised so that cost per every unit of electricity generated is optimum.

The above requirements can be met with proper planning, optimum design and reconfiguration of the system which leads to some incremental cost. But this incremental cost can be easily repaid through improved efficiency, reliability and flexibility of the system in the long run, lower monitoring requirements and improved battery life and lower replacement costs. For example, changing the configuration (moving from series to parallel configuration), adds more flexibility and enhances the probability of supply of power, even in poor weather and local operational conditions.

In a parallel design, the inverter is split into two as per the day load and night load requirement (Fig. 8). One inverter is a grid tied inverter which is responsible for feeding power directly from the solar PV to the day time loads. The other inverter is responsible primarily for (a) charging the battery (b) and providing the reference grid to the grid tied inverter for feeding power. In this configuration, the grid tied inverter can continue feeding power even if the battery goes into the fully discharged condition (low voltage disconnect point). Hence such an arrangement adds reliability into the system.

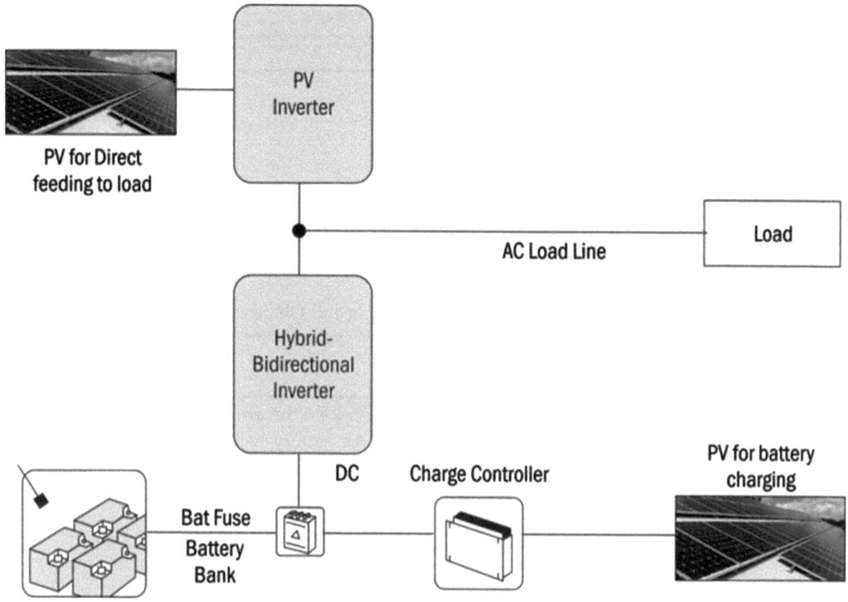

Fig. 8 Parallel configuration of solar photovoltaic system. *Source* TERI

Such inverters are selected in such a way that grid integration and integration of additional renewable energy sources at a later stage are also possible. In addition to this, these inverters are made with remote monitoring features through which daily output and utilisation data can be recorded remotely. Such data is especially useful for planning of future systems in areas where cloudy and rainy weather conditions are prevalent for long periods of time.

At the Rajanga site, a parallel configuration of inverters has been utilised as most of the livelihood generation activities (which operate during the day) are located in Rajanga and hence a robust system which can cater adequately to both day and night loads without putting strain on the battery is essential. Additionally, the loads themselves have been segregated such that the household loads are all on one feeder and the community and livelihood loads are on a separate feeder. In case the battery is in a heavily discharged state, the system will automatically cut-out of the community/livelihood loads through a DC contactor placed at the origin of the feeder.

Distribution network and water pumping facility. For each power plant, an independent power distribution network within the village has been planned. In the villages of Baguli, Rajanga Hamlet and Chadoi, this distribution network caters only to household and productive loads (if present). In Rajanga and Kanaka, the distribution network also connects to a water pump, located at some distance from the households. Rather than adopting a stand-alone system for the water pump, it has been connected to the main power plant in the village through a distribution line. Although the distribution line contributes to additional cost, such a connection has

Table 5 Day and nighttime energy requirements per village

Site	Day load (kW h)	Night load (kW h)
Rajanga	11.67	6.07
Kanaka	7.66	6.88
Baguli	0.30	5.84
Chaddoi	Nil	0.58
Rajanga Hamlet	Nil	0.58

Source Authors' compilation

Table 6 Design of 5 independent solar power plants in the Dhenkanal cluster

Site	AC/DC	Plant design	Total length of the distribution line (m)[a]
Rajanga	AC	6 kWp SPV power plant, 6 kVA inverter (grid tied), 48 V 500 Ah battery bank	2100
Kanaka	AC	5 kWp SPV power plant, 5 kVA inverter, 48 V 600 Ah battery bank	1400
Chadoi	AC	2.5 kWp SPV power plant, 2 kVA inverter, 48 V 500 Ah battery bank	1100
Baguli	DC	400 Wp, 24 V 200 Ah battery bank, 15 A charge controller	300
Rajanga Hamlet	DC	400 Wp, 24 V 200 Ah battery bank, 15 A charge controller	300

Source Authors' compilation
[a] The length of distribution line along the longest feeder has been kept within 1 km to keep the technical loss at the minimum

been preferred to (i) cater to any future requirements of increase in pump size and (ii) cater to inrush current required by the pump when it is switched on.

Provision for future grid interconnection. At the Rajanga site (one of the three AC mini-grid sites), a grid tied inverter has been installed. Regular solar inverters have been installed at the other two AC mini-grid sites to keep the project cost within reasonable limits (considering the higher costs of grid tied inverters). The grid tied inverter provision has been built into the largest and most accessible (by road) village, keeping in mind the possibility that some years in the future, the grid might reach this village. In case of grid extension, it is expected that the households will take grid connection and in such a case, the electricity can then be fed to the 11 kV grid, using the grid tied inverter and the VEC can continue to earn revenue from the DISCOM. The non-grid tied inverters can continue to operate on a stand-alone basis to support the productive and community load in the villages using the solar power capacity. They, however, have provision to get supply from the grid to charge the batteries with preference to solar charging.

Using standardised design methodologies and factoring in the considerations mentioned in this section, the day and nighttime energy requirements and final designs of the five systems implemented in the Dhenkanal sites in presented in Tables 5 and 6 respectively.

The system has now been installed, tested and commissioned. Box 3 provides some considerations used in this process.

Box 3 Installation and Commissioning of systems

A few important points considered by the project team during the installation and commissioning process are mentioned below:

- Following the guidelines of the Indian Electricity Act on safety and electrical supply [8] a comprehensive safety check has been conducted by a certified electrical inspector.
- 72 h (or 3 days) of complete trial run has been conducted to evaluate system performance during both day (charging cycle) and night (discharging cycle)
- Mobilising labour early in the project is a practical way to involve the local community, solicit their participation and also reduce implementation costs.

5 Concluding Remarks

The aim of this chapter was to present a case study of mini-grid implementation, based on a standardised framework. However, while working within this standardised framework, we have also attempted to highlight aspects where innovation is required, so that the final product is customised to suit the needs of the end-user, the availability of technology in the market, site-specific boundary conditions owing to weather, transport, user community characteristics and so on.

While the design and implementation of the project is complete, the impacts of these innovations are yet to be assessed. Under the OASYS project, our attempt over the next one year after project commissioning is to assess the performance of the systems from both technical and institutional points of view. Further research output on these aspects will be published as part of the OASYS project to provide actual field level data and analysis, which we believe will add further value to the work being done on mini-grids by practitioners across the globe.

Acknowledgement The authors would like to thank the entire team of professionals from TERI and our partner organisation IRADA, who have tirelessly contributed to the completion of the demonstration project in Dhenkanal district of Odisha. We are especially grateful to Mr. Joy Daniel Pradhan for his presence on the field and constant engagement with the village community. We would also like to thank Mr. Sudhakar Sundaray, Research Associate, TERI for his contribution to the technical design of the systems, Punam Energy Systems Private Limited, for their timely installation and commissioning of the systems based on our design and Ms. Apoorva Mathur, Research Intern in TERI, for her assistance with diagram and editing. We also acknowledge the support provided by EPSRC/DFID and Rural Electrification Corporation Limited to meet the capital cost of installation of the solar mini-grids in the villages. Authors of the reference materials are also gratefully acknowledged.

Annexure A

Detailed List of Project Implementation Activities

Project Development and Pre-Installation

Site selection

- Identification of the sites for the intervention: includes collecting information on electrification status of the State (such as Odisha in this case), progress of rural electrification programmes over the last 3–5 years, distance/remoteness of the sites from district headquarters or other well-connected towns, distance of electricity grid from the sites, terrain and climactic conditions.

Feasibility studies and surveys

- Field visit and primary survey for collecting household information, existing and future load scenarios, key livelihoods and other sources of income and site-specific boundary conditions
- Secondary data collection and its authentication
- Stakeholders consultation meetings: with village residents, prominent members of the community, locally active NGOs, local government officials
- Electricity demand assessment for household, street lighting, agriculture, commercial and industrial segments
- Energy Resource assessment (such as hydro, solar, biomass, wind, biogas, etc.)
- Electricity linked livelihood assessment (examples include water pumping, grinding, food processing, etc.)
- Market assessment and surveys for products and services resulting from electrification
- Cost-benefit analysis of proposed interventions.

Project development activities

- Industry surveys to gauge maturity, availability and cost of different technology options
- Techno-economic assessment and tariff determination
- Contracting with local partners
- Inter-agency co-ordination, clearances, application, approvals
- Procurement of land and infrastructure for power plants, community hall, etc.

Establishing the institutional setup

- Formation of a local electricity project governance body, based on specific site conditions
- Drafting guidelines and responsibilities of this local institution
- Ensuring sustained involvement of this local institution through regular meetings and training programmes in order to transfer decision-making responsibilities to the local institution.
- Communication of installation charges, tariffs and management of electricity project funds through a transparent process to the entire consumer group.
- Based on specific site conditions and project requirements: required: capacity building for financial management, awareness on benefits of clean electricity use (as compared to existing fuels such as kerosene), involvement of NGOs and other local bodies for synergizing energy and livelihoods.

Design, Procurement, Installation and Commissioning

Project design

- Assessment of resource options such as hydropower, biomass, small wind and solar photovoltaic among others to arrive at a feasible technology option
- Assessment of the public distribution network requirements (length of feeders and total distribution network length)
- Estimation of project costs including costs of the power plant, distribution network, household wiring, land, civil construction, transport and professional costs
- Preparation of the Detailed Project Report (DPR) which contains all the information collected above (resources, demands, institutions) and the design options and costs.

Procurement

- Preparation of a detailed Bill of Materials (BoM) for the power plant and other appliances being considered for the project
- Identification of suitable vendors/suppliers of renewable energy-based power systems (and other components of the project such as distribution and civil construction) and soliciting of quotations against the Bill of Materials.
- Vendor selection possibly using quality and cost-based selection process
- Quality assurance through inspection of previously established projects
- Site visit for finalisation of design and costs with the selected vendor.

Installation and commissioning

- Mobilisation of labour: crucial step in remote areas where a paucity of skills labour for activities such as civil work exists and costs of sourcing labour from towns and cities are high
- Civil works and site preparation for installation of distribution network
- Installation of the power plants and associated distribution network
- Technical and safety checks
- Trial run and commissioning of the system.

Post-Commissioning and Sustaining the Project

Operation, maintenance and monitoring

- Hands-on training on operation, maintenance and repair
- Organising the supply of spare parts
- Record-keeping for faults, energy generation and consumption and revenue
- Handing-over of the commissioned system to the local institution.

Business development

- Integration with complementary development schemes
- Skill development for income generation
- Raw material procurement for secondary activities
- Market development for finished goods.

Annexure B

Assessment Framework

Category	Sub-categories	Significance
Geographic and administrative information	Exact location and coordinates of the sites	For estimating the availability of resources such as solar, hydro or biomass energy from renewable energy atlases Design of the electricity distribution system using software

(continued)

(continued)

Category	Sub-categories	Significance
	Terrain and distance of the sites from nearest block headquarters or large town. Nature and quality of roads and impact of weather conditions on access to sites	Assessment of transport requirements, especially in the context of transport of heavy materials such as power plant equipment Estimation of project completion time and impact of weather on the same
	Village mapping—location of individual households, schools, hospitals, local government offices, agricultural land, water bodies, fallow land, forests, etc.	Design of power plant and distribution system Availability of space for construction of power plant Assessment availability of government services and natural resources
	Type of local government and current and proposed development schemes (government or NGO led)	To understand potential partnerships and plan for synergies with existing or forthcoming government/NGO programmes
Electricity/energy scenario	Grid electrification status of the village and hamlets. Quantity and quality of power, number of connections, potential for electrification	Often the main village is electrified but the hamlets are not. Although the electrical lines may have reached the village, it is important to assess the quantity and quality of power being received and the estimated time by which the grid will reach the sites, if it has not already
	Alternative sources of energy/electricity—existing decentralised renewable sources in the form of solar lanterns, solar home systems or others and Diesel Gensets, kerosene lamps, batteries, etc. And the quantity and cost of each option on a per household basis	These sources will provide the assessor information on current usage patterns and also the costs being paid for availing the services. These costs can be treated as benchmarks for project design
	Types of loads, household, commercial and agricultural loads currently in use	To assess which existing loads need to be provided with electricity (in case they are operating on diesel or other source currently) and the potential load profile of the area

(continued)

(continued)

Category	Sub-categories	Significance
Socio-economics	Composition of community, community leadership.	In the case where the power plant is to be community managed or community involvement is important, it is important to understand community dynamics and leadership, to better facilitate an intervention in the area and minimise risk of conflict Even for privately operated power plants, such data is essential before proposing an intervention in the area to ensure cooperation from the community as a whole
	Income profiles, key occupation, type and area of houses, household size, land, livestock and other capital ownership, education, proximity of market and other commercial establishments, key agricultural crops grown, other livelihood activities such as NTFP collection and labour	This data is key to identification of potential livelihood activities that can either be started or be scaled up
	Access to finance	Rural communities often do not get easy access to finance owing to low incomes and unavailability of documentation of assets. Power projects that require investment would need to assess the availability of finance either through FIs or MFIs in the region before making an investment decision.
Demand assessment	Duration of time for which lighting and other household services are required Daily, monthly and yearly requirement of commercial and agricultural equipment	Daily household load patterns in rural areas are general uniform throughout the year and the assessment should also aid in estimation of the cap on energy consumption per household and the prospective loads that could be added over the next few years For larger loads such as commercial and agricultural loads, accurate data on time, day and season of usage can lead to a more optimised system design and higher capacity utilisation factor of the power plant through intelligent load management

(continued)

(continued)

Category	Sub-categories Significance
Other factors which contribute to understanding the socio-economic and political dynamics of the community	Access to markets or other avenues for sale of produce. Involvement of local middle-men or organised agents in the collection of produce from the sites. Availability of modern communication facilities including cellphones and internet Existing community-based initiatives such as water conservation, forestry, etc. Extent of involvement of local governing institutions and NGOs Presence of local youth for employment at the power plant

References

1. Ailawadi VS, Bhattacharyya SC (2006) Access to energy services by the poor in India: current situation and need for alternative strategies. Nat Resour Forum 30(1):2–14
2. Alvial-Palavicino C, Garrido-Echeverría N, Jiménez-Estévez G, Reyes L, Palma-Behnke R (2011) *A* methodology for community engagement in the introduction of renewable based smart microgrid. Energy Sustain Dev 15:314–323
3. Bhattacharyya SC (2013) Rural electrification experience from South-east Asia and South America. Springer, London, pp 187–226
4. Bhattacharyya SC (2006) Energy access problem of the poor in India: Is rural electrification a remedy? Energy Policy 34(18):3387–3397
5. Chakrabarti S, Chakrabarti S (2002) Rural electrification programme with solar energy in remote region—a case study in an island. Energy Policy 30:33–42
6. Chaurey A, Krithika PR, Palit D, Rakesh S, Sovacool BK (2012) New partnerships and business models for facilitating energy access. Energy Policy 47:48–55
7. DFID (2002) Energy for the poor: Underpinning the millennium development goals. Department for International Development, London
8. Electricity Act (2003) Section 53: provisions relating to safety and electricity supply. Ministry of Power, Government of India. http://powermin.nic.in/acts_notification/electricity_act2003/distribution_electricity.htm. Accessed 12 Jan 2014
9. ESMAP (2002) Rural electrification and development in the Philippines: measuring the social and economic benefits. Joint UNDP/World Bank Energy Sector Management Assistance Program (ESMAP) Report. The World Bank: Washington, DC
10. Hiremath RB, Kumar B, Balachandra P, Ravindranath NH, Raghunandan BN (2009) Decentralised renewable energy: scope, relevance and applications in the Indian context. Energy Sustain Dev 13:4–10
11. IEA (2012) World energy outlook, 2012. International Energy Agency, Paris
12. IEEE (2007) IEEE guide for array and battery sizing in stand-alone photovoltaic (PV) systems, IEEE standards 1562-2007, 4–9
13. Imai K, Palit D (2013) Impacts of electrification with renewable energies on local economies: the case of India's rural areas. Int J Environ Sustain 9(2): 1–18
14. Iyer C, Sharma R, Khanna R, Laxman AV (2010) Decentralised distributed generation for an inclusive and low carbon economy for India. India Infrastructure Report

15. Krithika PR, Debajit Palit (2013) Participatory Business Models for Off-Grid Electrification. Rural Electrification Through Decentralised Off-grid Systems in Developing Countries. Springer, London, pp 187–225
16. Kumar A, Mohanty P, Palit D, Chaurey A, (2009) Approach for standardization of off-grid electrification projects. Renew Sustain Energy Rev 13:1946–1956
17. Laufer D, Schäfer M (2011) The implementation of solar home systems as a poverty reduction strategy—a case study in Sri Lanka. Energy Sustain Dev 15(3):330–336
18. Lahimer AA, Alghoul MA, Yousif F, Razykov TM, Amin N, Sopian K (2013).Research and development aspects on decentralized electrification options for rural household. Renew Sustain Energy Rev 24: 314–324
19. Mahapatra S, Dasappa S (2012) Rural electrification: optimising the choice between decentralised renewable energy sources and grid extension. Energy Sustain Dev 16:146–154
20. Ministry of Panchayati Raj (2007) Backward Regions Grant Fun: Programme Guidelines. http://www.nird.org.in/brgf/doc/BRGFFINALGUIDELINES.pdf. Accessed 13 Nov 2013
21. Ministry of Power (2009) Guidelines for village electrification through Decentralized Distributed Generation (DDG) under Rajiv Gandhi Grameen Vidyutikaran Yojana in the XI plan—scheme of rural electricity infrastructure and household electrification; Ministry of Power, Government of India
22. Ministry of Environment and Forest (MoEF) (2011). Guidelines for linear infrastructure intrusions in natural areas: roads and power lines. National Board for Wildlife. MoEF, Government of India
23. Mishra A, Sarangi GK (2011) Off-grid energy development in India: an approach towards sustainability, OASYS Working Paper 12. http://oasyssouthasia.dmu.ac.uk/docs/oasyssouthasia-wp12-dec2011.pdf. Accessed 25 Dec 2013
24. Mondal AH, Klein D (2011) Impacts of solar home systems on social developments in rural Bangladesh. Energy Sustain Dev 15:17–20
25. NASA (2008) Surface meteorology and solar energy. Website, last update 26 Mar 2008. https://eosweb.larc.nasa.gov/sse/RETScreen/. Accessed 15 Jan 2014
26. NRECA (2002) Economic and social impact evaluation study of the Bangladesh rural electrification program. NRECA International Ltd. Bangladesh: Dhaka
27. Palit D (2011) Performance assessment of biomass gasifier based power generation systems implemented under village energy security program in India. In: Proceedings of the international conference advances in energy research, Indian Institute of Bombay, Mumbai
28. Palit D, Sovacool BK, Cooper C, Zoppo D, Eidsness J, Crafton M, Clarke S (2013) The trials and tribulations of the Village Energy Security Programme (VESP) in India. Energy Policy 57:407–417
29. Palit D, Sharma KR, Sundaray S (2013) Cluster approach for effective decentralization in off-grid energy project: a case study from Dhenkenal district, Odisha. In: PC Ghosh (ed) 4th international conference on advances in energy research—book of proceedings, Indian Institute of Technology, Mumbai, p 1002–1011. http://www.ese.iitb.ac.in/icaer2013/data/ICAER.ConferenceProceedings.pdf. Accessed 15 Feb 2014
30. Palit D (2013) Solar energy programs for rural electrification: experiences and lessons from South Asia. Energy Sustain Dev 17(3):270–279
31. Schillebeeck SJD, Parikh P, Bansal R, George G (2012) An integrated framework for rural electrification: adopting a user-centric approach to business model development. Energy Policy 48:687–697
32. Sovacool BK, Drupady IM (2011) Summoning earth and fire: the energy development implications of Grameen Shakti in Bangladesh. Energy 36(7): 4445–4459
33. Ulsrud K, Winther T, Palit D, Rohracher H, Sandgren J (2011) The Solar Transitions research on solar mini-grids in India: Learning from local cases of innovative socio-technical systems. Energy Sustain Dev 15:293–303

34. United Nations (2012) The Secretary-General's High-level Group on Sustainable Energy for All. http://www.un.org/wcm/webdav/site/sustainableenergyforall/shared/Documents/SE%20for%20All%20-%20Framework%20for%20Action%20FINAL.pdf. Accessed 5 Jan 2014
35. Wijayatunga PDC, Attalage RA (2005) Socio economic impact of solar home systems in rural Sri Lanka: a case study. Energy Sustain Dev 9(2):5–9
36. Yadoo A (2012) Delivery models for decentralised rural electrification: case studies in Nepal, Peru and Kenya. International Institute for Environment and Development, London

Renewable Energy-Based Mini-Grid for Rural Electrification: Case Study of an Indian Village

Rohit Sen and Subhes C. Bhattacharyya

Abstract Although off-grid electrification has become a cost-effective and convenient option for many non-electrified areas, generally stand-alone individual options receive greater attention; and when mini-grid-based solutions are considered, traditionally a single technology-based limited level of supply is often considered, without paying attention to reliable round-the-clock supply of electricity. This chapter considers a hybrid combination of renewable energy technologies (RETs) as an alternative to grid extension for remote areas. Applying HOMER software, this study presents an analysis for choosing the best hybrid RET system for an Indian village and compares the result with conventional grid extension. It provides a systematic load demand analysis of the village, simulates optimal sizing of a hybrid system, calculates the economical distance limit (EDL) beyond which the use of the grid extension is not cost-effective and shows that the use of decentralised RET systems at an off-grid location can be a relevant option. HOMER results show that the solution is sustainable and techno-economically viable and environmentally sound.

Abbreviations

COE	Cost of Energy
Km	Kilometre
EDL	Economical Distance Limit
RET	Renewable Energy Technology
RES	Renewable Energy Sources
GHG	Green House Gases
LCC	Life-Cycle Cost

R. Sen
TERI—The Energy and Resources Institute, New Delhi, India
e-mail: rohit.sen@teri.res.in

S. C. Bhattacharyya (✉)
Institute of Energy and Sustainable Development, De Montfort University, Leicester, UK
e-mail: subhesb@dmu.ac.uk; subhes_bhattacharyya@yahoo.com

S. C. Bhattacharyya and D. Palit (eds.), *Mini-Grids for Rural Electrification of Developing Countries*, Green Energy and Technology,
DOI: 10.1007/978-3-319-04816-1_9,

LUCE	Levelised Unit Cost of Electricity
NPC	Net Present Cost
O&M	Operation and Maintenance
BET	Bioenergy Technology
T&D	Transmission and Distribution
SPV	Solar Photovoltaics
BDG	Biodiesel Generator
SHP	Small Hydropower
B100	100 % Pure Biodiesel
DG	Diesel Generator
MNRE	Ministry of New and Renewable Energy, India

1 Introduction

Rural electrification is relatively costly compared to electrification of urban areas. At any off-grid location, the delivered cost of electricity from the grid depends on four factors [18][1]:

(a) Cost of electricity (COE) generation from the central power plant.
(b) Cost of transmission and distribution (T&D) of electricity through the new network lines.
(c) Transmission and distribution (T&D) loss—(technical and non-technical).[2]
(d) Load factor.

Load factor plays an important role in the rural areas as there is a limited commercial and industrial load and most of the residential demand occurs in the evening, thereby creating an uneven pattern of electricity use. Moreover, Nouni et al. [18] report that the average cost of grid extension in remote rural areas varies from $8,000 to $10,000 for every kilometre and rapidly increases up to $22,000/km for extremely difficult terrains. Rural electrification is almost seven to ten times more expensive than that in urban areas.

Decentralised electricity generation with the help of renewable energy technologies (RET) has received considerable attention in recent years. RETs are environment-friendly and beneficial from the energy supply security perspective. In fact, the increasing costs of fossil fuels on the one hand and the gradual decrease in the cost of RET-based systems on the other, as the technology is maturing, have made the utilisation of RET attractive for rural electrification [13].

[1] This chapter is an extended version of Sen and Bhattacharyya [20].

[2] There is distribution of electricity in a mini-grid but often the distance covered is relatively small. Although there will be some distribution loss in the mini-grids, it is likely to be lower than the case where electricity is transported over long distances at low voltage.

However, the efforts in using renewable energies have often focussed on single technologies. For example, solar home systems (SHS), solar photovoltaic systems and micro-hydropower have been widely used, but such options are often unable to cater to consumers' needs adequately and reliably due to limited resource availability arising from variability of resources. Reliance on a single technology generally results in an oversizing of the system, thereby increasing the initial costs. Also these systems were mostly designed for a smaller load or for a limited duration of supply of electricity from 4 to 6 h/day. A hybrid system design can overcome the intermittent nature of renewable energy sources (RES), component oversizing issue and enhance reliability of supply. Depending on the availability of RES and the electricity need, a hybrid system would ensure electricity for larger loads and longer duration of supply (for 10–12 h generally but in some cases, it can be designed for 24 × 7 supply). A hybrid system requires higher initial investment compared to a single resource system, but can offer higher reliability and better system utilisation. Yet, hybrid systems have received limited attention and hardly any work has considered the issue of reliable supply of electricity in a rural context.

The purpose of this study is to find the best combination of RET from the available resources in a given village location that can meet the electricity demand in a reliable and sustainable manner and to analyse whether such a hybrid option is a cost-effective solution or not. To achieve this objective, we use an example of an Indian village, estimate the potential demand, identify the available resources, model electricity generation based on multiple combinations of RETs with the application of HOMER software, select the best option based on the COE generation and then compare these performance indicators to grid extension related costs. Our choice of the tool is influenced by its popularity, ease of use and flexibility. Despite our reliance on HOMER, our contribution arises from four novel features: (1) most of the studies in the past have considered wind turbines, solar power and diesel technologies, whereas we have considered four renewable technologies, namely micro-hydro, solar PV, wind turbines and biodiesel thereby extending the hybrid technology combinations; (2) the reliability of supply, which has not received adequate attention in the literature, is considered as a main supply objective; (3) we have included productive use of electricity in commercial and agricultural activities in addition to domestic energy demand, thereby enlarging the scope of the study; and (4) we have gone beyond a typical HOMER application by considering pre- and post-HOMER analysis (discussed in more detail below).

The organisation of the chapter is as follows: Sect. 2 presents a review of related studies; Sect. 3 briefly presents HOMER, Sect. 4 presents the case study and results obtained from the study. Section 5 then presents the post-HOMER analysis, while concluding remarks are presented in Sect. 6.

2 Literature Review

The purpose of the literature review presented here is (1) to review some relevant literature on off-grid electricity access using RET, (2) to provide evidence of knowledge gap that justifies the need for this work; and (3) to provide support for the methodology used in the study and is a source of information for comparison, triangulation and referencing. Accordingly, the following sub-sections present the relevant reviews.

2.1 Review of Literature on Off-Grid Electrification

Nouni et al. [18] conducted a study to identify appropriate locations for decentralised electrification in India using the levelised cost of supply as the indicator. The study analysed the cost of electricity supply to villages located within a radius of 5–25 km from an existing 11 kV distribution line and compared this with the decentralised options. An earlier study by Sinha and Kandpal [21] also analysed the factors influencing the cost of decentralised electricity supply. They also suggested that decentralised options can be cost-effective in villages further away from the grid and where the demand is limited.

Mahapatra et al. [12] have analysed domestic lighting in rural areas and looked for alternatives to traditional kerosene lamps and found that modern bioenergy systems (for the whole village) or SPV (for each house) can be a good, reliable option. Nouni et al. [17] indicate that modern bioenergy systems such as biomass gasifiers are becoming popular in India due to their cost-effectiveness and high load factor. Biomass gasification can be used not only for domestic electricity purposes, but also for small-scale industry and has proved suitable for supplying power to a cluster of villages or a cluster of houses within a village [17, 19].

Chakrabarti and Chakrabarti [3] conducted a case study on 'Sagar Dweep' island in West Bengal, India, for which they considered environmental and socio-economic factors and found that solar PV is the best option in this case, because of the high cost of transmitting electricity from conventional sources to extremely remote locations. The solar PV was also advantageous from a sociological point of view, leading to a significant improvement in commerce, trade and education and more participation of women in community activities. The environmental benefit gained resulted from the fact that externality costs which using fossil fuels would have caused were not incurred with the use of solar PV [3, 22].

2.2 Review of Studies on HOMER

HOMER (Hybrid Optimisation Model for Electric Renewables) developed by NREL (National Renewable Energy Laboratory, USA) appears repeatedly in the literature as a preferred tool. It can handle a large set of technologies (including PV, wind, hydro, fuel cells and boilers), loads (AC/DC, thermal and hydrogen), and can perform hourly simulations. HOMER is an optimisation tool that is used to decide the system configuration for decentralised systems. It has been used both to analyse the off-grid electrification issues in the developed as well as developing countries. In the case of developed countries, often advanced fuel systems such as hydrogen are considered. Examples of such studies include the following Khan and Iqbal [10] who investigated the feasibility of a hybrid system with hydrogen as energy carrier in Newfoundland, Canada; Barsoum and Pandian [1] who studied a hybrid solar-hydrogen system; Karakoulidis et al. [9] who considered a hybrid solar-diesel and fuel cell system; and Giatrakos et al. [5] a hybrid system for a Greek island.

For developing countries, a large number of studies exist and a detailed review of this literature is beyond the scope of this chapter.[3] Instead, we focus on a selected set for our purpose. Givler and Lilienthal [6] conducted a case study of Sri Lanka where they identified when a PV–diesel hybrid becomes cost-effective compared to a stand-alone small SHS (50 W PV with battery). This study considers an individual household base load of 5 W with a peak of 40 W, leading to a daily load average of 305 Wh. Through a large number of simulations, the study found that the PV–diesel hybrid becomes cost-effective as the demand increases. However, this study only focuses on the basic needs and does not include productive use of energy.

Munuswamy et al. [14] compared the COE from fuel cell-based electricity generation against the cost of supply from the grid for a rural health centre in India, applying HOMER simulations. The results show that beyond a distance of 44 km from the grid, the cost of supply from an off-grid source is cheaper. This work, however, just considered the demand of a rural health centre and was not part of any traditional rural electrification programme.

Hafez and Bhattacharya [7] analysed the optimal design and planning of a renewable energy-based micro-grid system for a hypothetical rural community where the base load is 600 kW and the peak load is 1,183 kW, with a daily energy requirement of 5,000 kWh/day. The study considers solar, wind, hydro and diesel resources for electricity generation. Although the study considers electricity demand over 24 h, the purely hypothetical nature of the assumptions makes the work unrealistic for many off-grid areas of developing countries.

Lau et al. [11] analysed the case of a remote residential area in Malaysia and used HOMER to analyse the economic viability of a hybrid system. The study uses a hypothetical case of 40 households with a peak demand of 2 kW per household.

[3] See Analytical frameworks and an integrated approach for mini-grid-based electrification for a detailed review of the relevant literature.

Table 1 Selected examples of hybrid technology analysis using HOMER

Reference	Technology application	Country of application	Supply duration/type
Givler and Lilienthal [6]	PV–battery–diesel	Sri Lanka	Basic needs
Hafez and Bhattacharya [7]	PV, wind, hydro, diesel, battery	Hypothetical	24 h service but unrealistic demand profile for a rural area of developing countries
Lau et al. [11]	PV–diesel hybrid	Malaysia	24 h service but uses a high demand profile for a rural area and does not use any productive load
Himri et al. [8]	Wind–diesel hybrid	Algeria	Adding wind turbine to an existing diesel-based supply; limited technology options
Nandi and Ghosh [15]	Wind–PV–battery	Bangladesh	Solar and wind hybrid; no productive demand
Nfah et al. [16]	PV, micro-hydro, LPG generator, battery	Cameroon	Diesel as main generator supplemented by PV and micro-hydro, load based on grid-connected urban households of Uganda was used
Bekele and Palm [2]	PV–wind hybrid	Ethiopia	PV and wind hybrid, randomised load profile from hypothetical load data

Source Sen and Bhattacharyya [20]

The peak demand is 80 kW and the base demand of around 30 kW is considered in the analysis. Although such high rural demand can be typical for Malaysian conditions, this might not be true for others. The study also does not consider any productive use of electricity.

Similar cases are presented in other studies as well. For example, Himri et al. [8] present a study of an Algerian village; Nandi and Ghosh [15] discuss the case of a Bangladeshi village, while Nfah et al. [16], Bekele and Palm [2] provide case studies of Cameroon and Ethiopia, respectively. Table 1 summarises the technology choices, demand focus and country of application of these studies.

2.3 Knowledge Gap

The above review shows the popularity of HOMER as a tool to analyse decentralised electricity supply systems. However, most of the studies do not consider electricity demand in rural areas carefully. As the optimal system configuration is obtained to meet the demand, demand analysis plays an important role. Most of the studies also focus on a limited level of supply and do not often consider the productive applications of electricity. In addition, while technology choices are dependent on local conditions, it is possible to investigate alternative combinations more imaginatively. For example, we have considered the biofuel option as an alternative to conventional diesel to increase the portfolio of RES. Finally, studies also limit their scope to techno-economic analysis and do not consider the business issues or practical considerations related to their implementation. In the absence of such considerations, most of the studies remain theoretical in nature. This chapter tries to bridge the above knowledge gaps and presents an application of HOMER by including a pre- and post-analysis to extend the scope of the work and knowledge base.

3 Methodology

3.1 Introduction

As mentioned in "Introduction", this study uses the HOMER software package for designing micro-power systems but complements it by undertaking pre- and post-HOMER analyses. This is indicated in Fig. 1. In the pre-HOMER analysis phase, a detailed assessment of the village load, site layout and available resources in the selected village is conducted. This is carried out outside HOMER and data is fed into the software. In the HOMER analysis the hybrid RET system is designed, followed by a techno-economic analysis. It compares a wide range of equipment with different constraints and sensitivities to optimise the system design. The analysis is based on the technical properties of the system and the life-cycle cost (LCC) of the system. The LCC comprises the initial capital cost, cost of installation and operation costs over the system's life span. HOMER performs simulations to satisfy the given demand using alternative technology options and resource availability. Based on the simulation results, the best suited configuration is selected. In the post-HOMER phase, the business-related analysis is performed to a limited extent (see Sect. 5), but an example of the extended business analysis is presented in "From SHS to mini-grid based off-grid electrification: A case study of Bangladesh".

We have considered a combination of the following technologies, namely small hydropower (SHP), wind turbines, solar PV (SPV) systems, batteries and a biodiesel generator (BDG) for backup (see Fig. 2 for a schematic system configuration diagram). In the hybrid system, the demand from the village is AC-coupled; the SHP and the BDG are connected to the AC side of the network, whereas the SPV, wind turbine and the batteries are connected to its DC side. Usually a conventional backup diesel generator (DG) is used to supplement the RE system for peak loads and for periods with resource constraints. In this study, a BDG (B100) is used, making the whole system renewable, clean and carbon neutral system, not only for the purpose of electricity generation but also for working effectively towards GHG emissions mitigation by not burning any fossil fuels. This makes the study different from others.

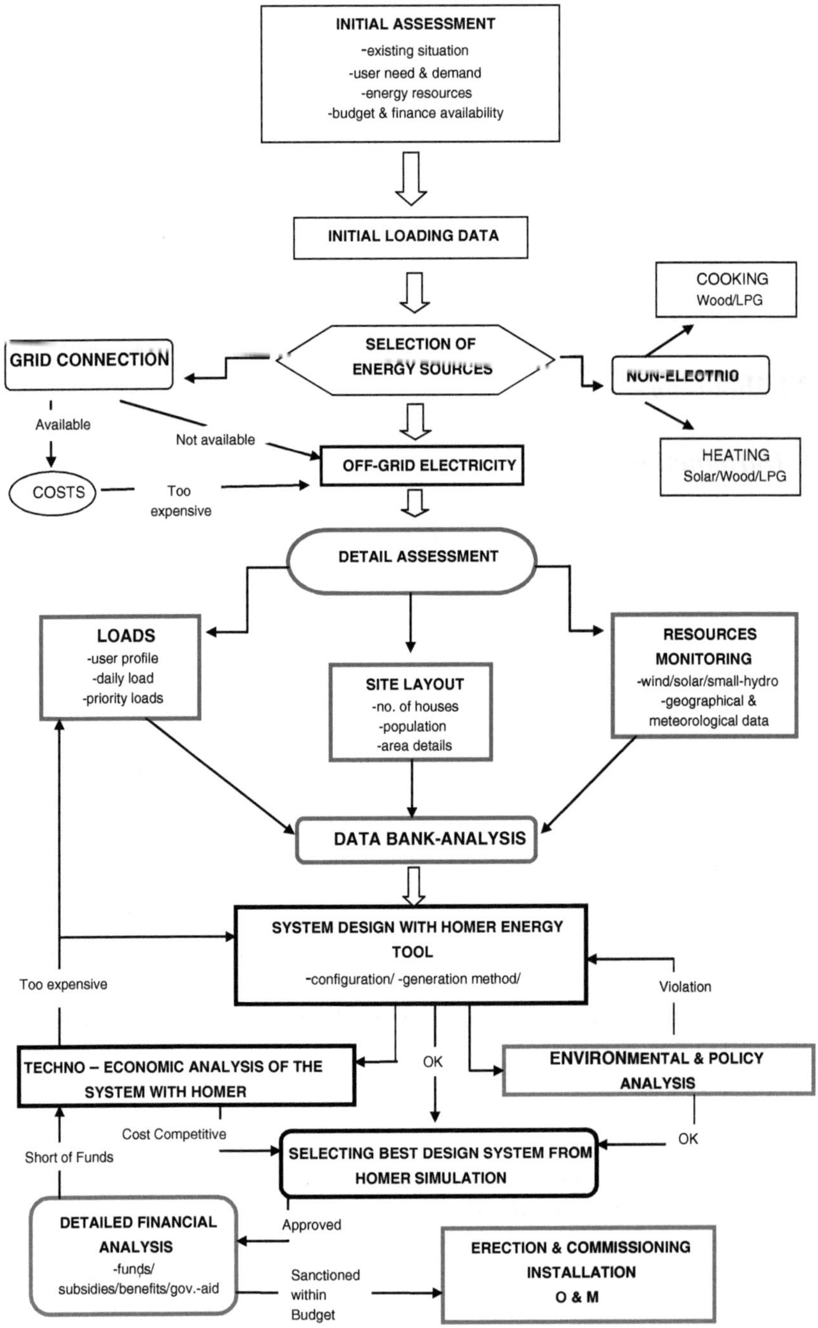

Fig. 1 Framework of analysis

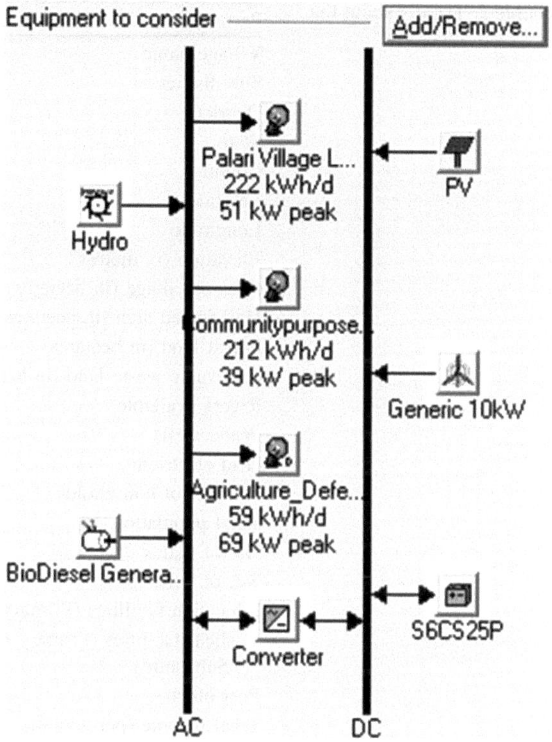

Fig. 2 Design of the selected RETs for the hybrid system

3.2 System Modelling

The selected off-grid remote rural village for this study is Palari, a small village in Bastar district in the Indian state of Chhattisgarh. The details of the village are listed in Table 2. The nearest town is Kondagaon, which is about 15–20 km away, both Kondagaon and Palari are located in close proximity to the national highway 43 (NH-43). The area around the village is partially hilly with flat plains constituting the rest. The village has water and drinking water facilities in the form of water wells and hand pumps. The village has no access to grid electricity, which offers an opportunity for off-grid electrification of the village.

3.2.1 Village Load Assessment

In a remote rural village the demand for electricity is not high compared to urban areas. Electricity demand is for domestic use (for appliances such as radio, compact fluorescent lamps, ceiling fans and table fans), agricultural activities (such as water pumping), community activities (such as in community halls, schools and clinics) and for rural commercial and small-scale industrial activities (such as cold storage, small milk processing plants and cottage industries).

Table 2 Details about the selected village

Particulars	Details
Village name	Palari
Sub-district	Kondagaon
District	Bastar
State	Chhattisgarh
Country	India
Latitude	19° 635′′ N
Longitude	81° 672′ E
Elevation (in metres)	587
Area of village (in hectares)	370
Unirrigated area (in hectares)	>200
Forest land (in hectares)	14
Cultivable waste land (in hectares)	7
Rivers available	1
Water wells	1
Grid electricity	0
Number of households	304
Total population	1,624
No. of males	764
No. of females	860
Education facilities (Primary School)	1
Medical facilities (Primary Health Subcentre)	1
Post office	1
Total income (per annum)	Rs. 1,75,100/$ 3,892
Total expenditure (per annum)	Rs. 1,13,400/$ 2,520

Source Census 2001. (http://www.censusindia.gov.in/)

In the pre-HOMER study, the village energy load requirement is carefully estimated considering existing load profile data available in state government records for similar rural areas. We have also consulted published literature on Indian villages and triangulated with expert opinions and personal judgement. As the energy requirements in the village vary from season to season, we have estimated demand separately for two distinct seasons prevailing in this area, namely summer (April to October) and winter (November to March) considering the appliance holding and use patterns for households, potential commercial activities and energy use in productive applications. Table 3 provides the summary of estimated demand for summer and winter seasons. Clearly, the demand estimation is a crucial element of the entire system design, and further improvement is possible here by incorporating social information of the users as well as their preferences. Moreover, demand management scenarios can also be considered but for simplicity, this was not included.[4]

[4] Alternative scenarios of rural electricity demand are presented for the Bangladesh study in From SHS to mini-grid based off-grid electrification: A case study of Bangladesh.

Table 3 Estimated electricity demand for Palari village

S. No.	Load	No. in use	Power (Watts)	Summer (April–Oct.)		Winter (Nov.–March)	
				Hrs/day	Watt-hrs/day	Hrs/day	Watt-hrs/day
	Domestic purposes						
1	Low-energy lights (CFL)	1	20	6	120	7	140
2	Low-energy lights (CFL)	1	20	6	120	7	140
3	Low-energy lights (CFL)	1	11	5	55	6	66
4	Radio	1	10	3	30	4	40
5	Ceiling fan	1	30	15	450	0	0
6	Table fan	1	15	9	135	**0**	0
	Total				**910**		**386**
A	**No. of houses**	**304**			**276640**		**117344**
	Industrial/commercial/community purposes						
1	Shops	10	500	8	40000	7	35000
2	Community centre	1	1000	8	8000	6	6000
3	Small manufacturing units	5	3000	12	180000	10	150000
4	Street lights (CFL)	5	30	10	1500	12	1800
B	**Total**				**229500**		**192800**
	Agriculture & irrigation purposes						
1	Water pump	8	745	5	29824	3	17894
2	Irrigation pump	4	1490	6	35789	4	23859
3	Well	1	745	4	2982	2	1491
C	**Total**				**68595**		**43244**
	Medical centre						
1	Low-energy lights (CFL)	4	20	4	320	6	480
2	Ceiling fan	4	30	6	720	0	0
3	Refrigerator	1	600	20	12000	16	9600
D	**Total**				**13040**		**10080**
	School						
1	Compact fluorescent lights	5	20	2	200	4	400
2	Ceiling fan	2	30	6	360	0	0
3	Computer (desktop)	1	300	2	600	2	600
4	Television	1	100	2	200	2	200
E	**Total**				**1360**		**1200**

Note We have assumed energy efficient appliances such as fans and lights

Knowing that the load factor in such locations tends to be poor, some demand has been distributed strategically over the 24-h period to improve the system load factor to attain a flatter load curve, system optimisation and to meet all possible load demand without increasing the capacity of the hybrid power plant. The village load has been divided into three important categories:

(1) Primary Load 1—This includes the domestic load, medical centre and school demand. The load demand is approximately 222 kWh/day and 51.2 kW peak. It has a load factor of 0.181 (see Fig. 3).

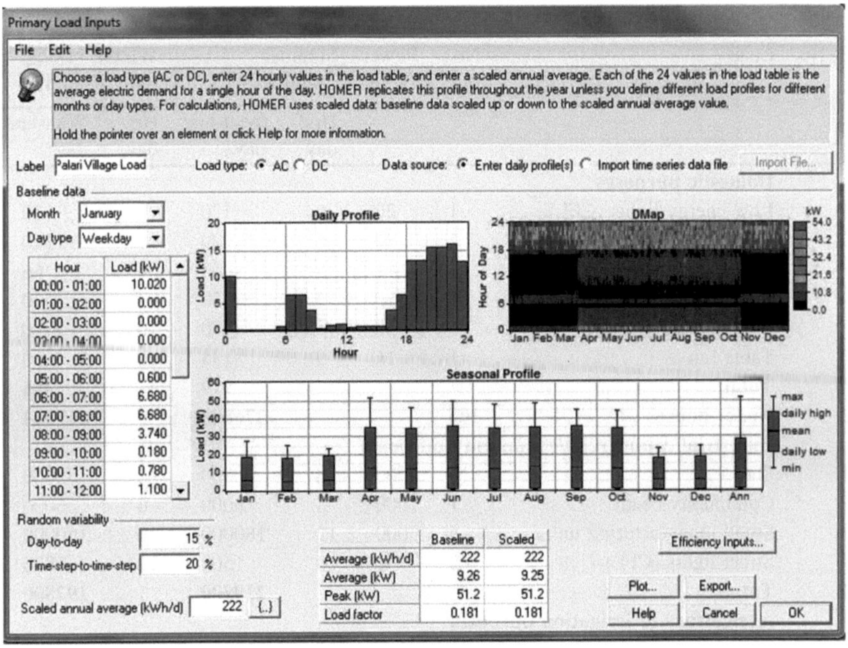

Fig. 3 Load profile of primary load 1

(2) Primary Load 2—This includes the demand for the community centre, shops, local business and small manufacturing units. It is approximately 212 kWh/day and 39.4 kW peak. It has a load factor of 0.224 (see Fig. 4).
(3) Deferred Load—This includes the agricultural load of the village. The scaled annual average deferred load is 58.6 kWh/day and has a peak load of 68.6 kW. It has a storage capacity designed for 30 kW and is also connected to the AC side.

The load assessment was done in Excel worksheet, using customised data templates for this purpose. This is an area of further work which can produce a generic pre-HOMER tool for wider applications.

3.2.2 Resources Assessment

We have considered solar, wind, micro-hydro and biodiesel resources in this simulation. The resource assessment is presented below.

The solar resource used for Palari village at a location of 19° 59′ N latitude and 81° 59′ E longitude was taken from NASA Surface Meteorology and Solar Energy website.[5] The annual average solar radiation was found to be 5.17 kWh/m^2/day,

[5] http://eosweb.larc.nasa.gov/sse/

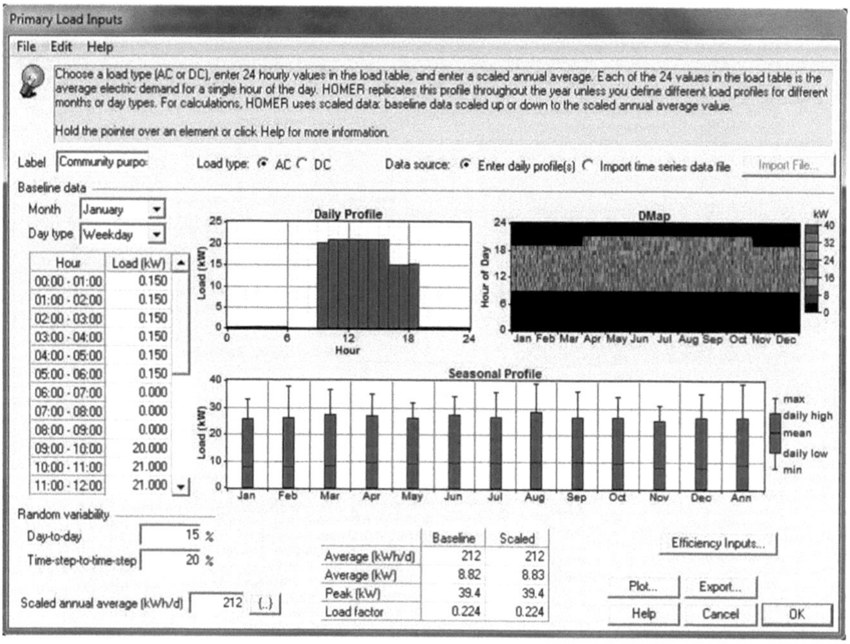

Fig. 4 Load profile of primary load 2

and the average clearness index was found to be 0.548. The solar radiation is available throughout the year; therefore, a considerable amount of PV power output can be obtained (see Fig. 5).

The monthly average wind resource data from an average of 10 years was taken from the above NASA resource website based on the longitude and latitude of the village location. The annual average wind speed for the location is 3.5 m/s with the anemometer height at 50 m. The wind speed probability and average monthly speed throughout the year is also observed. It shows that there are 15 h of peak wind speed. The wind speed variation over a day (diurnal pattern strength) is 0.25 and the randomness in wind speed (autocorrelation factor) is 0.85 (see Fig. 6).

The Alternate Hydro Energy Centre (AHEC), India, has identified a number of locations for the development of Small Hydropower projects in Chhattisgarh.[6] One of the projects, Kondagaon on Narangi River near Palari village, is identified with a potential output of 500 kW at 5 m head. The monthly average flow has been carefully estimated based on the average precipitation, average temperatures and topography of the region. The residual flow was assumed to be 4,000 l/s. The flow in the river drops from September to May and rapidly increases up to 58,000 l/s in

[6] Chhattisgarh SHP Projects Identified by AHEC—CREDA.pdf, Available from http://www.credacg.org/6-_Chhattisgarh_SHP_Projects_Identified_by_AHEC.pdf (Assessed on 10/12/2012).

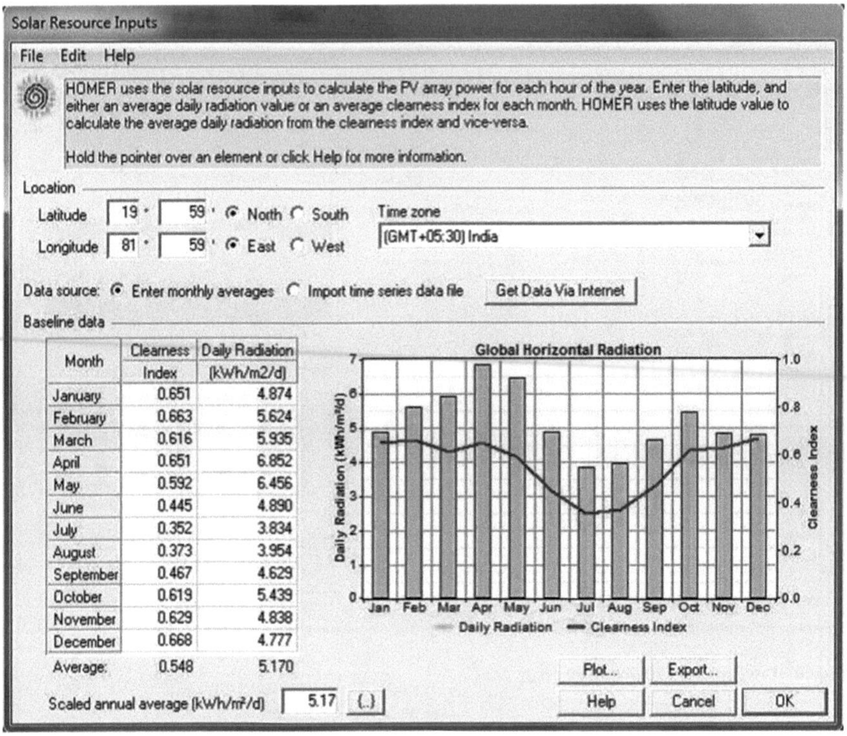

Fig. 5 Solar energy profile at the selected village

August due to heavy rainfall in the area. Hence, power generation from the hydro source varies depending on the water availability during the year (see Fig. 7).

Biodiesel is a biofuel predominantly made from vegetable oil and sometimes animal fat. Biodiesel can be used to run diesel engines with minor engine modifications (if required). In India, with the help of extensive agricultural research, Jatropha Curcas oilseed was chosen as main feedstock for the biodiesel production in India's Biodiesel programme. The most commonly available are B20 (containing 20 % of biodiesel and 80 % petroleum diesel in the blend), and B100 which is pure biodiesel. Biodiesel has a shelf life of 6 months, after which it has to be tested again.

As there is a biodiesel plant in the neighbouring city of Raigarh, it is assumed that the fuel will be available from this plant. The fuel price is considered to be 0.6 $/l. The current market price of biodiesel in India is about 0.58 $/l,[7] though it varies regionally due to tax and other costs.

[7] http://www.commodityonline.com/news/Raise-price-for-bio-diesel-from-jatropha–India-ministry-35338-3-1.html (Accessed on 08/06/2011).

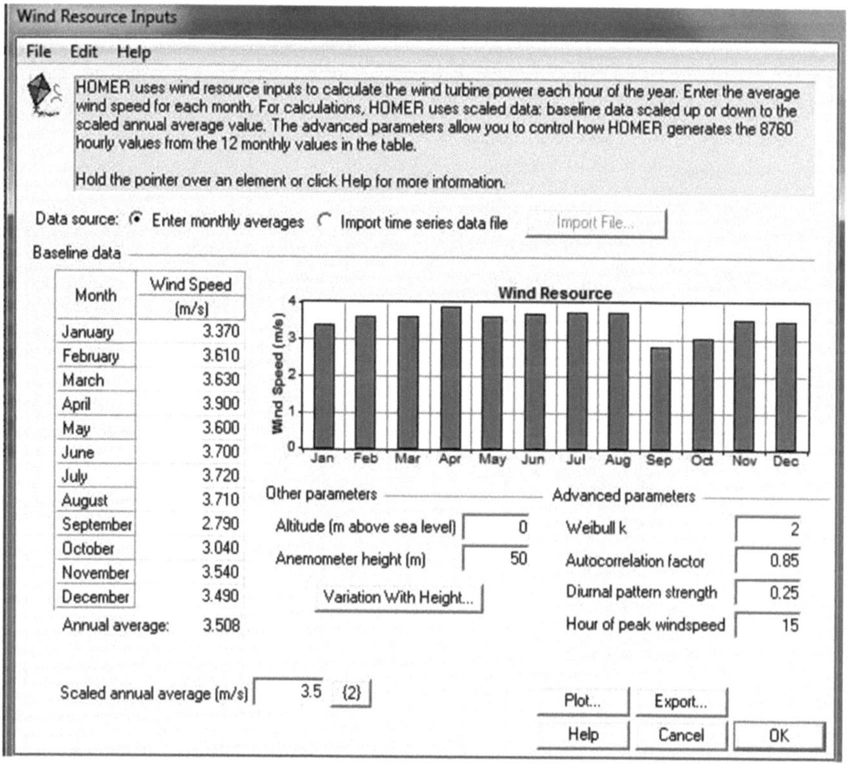

Fig. 6 Wind energy resource at the selected village

3.2.3 Components Assessment

In this HOMER analysis, solar PV, wind turbines and run-off river hydropower are the intermittent resources and the biodiesel is kept for backup. Batteries and converter are for storing and converting electricity, respectively. The grid connection in this study is only used as a comparison for the analysis and determination of the economic distance to grid (EDL). The performance and cost of each of the system's components is a major factor for the cost results and the design.[8]

The SPV panels are connected in series. The power generated by SPV is more than that from wind turbines at this location due to better solar insolation. The capital cost and replacement cost for a 1-kW SPV are taken as $6,000 and $5,000, respectively.[9]

---

[8] The components' technical and cost parameters for this study are based on data collected from the Ministry of New and Renewable Energy Sources (Government of India), the Energy and Resources Institute (TERI) in India, previous published literatures, information from personal sources of Indian manufactures and expert opinion.

[9] Prices in 2010–2011 were considered and include the cost of charge controller and power optimiser.

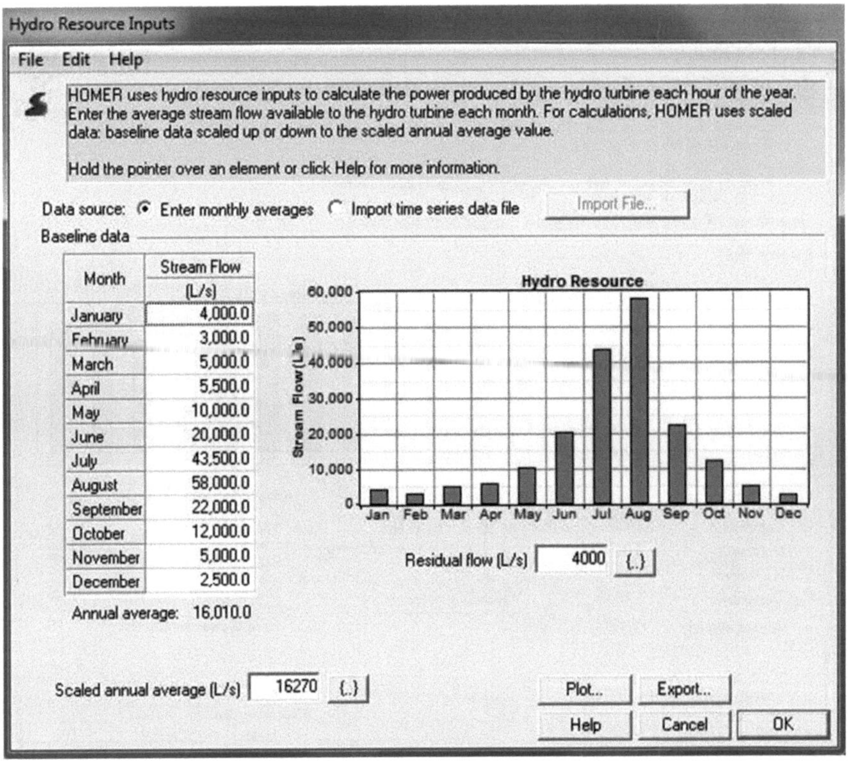

Fig. 7 Small hydro resource potential in the village

As there is very little maintenance required for PV, only \$10/year is taken for O&M costs. Like for all other components considered in the following paragraphs, the costs per kW considered include installation, logistics and dealer mark-ups. The SPV is connected to the DC side of the system and has a lifetime of 20 years. The de-rating factor considered is 90 % for each panel to approximate the varying effects of temperature and dust on the panels. The panels have no tracking system and are modelled as fixed titled south at 19° 59′ N latitude of the location with the slope of 45°.

A generic 10 kW horizontal axis wind turbine is considered. The amount of electricity generated by the wind turbine greatly depends on the availability of and variations in the wind speed. The G10 wind turbine selected gives a 10 kW of DC output. The cost of one unit is taken as \$40,000, while the replacement and the maintenance cost are considered to be \$32,000 and \$200/year, respectively. The wind turbine has a hub height of 25 m and a lifetime of 25 years.

The SHP is designed for a power output of 30 kW depending on the village load. The turbine is designed for a net head available of 5 m and has a design flow of 815 l/s. The turbine efficiency is 75 % and has a pipe head loss of 5.68 %. The SHP gives an AC output and has a lifetime of 25 years. The capital cost for a

30-kW SHP is taken as $42,000, while the replacement cost and O&M cost are considered to be $35,000 and $4,000, respectively.

The capital cost, replacement cost and O&M costs of a 1-kW BDG are taken as $1,200, $1,000 and $1.03/h, respectively.[10] A normal old DG can be used as well, but it might need certain modifications. The per kW costs are for a new modern DG that can be used for biodiesel as fuel as well and include the costs of installation, logistics and dealer mark-ups. The generator is connected to an AC bus with a lifetime of 15,000 operating hours. The minimum load is taken as 30 % of the capacity; moreover, HOMER requires the partial load efficiency to simulate this component. HOMER calculates the total operating cost of the generator based on the amount of time it has to be used in a year.

Batteries are used as a backup in the system and to maintain a constant voltage during peak loads or a shortfall in generation capacity. The battery chosen for this study is Surrette 6CS25P. It is a 6-V battery with a nominal capacity of 1,156 Ah (6.94 kWh). It has a lifetime throughput of 9,645 kWh. The capital cost, replacement cost and O&M costs for one unit of this battery were considered as $1,000, $800 and $50/year, respectively.[11] HOMER models the batteries on charging and discharging cycles.

The capital cost, replacement cost and O&M costs of the converter for 1 kW systems were considered as $700, $550 and $100/year, respectively following Deshmukh and Deshmukh [4]. The lifetime of the converter is taken as 15 years, whereas the inverter and rectifier efficiencies are taken as 90 % and 85 %, respectively.

Figure 8 provides the list of components used in the search for the optimised configuration. HOMER considers the entire set to find the least cost option that meets the demand of the village.

3.2.4 Sensitivity of Inputs

The key variables for the micro-power system are, however, often uncertain. This is a major problem that needs to be overcome in the designing of the system. Here the uncertainties in the RES (wind, hydro, solar and biomass) have been taken into account. The sensitivity options help us analyse how changes in an optimisation variable affect the outcome. The alternative values used for some variables are: three biofuel prices 0.60, 0.714 and 0.804 in $/l; two wind speeds of 3.5 and 5.0 m/s and two design flow rates for SHP 815 and 0.1 l/s.

[10] The prices considered are an interpolation of data (quotations) obtained from local Indian manufactures and distributors.

[11] The prices considered are an interpolation of data (quotations) obtained from local Indian manufactures, distributors and previous published literatures.

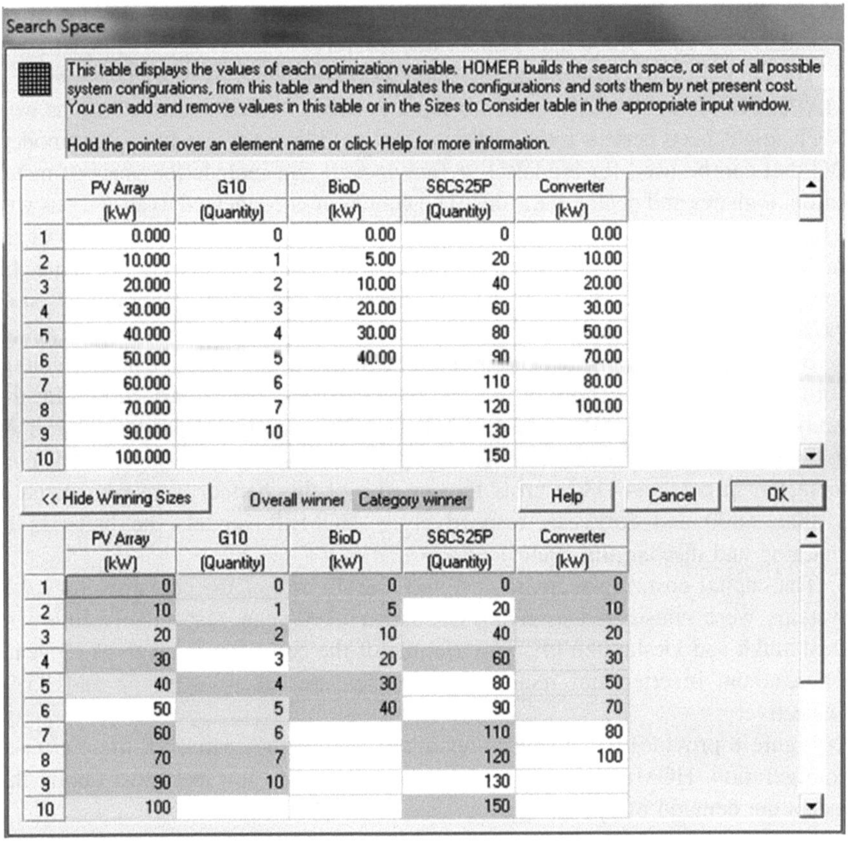
Search Space

This table displays the values of each optimization variable. HOMER builds the search space, or set of all possible system configurations, from this table and then simulates the configurations and sorts them by net present cost. You can add and remove values in this table or in the Sizes to Consider table in the appropriate input window.

Hold the pointer over an element name or click Help for more information.

	PV Array (kW)	G10 (Quantity)	BioD (kW)	S6CS25P (Quantity)	Converter (kW)
1	0.000	0	0.00	0	0.00
2	10.000	1	5.00	20	10.00
3	20.000	2	10.00	40	20.00
4	30.000	3	20.00	60	30.00
5	40.000	4	30.00	80	50.00
6	50.000	5	40.00	90	70.00
7	60.000	6		110	80.00
8	70.000	7		120	100.00
9	90.000	10		130	
10	100.000			150	

<< Hide Winning Sizes Overall winner Category winner Help Cancel OK

	PV Array (kW)	G10 (Quantity)	BioD (kW)	S6CS25P (Quantity)	Converter (kW)
1	0	0	0	0	0
2	10	1	5	20	10
3	20	2	10	40	20
4	30	3	20	60	30
5	40	4	30	80	50
6	50	5	40	90	70
7	60	6		110	80
8	70	7		120	100
9	90	10		130	
10	100			150	

Fig. 8 Variable ranges used for optimisation

3.2.5 System Dispatch

Apart from choosing the different components, HOMER has to make various decisions regarding the working of these components, such as whether power should be sold to or bought from the grid, whether batteries should be charged or discharged, which generators should operate and at what power level, and the like. Furthermore, all of these decisions have to be taken for every hour of the day.

Micro-power systems face challenges arising from the variability of the power output of RES and the possibility that the electric load demand might suddenly increase to surpass the operating capacity (which will lead to an outage). Therefore, every system needs an operating reserve, which is surplus generating capacity to provide a safety gap and be able to respond to these problems. The amount of a system's operating reserve always equals to its operating capacity minus the electrical load. The operating reserve adds to the cost as higher capacity has to be installed but it is appropriate to maintain some extra capacity. Similarly, HOMER follows a load priority system, where it takes a separate set of decisions on the allocation of electricity produced by the system. HOMER serves the electricity produced from an AC source to an AC load first, the same goes for DC. In this study, as all loads are of AC type, HOMER serves primary load 1 and primary load 2 followed by the deferred load.

3.3 Economic Modelling

As HOMER aims to minimise the total net present cost (NPC) both in finding the optimal system configuration and in operating the system, economics play a crucial role in the simulation. The main difference between renewable and non-renewable resources regarding the costs is that non-renewable resources usually have low capital and high operating costs, whereas for RES the costs are generally distributed in the opposite way: After a considerable investment in the beginning, the system can be operated at a comparatively low cost. To be able to compare the economics of numerous different system configurations with a varying share of renewable and non-renewable energy sources, HOMER has to take into account both the operating and capital costs. Since the life-cycle cost comprises all costs incurred during the system's life span, it considers these factors and therefore is the appropriate parameter to compare the economics of different configurations. The indicator chosen to compare the different configurations' economics is the life-cycle cost (LCC), and the total net present cost is taken as the economic figure of merit. All economic calculations are in constant dollar terms.

3.3.1 Economic Inputs

The project's lifetime is considered to be 25 years and an annual discount rate of 10 % is used in the analysis. The system fixed capital cost is considered to be $10,000 for the whole project, and the system fixed O&M cost is estimated to be $500/year for the project lifetime.[12] The system fixed capital costs include various civil constructions, logistics, labour wages, required licences, administration and government approvals and other miscellaneous costs.

3.3.2 The Grid

In this study the grid is used as standard benchmark by HOMER, to be compared with the technical and cost parameters of the off-grid hybrid RETs system.

[12] These costs are an interpolation from previous literature, estimates from TERI's existing projects and quotations from local Indian civil contractors.

Therefore, the cost of grid extension is taken in the analysis to see whether a grid extension is viable or an off-grid system is more appropriate. The capital cost of grid extension per kilometre for the Palari village terrain is considered to be $8,000/km [18]. The annual O & M cost per kilometre is considered to be $1,500/year/km,[13] and the grid power price is assumed from an interpolation as $0.44/kWh [9].

3.3.3 System Constraints

If a small amount of load is allowed to go unserved, this can significantly enhance the economic performance of the hybrid RET system. If the system can be designed to be down for a short period of time in a year or if some load can be shed, this can help in reducing the battery bank and hence significantly reduce the capital cost of the hybrid system. The maximum annual capacity shortage is set at 3 % and the minimum renewable fraction is taken as 50 %. An operating reserve of 10 % of the hourly load is also maintained.

3.3.4 System Control

Two different types of control strategies are modelled by HOMER. With the load-following strategy, the generator is only allowed to generate power to meet the load at a given time. In contrast to this, in the cycle charging strategy the generator is allowed to operate up to maximum power to charge the batteries and serve the load at the same time. In this study, both the strategies have been selected to see which is more suitable in the given constraints of the system. The set point charge for the batteries is kept at 80 %.

3.3.5 Simulation

HOMER performs the simulation for a number of prospective design configurations. After examining every design, it selects the one that meets the load with the system constraints at the least life-cycle cost. HOMER performs its optimisation and sensitivity analysis across all mentioned components and their resources, technical and cost parameters, and system constraints and sensitivity data over a range of exogenous variables. The competitiveness of the best suited hybrid RET system for rural electrification is compared with the conventional option of grid extension, based on the COE for both options, and based on this the economical distance limit (EDL) is determined. The cost of low tension distribution lines within villages has been excluded, since it is the same in all the cases.

---

[13] The costs are an estimated assumption based on the interpolation from published literatures.

4 Results and Discussion

This section presents the results of our analysis. First, the optimisation results are presented, which is followed by the outcomes of our sensitivity analysis. The economic and environmental aspects are also considered.

4.1 Optimisation Results

For the off-grid electrification of Palari village, various combinations of hybrid systems have been obtained with SPV, wind turbines, SHP, batteries and convertors from the HOMER optimisation simulation. Figure 9 presents a screen shot containing the summary of simulation outcomes.

All possible hybrid system configurations are listed in ascending order of their total net present cost. The best possible combination of SPV, SHP, a BDG and batteries is highlighted in blue, and the next best possible combination, marked with a red-coloured box, includes the SPV, a wind generator, SHP and a biodiesel generator. The blue highlighted combination is able to fully meet Palari's load demands at the lowest possible total net present cost.

The optimal combination of RET system components for our case study is a 20-kW PV array, 30 kW SHP, 10 kW BDG, 40 Surrette 6CS25P batteries, 20 kW inverter and a 20-kW rectifier with a dispatch strategy of cycle charging.[14] No wind turbine is selected at this site[15] (see Fig. 10). This system is considered at 3.5 m/s of wind speed, $0.6/l of biodiesel cost and 815 l/s of design flow rate for the SHP. The total net present cost, capital cost and the COE for such a hybrid system are $673,147, $238,000 and $0.420/kWh, respectively. The COE of $0.420/kWh from this hybrid system is cheaper than that of $0.44/kWh from grid extension as considered for this study. Therefore, grid extension does not appear to be a viable option to meet the village load. But if the COE from the grid supply falls below $0.420/kWh, grid extension becomes viable.

Figure 11 shows the monthly distribution of the electricity produced in kW by the SPV, SHP and BDG. From December to February, the biodiesel generator is mostly used combined with SPV as hydropower is unavailable due to low flow in the river. Also, from June to August the peak load is met by SPV and BDG.

It is evident from Fig. 11 and Table 4 that small hydro station dominates the electricity output in this case. The SHP operates at full load for 9 months and produces 186,649 kWh/year, achieving a capacity factor of 71 %. At this level of operation, the levelised cost of hydro-only system becomes just 4.62 cents/kWh.

---

[14] In cycle charging mode of operation, the generator charges the battery with surplus power each time the generator operates.

[15] All generators can be co-located at the SHP power house to reduce the length of the distribution system for carrying power from the village.

Sensitivity Results | Optimization Results

Sensitivity variables

Wind Speed (m/s) 3.5 | BioDiesel-(B100) Price ($/L) 0.6 | Design Flow Rate (L/s) 815

Double click on a system below for simulation results.

Categorized

PV (kW)	G10	Hydro (kW)	BioD (kW)	S6CS25P	Conv. (kW)	Disp. Strgy	Initial Capital	Operating Cost ($/yr)	Total NPC	COE ($/kWh)	Ren. Frac.	Capacity Shortage	Diesel-(B10 (L)	BioD (hrs)
20		30.0	10	40	20	CC	$ 238,000	47,939	$ 673,147	0.420	0.90	0.02	13,646	2,663
40	1	30.0	5	60	30	CC	$ 419,000	31,821	$ 707,837	0.444	0.96	0.03	5,865	2,334
70		30.0		110	30	CC	$ 603,000	20,395	$ 788,124	0.493	1.00	0.03		
		30.0	20	60	20	CC	$ 150,000	72,843	$ 811,198	0.497	0.81	0.00	23,039	2,208
70	1	30.0		110	30	CC	$ 643,000	21,330	$ 836,615	0.523	1.00	0.03		
	1	30.0	20	60	20	CC	$ 190,000	72,094	$ 844,404	0.517	0.82	0.00	22,461	2,147
20		30.0	20		10	LF	$ 203,000	105,612	$ 1,161,647	0.719	0.85	0.02	27,302	3,803
20	1	30.0	20		10	LF	$ 243,000	104,274	$ 1,189,499	0.736	0.85	0.02	26,778	3,714
	2	30.0	20		10	CC	$ 163,000	120,866	$ 1,260,110	0.786	0.79	0.03	33,489	4,292
		30.0	30			LF	$ 88,000	176,146	$ 1,686,882	1.036	0.76	0.01	44,413	4,446
100			20	180	50	CC	$ 849,000	125,094	$ 1,984,485	1.230	0.73	0.02	35,268	3,602
160	1		10	260	70	CC	$ 1,331,000	75,422	$ 2,015,606	1.266	0.93	0.03	12,738	2,702
260	4			300	70	CC	$ 2,079,000	49,974	$ 2,532,612	1.594	1.00	0.03		
100	1		40		60	LF	$ 733,000	321,640	$ 3,652,537	2.267	0.61	0.02	81,351	5,971
100			40		50	LF	$ 693,000	327,234	$ 3,663,319	2.275	0.61	0.02	82,991	6,097

Fig. 9 Summary of simulation results

Cost summary

Total net present cost	$673,147
Levelized cost of energy	$0.420/kWh
Operating cost	$47,939/yr

System architecture

PV Array	20 kW
Hydro	30 kW
Biodiesel Generator	10 kW
Battery	40 Surrette 6CS25P
Inverter	20 kW
Rectifier	20 kW
Dispatch strategy	Cycle Charging

Electrical

Component	Production (kWh/yr)	Fraction
PV array	34,439	14%
Hydro turbine	186,649	76%
Biodiesel Generator	25,294	10%
Total	246,382	100%

Fig. 10 Optimal least cost hybrid system for the case study

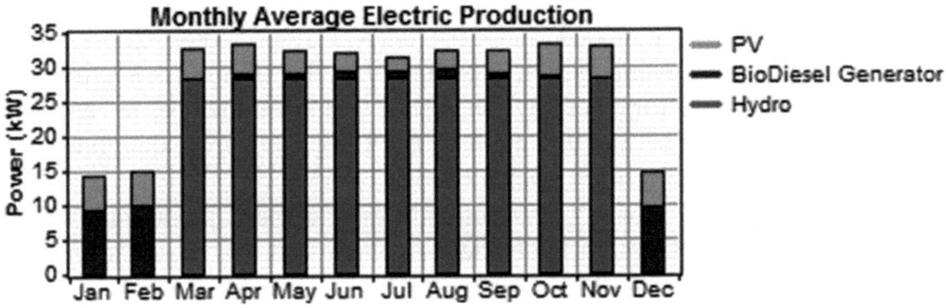

Fig. 11 Monthly average electricity production from the best hybrid configuration system

Only during the winter months when water is inadequate, biodiesel plant becomes the dominant producer. For the selected system the biodiesel plant operates for 2,663 h (capacity factor 28.9 %), produces 25,294 kWh/year and consumes 13,646 l of biofuel. However, this is a costlier option than hydropower and the marginal COE from the biodiesel plant is $0.21/kWh. The penetration of solar energy reduces the biodiesel output, particularly outside winter months. The solar panels produce 34,439 kWh/year, operating for 4,357 h (or recording a capacity factor of 19.7 %). The levelised cost of solar electricity turns out to be $0.415/kWh.

Table 4 Techno-economic details of the three hybrids system configurations

Configurations	Unit	Best hybrid	Second best hybrid	Third best hybrid
Wind speed	m/s	3.5	3.5	5
Biodiesel-(B100) price	$/L	0.6	0.6	0.6
Design flow rate	L/s	815	0	0
Solar PV	kW	20	100	120
Wind turbine (G10 kW)	no.	0	0	6
Hydro	kW	29.98	0	0
Bio-D	kW	10	20	10
Batteries—Surrette 6CS25P		40	180	200
Converter	kW	20	50	70
Dispatch strategy	no.	CC	CC	CC
Total capital cost	$	2,38,000	8,49,000	12,31,000
Total NPC	$	6,73,147	19,84,485	18,98,258
Total annual capacity cost	$/yr	26,220	93,533	1,35,617
Total annual replacement cost	$/yr	3,623	14,232	17,815
Total O&M cost	$/yr	36,129	89,701	47,937
Total fuel cost	$/yr	8,188	21,161	7,759
Total annual cost	$/yr	74,159	2,18,627	2,09,127
Operating cost	$/yr	47,939	1,25,094	73,511
COE	$/kWh	0.42	1.23	1.192
PV production	kWh/yr	34,439	1,72,196	2,06,636
Wind production	kWh/yr	0	0	53,004
Hydro production	kWh/yr	1,86,649	0	0
Bio-D production	kWh/yr	25,294	63,719	22,950
Total electrical production	kWh/yr	2,46,382	2,35,916	2,82,589
AC primary load served	kWh/yr	1,55,444	1,56,290	1,54,027
Deferrable load served	kWh/yr	21,308	21,383	21,350
Renewable fraction	%	0.9	0.73	0.92
Capacity shortage	kWh/yr	4,393	2,805	5,522
Capacity shortage fraction	%	0	0	0
Unmet load	kWh/yr	3,056	2,144	4,440
Unmet load fraction	%	0.02	0.01	0.02
Excess electricity	kWh/yr	62,412	25,258	73,398
Biodiesel-(B100)	L/yr	13,646	35,268	12,932
Breakeven grid extension distance	km	−2.09	58.58	54.59

62,412 kWh/year of electricity which is 25 % of total electricity generated goes unused due to low demand and is fed to dump loads. This is particularly high in summer months when the hydro plant operates fully. This shows that this system has the capability in meeting the demand growth in the future. The demand can also be increased by serving the demand of other nearby villages, because as the demand increases the load factor increases and hence the cost per kWh will decrease.

Figure 12 shows the cash flow summary for the optimal system. The capital cost of the BDG makes up only 5 % of the system's total capital cost, whereas almost 50 % of the initial investments go to the SPV arrays. Once installed,

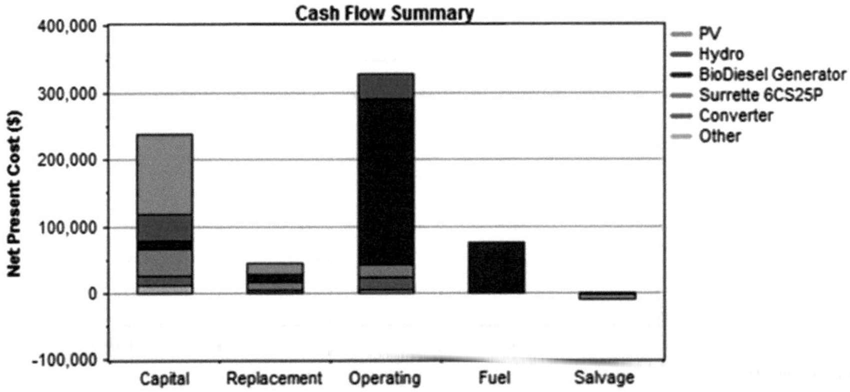

Fig. 12 Cash flow summary based on the selected components

however, SPV is cheap to maintain and operate compared to the BDG, which in the end is responsible for 51.5 % of the system's total annual cost of $74,159. Small hydro plant on the other hand is a relatively cheaper option and contributes less to the overall cost.

4.2 Sensitivity Results

Sensitivity analysis eliminates all infeasible combinations and ranks the feasible combinations taking into account the uncertainty of parameters. HOMER allows taking into account future developments, such as increasing or decreasing load demand as well as changes regarding the resources, for example fluctuations in the river's water flow rate, wind speed variations or the biodiesel prices. Here, various sensitive variables are considered to select the best suited combination for the hybrid system to serve the load demand.

If the hydropower option is not available, the second-best hybrid configuration will comprise 100 kW SPV, 20 kW BDG and 180 batteries and will have a COE of $1.230/kWh. This hybrid system has an economic distance to grid of 58.6 km and generates 25,258 kWh/year of excess electricity. This configuration can be used at off-grid locations where hydropower is not available. The third hybrid configuration adds wind turbines to the configuration when the wind speed increases from 3.5 to 5 m/s and no hydro is available. Hence with an increasing number of components, the capital cost and total NPC also increase. The third hybrid configuration of wind turbines, SPV, BDG and batteries has a COE of $1.192/kWh.[16] Table 4 shows all the relevant techno-economic details considering the best three hybrid system configurations considered on HOMER for this study.

[16] HOMER ranks options by net present value and not by cost of electricity.

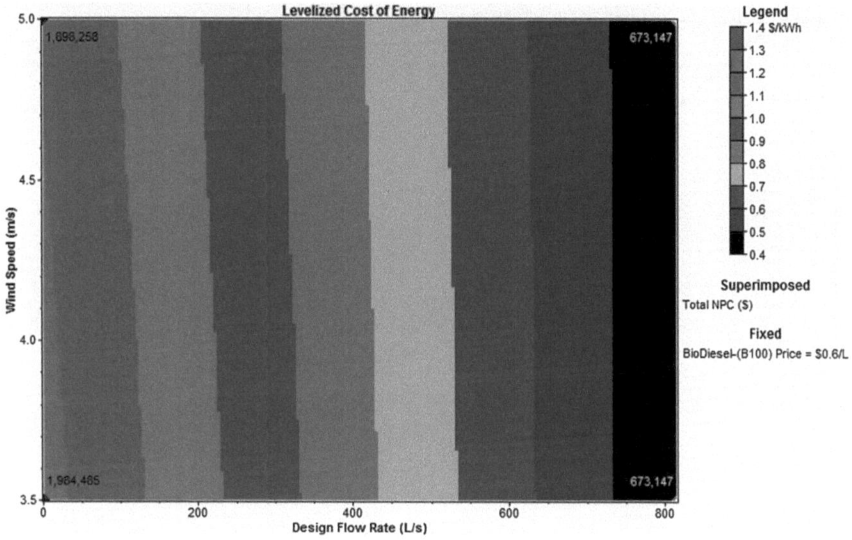

Fig. 13 Surface plot of cost of electricity

This shows that the system configurations without SHP tend to be more costly than other renewable technologies. Even if the wind speed increases, generating more electricity from the wind turbine, the system costs do not reduce (see Fig. 13). The surface plot for the levelised COE with total NPC superimposed is presented in Fig. 13. The biodiesel price is fixed at $0.6/l, the hydropower design flow rate is depicted on the x-axis, and wind speed variation on the y-axis. It can be observed that as the design flow rate increases, the power output from SHP increases and hence there is a reduction in the total NPC. As the total NPC decreases, the system's COE decreases as well. This shows that with a change in sensitivity variables the capacity of an individual component increases and hence the configuration of the system changes. Therefore, a hybrid system with SHP proves to be the cheapest option compared to other RETs due to lower capital cost of a small hydropower plant.

Figure 14 shows the result for the breakeven distance for grid extension (or EDL). It shows that the distance varies from a negative value to 60 kms depending on the total NPC and levelised COE. For the selected hybrid configuration, the EDL comes out be a negative value meaning that a decentralised RE hybrid system is a better option than the grid extension. It is clearly evident from the line graph that as the design flow rate increases with wind speed and biodiesel cost at a fixed value of 3.5 m/s and $0.6/l, respectively, the total NPC of the system decreases. At 100 l/s of design flow rate the EDL comes out to be 50 kms, and at 800 l/s the EDL comes out to be negative distance of −2 kms. Hence, the total NPC and levelised COE of a system determine the EDL with respect to the input parameters.

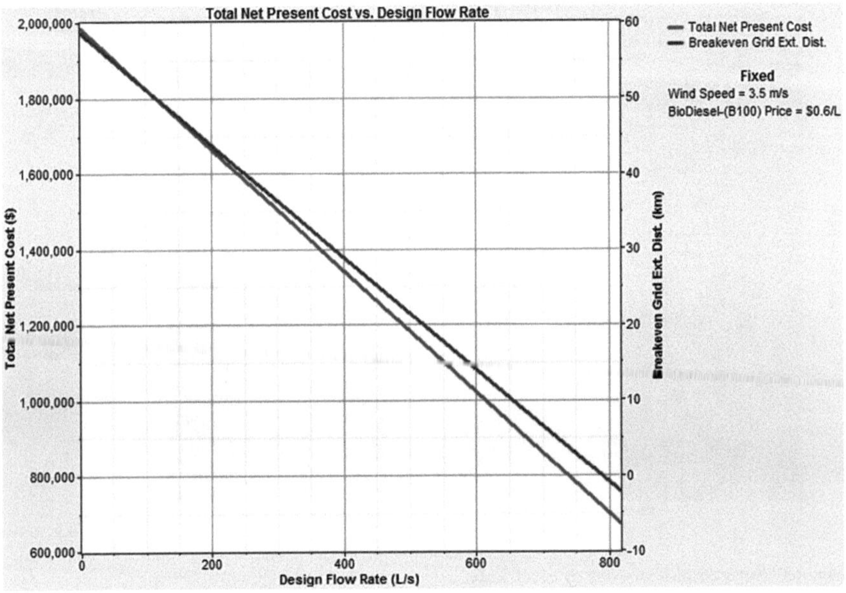

Fig. 14 Line graph for total NPC vs. design flow rate and breakeven grid extension distance

Table 5 Emission reduction

Pollutant	Emissions (kg/yr)
Carbon dioxide	33,832
Carbon monoxide	44.3
Unburned hydrocarbons	0.688
Particulate matter	4.68
Sulphur dioxide	0.576
Nitrogen oxides	894

4.3 Emissions

The optimal hybrid RET system would save 33,832 kg/yr of CO_2 over 1 year in operation compared to a central power generation plant or a stand-alone DG system. In addition, emission of particulate matters and nitrogen oxides will be reduced due to reliance on renewable energy systems (see Table 5).

Based on the above analysis, it can be concluded that a hybrid system becomes a viable option in an off-grid location in India. If small hydropower potential exists, it can offer economically attractive power supply. In the absence of small hydro potential, the cost of supply increases and interventions by the government may be required to make the investment socially desirable.

5 Post-HOMER Analysis

The optimal hybrid electricity supply system requires the development of 60 kW of power generating capacity (30 kW small hydro, 10 kW biodiesel and 20 kW solar PV) in a remote location and arrangements for necessary distribution to 304 households and other customers in the village. Although HOMER suggests technical feasibility of such a system and indicates the breakeven cost at which the investment can be recovered (see Table 4), the business dimensions are generally not covered. The post-HOMER analysis is required to develop a complete understanding of the business case. In the following paragraphs, some issues are highlighted.

The first question that arises relates to financing of the investment. For example, in the optimal case, an investment of $238,000 will be required for a 60-kW system (or an average of $400/kW approximately)—see Table 4. Although the investment volume is not large either for any conventional lender (such as banks) or for any utility investor, significant risks are involved in the investment. First, a part of the investment is not re-deployable (e.g. the investment for SHP). If the project does not succeed for any reason, the investment will be a sunk cost for the investor and will represent a bad investment. Second, the electricity market in the area is not developed and the assumptions related to the demand may not materialise, or may take longer to realise. This will adversely affect the cost recovery process. Third, the business environment may be affected by political, regulatory and governance challenges, thereby affecting such investments. Fourth, there are practical difficulties (e.g. availability of skilled manpower, managing supply logistics and poor transport facilities) that can add to costs, delay project delivery and reduce profitability of the projects. In such cases, appropriate incentives and support mechanisms will play an important role to attract investment and mitigate risks. In the state of Chhattisgarh, capital grant is available for solar, wind and biogas plants. Also, the state renewable energy agency is actively promoting off-grid electrification and has trained technicians who work in rural areas. Therefore, some risk mitigation is already underway.

A related issue that requires careful consideration is the choice of an appropriate business model for delivering the project. While a private investor brings expertise and innovative ideas, the cost of supply can be higher. Moreover, a private investor will essentially be profit-driven and unless the business case suggests profitability, it is unlikely that private investment will flow. On the other hand, state utility services have not been successful in providing electricity in the remote areas, and therefore, such agencies are unlikely to be interested in off-grid electricity delivery. A middle path may be found in the form of local co-operatives or private–public partnership projects where both the social dimension and the business-like approach are combined. In our particular case, the state renewable energy agency is proactive in building joint ventures and promoting private investment, but further work is required to decide a specific business model.

Clearly, the tariff issue will play a crucial role but remains a challenging task in the rural context due to the following factors. First, investors will be interested in recovering the investment over a shorter period of time and very few lenders will consider a loan period of 25 years. As the cost recovery period reduces, the cost of supply will increase, which may in turn make the project less attractive to the users. Second, the discount rate used for business decisions depends on the investor. For example, a private investor is likely to use a higher discount rate to reflect the cost of capital, riskiness of the investment and its desire to recover the investment quickly. On the other hand, state agencies or local communities may use a low discount considering the social nature of the investment. The tariff will accordingly depend on this. Third, the grid-based electricity supply in other areas may be subsidised and consumers in the off-grid area may expect similar tariff treatments. However, the cost of supply for the off-grid case may be quite different from the grid-based supply and the consumer base is significantly small. Therefore, there is very limited cross-subsidy potential in the off-grid case and unless there is direct subsidy support from the government, price parity with the grid-based supply can only jeopardise the viability of the project. In this particular case, the state allows subsidy for rural areas and the support is available for off-grid consumers at a fixed rate. While this may reduce the tariff burden on the consumer, it is unclear whether this level of subsidy is sufficient for cost recovery or not and whether the subsidy will be available over the entire life of the project.

Finally, the issue of regulating the off-grid supply through a mini-grid system as is envisaged in the case study requires careful consideration. The business will not function effectively unless the rules of the game are clearly laid out and the compliance with the rules is monitored through a supervisory system. As the consumers are likely to be poor and vulnerable, protecting them against any monopoly abuse, health and safety risks and other unfair treatments assumes greater importance. Simultaneously, the investor needs to be protected and encouraged to provide the desired level of service. However, unclear regulatory environment and lack of regulatory capacity can hinder developing such projects. This is an area of concern for the present case study where no specific regulatory arrangement exists for mini-grids.

6 Conclusion

Our search for a technically feasible and economically viable hybrid solution for off-grid electricity supply to a remote village such as Palari resulted in a least-cost combination of small hydropower, solar PV, biodiesel and batteries that can meet the demand in a dependable manner at a cost of \$0.420/kWh. Given the availability of small hydropower in this location, most of the electricity in the optimal solution comes from the hydro plant and it provides a cheap source of power to the locality. However, the system reliability cannot be ensured due to variable nature of water availability and lack of adequate water flow in winter unless other technology options are considered. The biodiesel plant and the solar PV plants contribute 10 and 14 %, respectively, to electricity generation but being costlier options than small hydropower, they raise the overall COE. If the small hydro plant is not available (or no hydro resources are available), the electricity demand can be met with a hybrid system comprising of solar PV, small wind turbines and biodiesel plants. But the COE supply will increase threefold, thereby making the system less attractive to users. Thus, three main lessons from this case study are: (1) where hydro potential exists, it is important to take advantage of the resource; (2) a combination of technologies improves supply reliability and hence makes better business sense; and (3) the cost of supply of renewable energy-based electricity may not always be a cost-effective option for remote applications unless appropriately supported by the government.

Although our work is based on a standard software HOMER, it goes beyond the conventional applications of the software by systematically considering the pre- and post-application phases. In the pre-HOMER stage, we have considered the local demand in detail and have included multiple types of users (residential, institutional, commercial, agricultural and industrial) and considered seasonal variation in the demand. As HOMER takes the demand as given and finds the least cost combination of supply options to meet the demand, realistic demand estimation assumes an important role. Our study contributes in this area by highlighting this aspect and incorporating a detailed demand analysis feature in the study. In the post-HOMER phase, we highlight the business-related dimensions that influence the project delivery. We have briefly considered the financing challenge, business model selection, tariff issue and the regulatory concerns. This is an attempt to go beyond the techno-economic analysis.

Surely, further work is required in both pre- and post-HOMER areas. We believe that a standard template can be designed for a systematic estimation of demand for off-grid areas and to capture the stakeholder perspectives. Even demand scenarios can be included to take the simulations to another level of iteration. Similarly, a systematic approach of considering the business case of the optimal solution and its delivery-related issues can enhance the overall appreciation of the micro-energy systems.

References

1. Barsoum N, Pandian V (2007) Balancing cost, operation and performance in integrated hydrogen hybrid energy system. In: Proceedings of the FIRST Asia international conference on modelling and simulation (AMS 07), IEEE Explore, Phuket, pp 14–18
2. Bekele G, Palm B (2010) Feasibility study for a sustainable solar-wind-based hybrid energy system for application in Ethiopia. Appl Energy 87(2):487–495
3. Chakrabarti S, Chakrabarti S (2002) Rural electrification programme with solar energy in remote region—a case study in an island. Energy Policy 30(1):33–42
4. Deshmukh MK, Deshmukh SS (2008) Modelling of hybrid renewable energy systems, Renew Sustain Energy Rev 12:235–249

5. Giatrakos GP, Tsoutsos TD, Mouchtaropoulos PG, Naxakis GD, Stavrakakis G (2009) 'Sustainable energy planning based on a stand-alone hybrid renewable energy/hydrogen power system: application in Karpathos Island, Greece'. Renew Energy 34:2562–2570
6. Givler T, Lilienthal P (2005) Using HOMER® Software, NREL's Micro power Optimization Model, To Explore the Role of Gen-sets in Small Solar Power Systems Case Study: Sri Lanka, Technical Report NREL/TP-710-36774. Available from http://www.osti.gov/bridge
7. Hafez O, Bhattacharya K (2012) Optimal planning and design of a renewable energy based supply system for microgrids. Renew Energy 45:7–15
8. Himri Y, Stambouli AB, Draoui B, Himri S (2008) Techno-economical study of hybrid power system for a remote village in Algeria. Energy 33(7):1128–1136
9. Karakoulidis K, Mavridis K, Bandekas D, Adoniadis P, Potolias C, Vordos N (2011) Techno-economic analysis of a stand-alone hybrid photovoltaic-diesel-battery-fuel cell power system. Renew Energy 36:2238–2244
10. Khan M, Iqbal M (2005) Pre-feasibility study of stand-alone hybrid energy systems for applications in Newfoundland. Renew Energy 30(6):835–854
11. Lau KY, Yousof MFM, Arshad SNM, Anwari M, Yatim AHM (2010) Performance analysis of hybrid photovoltaic/ diesel energy system under Malaysian conditions. Energy 35(8): 3245–3255
12. Mahapatra S, Chanakya H, Dasappa S (2009) Evaluation of various energy devices for domestic lighting in India: technology, economics and CO2 emissions. Energy Sustain Dev 13(4):271–279
13. Muneer T, Muhammad A, Saima M (2005) Sustainable production of solar electricity with particular reference to the Indian EconomyRn. Renew Sustain Energy Rev 9(5):444–473
14. Munuswamy S, Nakamura K, Katta A (2011) Comparing the cost of electricity sourced from a fuel cell-based renewable energy system and the national grid to electrify a rural health centre in India: a case study. Renew Energy 36:2978–2983
15. Nandi S, Ghosh HR (2010) Prospect of wind-PV-battery hybrid system as an alternative to grid extension in Bangladesh. Energy 35(7):3040–3047
16. Nfah EM, Ngundam JM, Vandenbergh M, Schmid J (2008) Simulation of off-grid generation options for remote villages in Cameroon. Renew Energy 33(5):1064–1072
17. Nouni M, Mallick S, Kandpal T (2007) Biomass gasifier projects for decentralized power supply in India: A financial evaluation. Energy Policy 35(2):1373–1385
18. Nouni M, Mallick S, Kandpal T (2008) Providing electricity access to remote areas in India: an approach towards identifying potential areas for decentralized electricity supply. Renew Sustain Energy Rev 12(5):1187–1220
19. Ravindranath N, Someshekar H, Dasappa S, Reddy C (2004) Sustainable biomass power for rural India: case study of biomass gasifier for village electrification. Curr Sci 8(7):932–941
20. Sen R, Bhattacharyya SC (2014) Off-grid electricity generation with renewable energy technologies in India: an application of HOMER. Renew Energy 62:388–398
21. Sinha C, Kandpal T (1991) Decentralized v grid electricity for rural India—the economic factors. Energy Policy 19(5):441–448
22. Ulsrud K, Winther T, Palit D, Rohracher H, Sandgren J (2011) The solar transitions research on solar mini-grids in India: learning from local cases of innovative socio-technical systems. Energy Sustain Dev 15:293–303

From SHS to Mini-Grid-Based Off-Grid Electrification: A Case Study of Bangladesh

Subhes C. Bhattacharyya

Abstract This chapter presents a local-level study of a village off-grid system in Bangladesh. Using the recent census information and household survey information, the study presents the current status of rural electrification in Bangladesh and indicates the characteristics of rural energy use. It then applies an integrated methodology that identifies the demand in the off-grid village context using six alternative scenarios that capture the residential demand by household type, commercial demand and productive demand. The techno-economic analysis of the optimal off-grid system architecture is then presented using HOMER software. Three energy resources are considered, namely solar energy, wind and diesel fuel. The optimal configuration suggested for the scenarios consists of diesel generators for the basic level of demand and PV-diesel hybrid for higher demand and reliable supply scenarios. The cost of electricity per kWh remains high for the basic level of supply and decreases as the system size increases. However, the capital and asset replacement costs increase considerably for bigger systems. The business case is then analysed for each scenario and it is found that grid price parity is impossible to reach with even full capital cost subsidy, indicating a significant amount of operating cost subsidy requirement that makes the larger systems financially unviable. Moreover, the small mini-grid system for the basic level of supply emerges as a cheaper option than providing the consumers with solar home systems. However, the monthly electricity bill will become unaffordable for most consumers when demand restrictions are removed. Accordingly, the chapter suggests a mini-grid-based electricity supply to provide the basic level of provision alongside productive energy use during off-peak hours as the starting point. If the business develops and demand improves, the system can be expanded subsequently using appropriate technology combinations.

S. C. Bhattacharyya (✉)
Institute of Energy and Sustainable Development, De Montfort University,
Leicester LE1 9BH, UK
e-mail: subhesb@dmu.ac.uk; subhes_bhattacharyya@yahoo.com

S. C. Bhattacharyya and D. Palit (eds.), *Mini-Grids for Rural Electrification of Developing Countries*, Green Energy and Technology,
DOI: 10.1007/978-3-319-04816-1_10,

1 Introduction

Being at the forefront of Solar Home System (SHS) dissemination in the world, Bangladesh holds a special place in any discussion on off-grid electrification. In this densely populated, low-income country of 152 million people (in 2012), the overall rate of electrification is reported at 56 % in 2011, thereby forcing about 40 % of the population to rely on kerosene for lighting purposes [2]. However, there exists a significant variation between the rural and urban areas. 80 % of the population resides in rural areas but only about 49 % of the rural population is electrified, whereas about 89 % of the urban population is said to be electrified [2]. Although SHS has been successfully introduced in the country, particularly by Grameen Shakti, it has reached only 4 % of the rural households and 0.5 % of the urban households so far [2]. The government aims to provide electricity to all by 2021 and although the strategy appears to consider both off-grid and grid extension options, the task looks increasingly challenging.

Although a lot of academic and other studies have analysed the case of Bangladesh, the literature focuses on two dimensions: the success of Bangladesh in introducing rural electrification through the rural electrification cooperatives (see [8] for example) and the success of Grameen Shakti in introducing SHS (see [7] for example). However, neither grid extension nor SHS has succeeded in ensuring universal electrification of the country and in the case of SHS the use of electricity for productive purposes has remained insignificant. A few other studies (e.g. Nandi and Ghosh [5], Mondal and Denich [4], Roy [6], among others) have considered the case of hybrid off-grid systems for rural electricity supply but their analysis remains limited to just techno-economic analysis using a simulation tool, namely HOMER. Most of these studies are hypothetical in nature, rely on representative households consuming identical levels of energy for a given period of time, use generic technology/ financial information and thus provide an overall understanding of the hybrid option. Although these provide useful information, such techno-economic analysis does not really indicate whether the service can be provided as a viable business, whether costs can be recovered through affordable tariffs and whether the investment can be mobilised and if so, under what conditions.

The purpose of this chapter is to argue for a transition from SHS to mini-grid-based off-grid power supply in Bangladesh through a comprehensive analysis of the business case for such a transition. The aim is to understand the needs of the rural communities and identify appropriate solutions based on local resource availability so that an affordable solution can be proposed that is financially viable and socially desirable. This work thus goes beyond the standard application of a simulation tool and adds value by bridging the above knowledge gap.

The chapter is organised as follows: Sect. 2 presents a review of electricity access and background economic conditions in Bangladesh; Sect. 3 presents the methodology used in this work; Sect. 4 presents the case study background information; Sect. 5 presents the techno-economic analysis of alternative scenarios using HOMER; and Sect. 6 presents the business case analysis. Finally concluding remarks are provided in the concluding section.

2 Review of Socio-Economic Background Information About Bangladesh

The Census of 2011 has produced significant new information about the population and its socio-economic features. In addition, a recent survey, the Household Income and Expenditure Survey of 2010, also provides very relevant information. This information is helpful in understanding the energy use, incidence of poverty and affordability of the consumers. In this section, a brief review of relevant information is provided.

2.1 Basic Socio-Economic Background

Bangladesh is located between 20°34′ and 26°38′ North Latitude and 88°01′ and 92°41′ East Longitude with a total surface area of 147,570 km^2. Administratively, the country is divided into seven divisions[1] which are further divided into 64 districts. Dhaka is the capital of the country.

According to the 2011 Census [2], there are 31.7 million households in Bangladesh of which 6.2 million are in urban areas and the remaining 25.5 million live in rural areas. Table 1 provides the distribution of population by division. This clearly shows that Dhaka division holds about one-third of the population and about 45 % of the country's urban population. Chittagong division holds another 17.5 % of the population and 19 % of the urban population. These two divisions account for about one-half of the country's population and 64 % of the urban population.

The Household Income and Expenditure Survey 2010 [1] provides information about household income and expenditure as well as distribution of income and expenditure. The average monthly income per household at current prices was estimated at taka 11,479 in 2010.[2] However, the country suffers from high income inequality. The poorest of the poor (bottom 5 % of the population by income) just accounted for 0.78 % of the income while the income accrued to the top 5 % of the households was 24.61 %. Seventy per cent of the income went to top 40 % of the households, leaving only 6 % of the income for the lowest 30 % of the households

[1] These are Barisal, Chittagong, Dhaka, Khulna, Rajshahi, Rangpur and Sylhet.

[2] In 2010, the exchange rate for Bangladeshi taka was as follows: 1 GBP = 105 taka and 1 US dollar = 75 taka.

Table 1 Household distribution by administrative divisions of Bangladesh (in millions)

Division	Rural	Urban	Total
Barisal	1.61	0.24	1.85
Chittagong	4.36	1.19	5.55
Dhaka	7.8	2.78	10.58
Khulna	3.11	0.59	3.7
Rajshahi	3.75	0.71	4.46
Rangpur	3.35	0.44	3.79
Sylhet	1.54	0.22	1.76
Total	25.53	6.17	31.7

Source BBS [2]

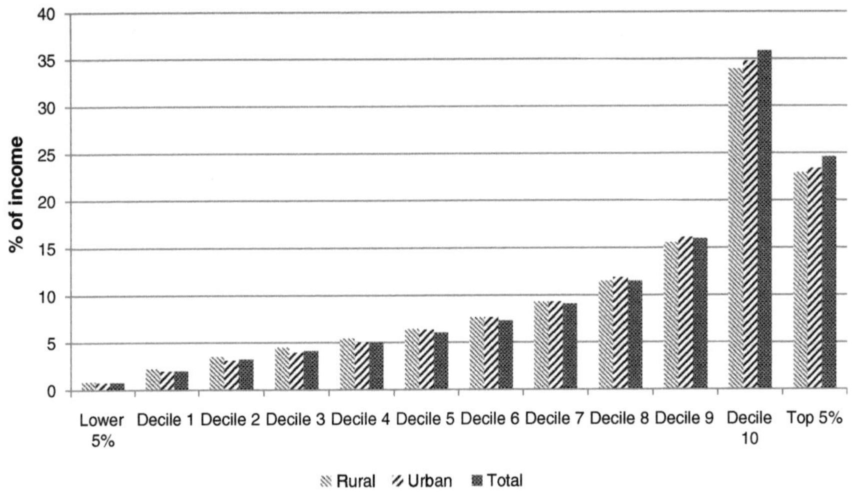

Fig. 1 Skewed income distribution in Bangladesh. *Data source* BBS [1]

[1]. The income distribution does not change much between urban and rural areas and the exponential nature of income growth for the households in the top deciles makes the income distribution skewed (see Fig. 1).

The country remains dependent on agriculture for household income, although commerce and the salaried class have emerged as other major income sources (see Fig. 2). There is however a significant difference between urban and rural households in terms of source of income. The salaried class and business activities form the main income source in urban areas, while agricultural activities and wage-based income are two income sources for rural areas.

The average monthly household expenditure was estimated at 11,200 taka at current prices in 2010. However, there is a significant difference in terms of urban and rural expenditure per household. The national average of urban household expenditure per month is 16,474 taka in 2010 while the rural household

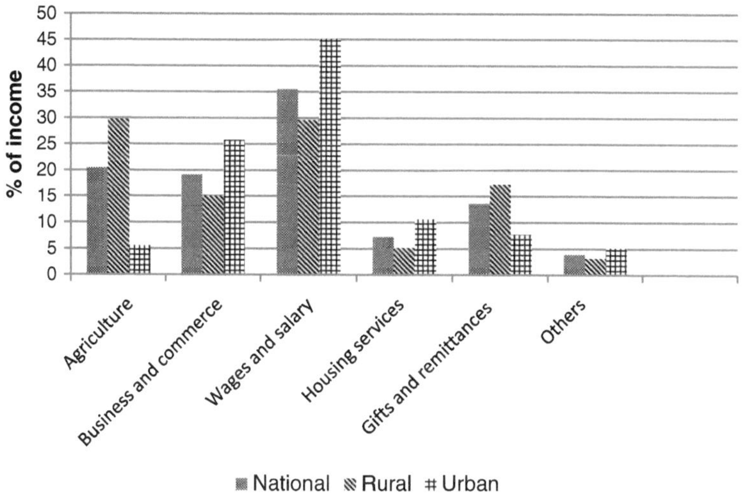

Fig. 2 Source of household income. *Data source* BBS [1]

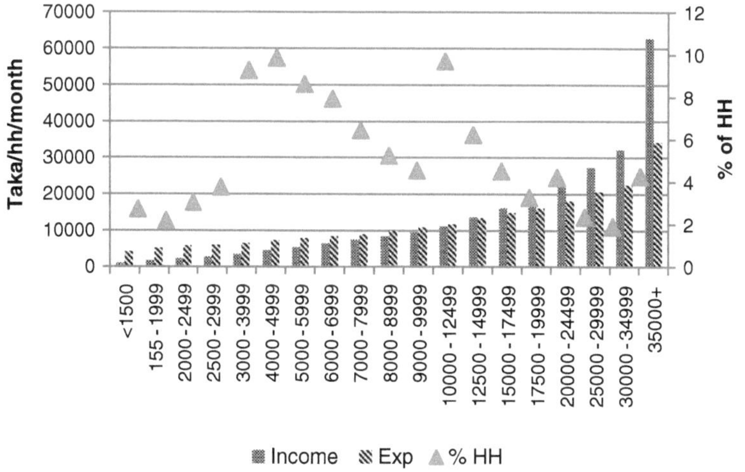

Fig. 3 Distribution of household income-expenditure at the national level. *Data source* BBS [1]

expenditure is 9,648 taka (or about 40 % less than the urban households). The expenditure also varies by income class of the houscholds (see Fig. 3). It becomes clear that more than 50 % of the national population has an insufficient household income to meet the household needs (i.e. they are living beyond their means). However, the situation is even worse in the rural areas where more than 70 % of the population is living beyond their means. This has important bearing on electricity access issues. On the other hand, the inequality in terms of expenditure is

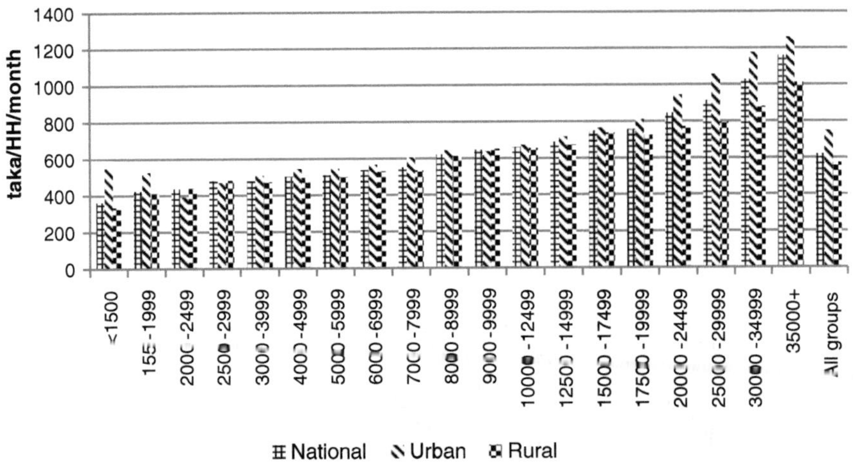

Fig. 4 Difference in spending on lighting and fuel by income categories. *Data source* BBS [1]

less severe compared to that of income. This perhaps suggests that the households in the top income decile have the potential to expand their consumption in the future following the rich lifestyles of the developed world. This is particularly important in urban areas where almost 9 % of the urban households come under the highest income group (more than 35,000 taka per month).

2.2 Energy-Related Expenditure by Income Categories

The survey also provides monthly expenditure on fuel and lighting. Although this does not indicate expenditure for lighting and cooking separately nor gives the quantity of fuel used, it provides insights into the consumer fuel preferences. On average, the monthly expenditure on fuel and lighting was 619 taka per household in 2010. This is about 5.5 % of the monthly household expenditure. There is some variation between urban and rural households: the monthly spending on lighting and fuel by an urban household was 746 taka (4.81 % of total expenditure), whereas a rural household spent 572 taka (5.95 % of total expenditure). However, the picture changes quite significantly for households in the lowest and highest income classes. The households in the lowest income group spend about 10 % of their income on fuel, whereas the richest group spends about 3–4 % of their budget on fuel. The poorest households typically spend about 30 % of what the richest households spend on fuel and lighting but the difference reduces in urban areas, where the poorest households spend almost 50 % of the amount that their richest counterparts devote to fuel and lighting. The other noticeable feature is the similarity in spending by the middle income groups irrespective of urban or rural

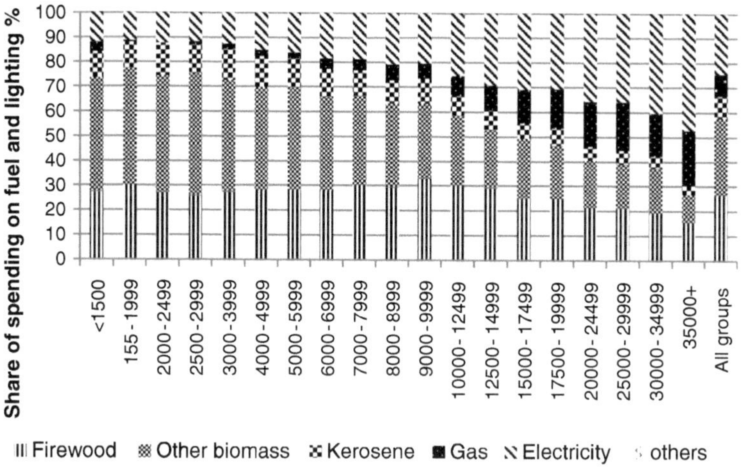

Fig. 5 Fuel mix by income categories at the national level. *Data source* BBS [1]

locations. This perhaps reflects the greater importance given to fuels and lighting by these consumer classes (see Fig. 4).

The fuel mix[3] also undergoes a major shift: the poor on average spend about 20 % of their fuel budget on modern fuels (coal, oil, gas and electricity) while the rich spend about 70 % on modern fuels (see Fig. 5). This clearly supports the 'energy ladder' concept where the preference for modern fuels improves as income increases. However, it is interesting to note here that modern fuels account for more than 50 % of the budget at a relatively high level of income of 17,500 taka per month per household.

Assuming that households use kerosene and electricity for lighting and appliance uses, the average spending on these fuels at the national level by income groups is shown in Fig. 6. On average, the monthly spending for lighting and appliance use is about 190 taka per household, about 30 % of which is spent on kerosene and the rest on electricity. However, the picture changes significantly at the lower income levels where more than 50 % of the lighting related expense is attributable to kerosene use, while for the rich the share falls to 7.5 %.

Electricity tariff in Bangladesh is regulated by the Bangladesh Energy Regulatory Commission and residential consumers are charged according to an increased block tariff. The prevailing tariff with effect from 1 March 2012 is 3.05 taka/kWh for the first 100 kWh, 4.29 taka for the next 300 kWh and 7.89 taka/kWh for any consumption above 400 kWh. At this tariff, the lowest income households can consume about 10 kWh per month with their monthly

[3] Strictly speaking, this is share of spending on fuel and lighting by different income categories. This leads to fuel mix under the assumption of a uniform price per unit of equivalent energy from each source.

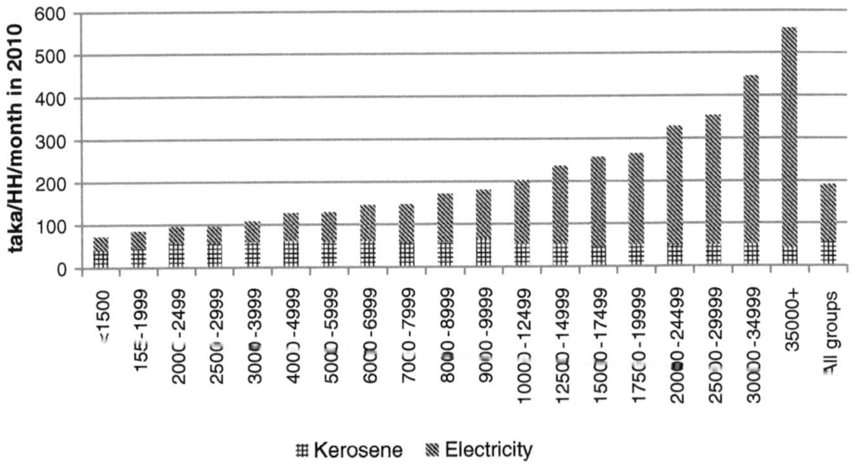

Fig. 6 Distribution of monthly household spending on lighting fuels. *Data source* BBS [1]

energy budget, while the highest income category households consume about 150 kWh per month. Even the high-end consumers appear to be consuming only modest amounts of electricity, which may be due to chronic shortage in electricity supply prevailing in the country.

However, the rural and urban differences in the spending by income categories are worth mentioning. The average spending for urban households increases to 321 taka/month/household, while for rural households this drops to 144 taka. More importantly, the share of spending on kerosene remains very high in rural households—68 % of the total spending on kerosene and electricity for the poorest income category and 25 % for the highest income category, with an average of 48 % for the rural households as a group. The inverse is true for urban households: the poorest families spend about 14 % of the expenditure on kerosene and electricity to purchase kerosene while the richest households spend about 1.5 % on kerosene.

A few observations/inferences can be drawn from the above. First, the economically weakest sections of the population spend proportionally higher shares of their income on energy-related services, thereby suffering energy poverty in an economic sense. Second, due to their high dependence on traditional energies, this category of households also derives high levels of social benefits of modern energy supply due to high potential for external cost reduction through improved health, better environmental conditions, reduction in drudgery and improved potential for human capacity development. This also justifies interventions on externality grounds, knowing that markets would not perform effectively in the presence of strong externalities. Third, the limited affordability of this category of households is a barrier to any transition to modern energy sources, which in turn requires careful design of the services to lower cost of supply and to link with rural economic development that in turn can reduce the incidence of poverty. Fourth,

Table 2 Access to electricity for lighting in each administrative division (percent of population)

Division	Rural	Urban	Total
Barisal	28.4	80.7	35.2
Chittagong	58	90.4	64.9
Dhaka	56.3	94.7	66.4
Khulna	51.2	85.1	56.6
Rajshahi	49.3	79.9	54.2
Rangpur	29.4	69.8	34.1
Sylhet	43.2	88.8	48.9
Total	48.8	88.7	56.6

Source BBS [2]

although affordability improves with higher income levels and electricity consumption improves with income, the level of electricity use for residential purposes appears to be limited. This has implications for business models for electricity supply.

2.3 Electricity Access and Energy Preference for Lighting

Table 2 provides the level of electricity access for lighting purposes in each administrative division of Bangladesh. The urban areas of all divisions have reported relatively high level of electricity access but the rural areas still confront with the challenge. Dhaka and Chittagong divisions have reported better access to electricity comparatively, while Barisal and Rangpur divisions have relatively low levels of electricity access.

At the district level, Dhaka clearly outperforms the rest with an overall electrification rate of 97 %. Almost 99 % of the urban households of the district have been electrified, while 92 % of the rural households are electrified. On the other hand, Lalmonirhat district with an overall electrification of 20 % has the lowest level of electrification in the country. In general, the urban areas are well electrified in most districts but because the urban population accounts for about 20 % of the country's population, the overall electrification rate remains highly influenced by the rural population that remains largely non-electrified in a large number of districts. It appears that proximity to centres of economic activities has positively influenced the electrification rate. For example, the capital region of Dhaka division or the commercial hub in the Chittagong division has benefited rural electrification efforts. However, there still remain pockets of relatively low electrified areas in these divisions. For example, Netrokona district in Dhaka division or Cox's Bazar in Chittagong division has an overall electrification of 30 %.

The frequency distribution of districts in terms of level of electrification is presented in Table 3 (details of district level electrification rate are presented in

Table 3 Frequency distribution of districts in terms of electrification level

Criteria	Rural	Urban	Total
<=20 % electrified	3	0	0
Between 20 and 30 %	10	0	7
Between 30 and 40 %	14	0	11
Between 40 and 50 %	16	1	19
Between 50 and 60 %	11	0	10
Between 60 and 70 %	5	2	8
Between 70 and 80 %	2	22	5
Between 80 and 90 %	1	27	1
Above 90 %	2	12	3
	64	64	64

Source Based on BBS [2]

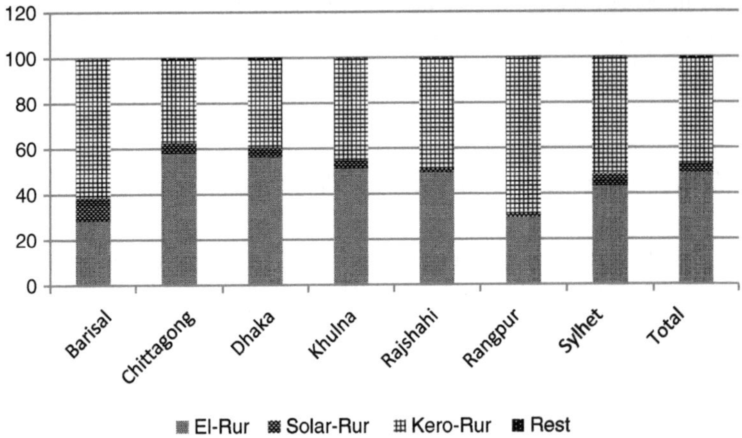

Fig. 7 Distribution of rural households in terms of lighting energy use. *Data source* [2]

Annex A.1). There are 41 districts with rural electrification between 30 and 60 % and 13 others with less than 30 % village electrification. Greater attention is required to develop a strategy for improving rural electrification of these areas to achieve universal electrification target by 2021.

As indicated earlier, limited access to electricity in rural areas has required continued reliance on kerosene for lighting and the lighting fuel mix differs significantly between urban and rural areas (see Figs. 7 and 8) of all divisions. Kerosene fills the gap in urban areas as well, where electricity is not available, but in some rural households solar energy has also emerged as a minor solution in most divisions (see Fig. 7). Rural areas of Barisal division has seen a higher level penetration of solar systems (about 10 % of household using them), which clearly stands out. However, the other division with a comparatively low electrification rate (namely Rangpur) has not seen such a growth in solar energy systems.

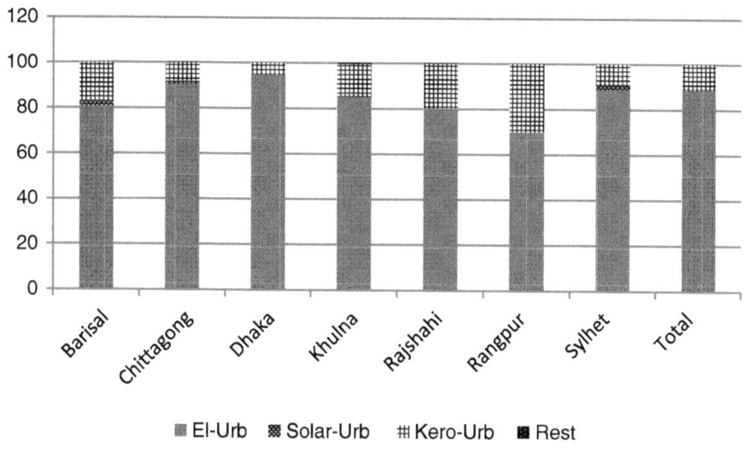

Fig. 8 Distribution of urban households in terms of lighting energy use. *Data source* BBS [2]

Any analysis of off-grid opportunities in the country needs to consider the above background. It is likely that no one solution will fit all conditions given the resource potential, size of the villages, income distribution and opportunities for productive applications. In the subsequent sections a systematic approach is used to identify the appropriate solutions under different conditions.

3 Methodology

Unlike most studies that focus on techno-economic feasibility of a given solution or alternative solutions, this study presents a multidimensional analysis covering the techno-economic, business and governance dimensions. Although techno-economic analysis still remains relevant, the work does not stop there. The outcome is further processed to consider the appropriate business delivery option, the conditions required to achieve such a delivery model and the regulatory governance system required to ensure business development and consumer protection. Moreover, given the diversity of local conditions that exist in reality, instead of using a stereotypical representative village or locality with fixed characteristics, we rely on scenarios of cases that capture different socio-economic conditions, stakeholder preferences, potential opportunities and alternative options. Thus our analysis aims to add value by expanding the knowledge frontier through a holistic analysis of off-grid systems.

The flowchart of the framework is presented in Fig. 9. The methodology is described in "Analytical Frameworks and an Integrated Approach for Mini-Grid-Based Electrification".

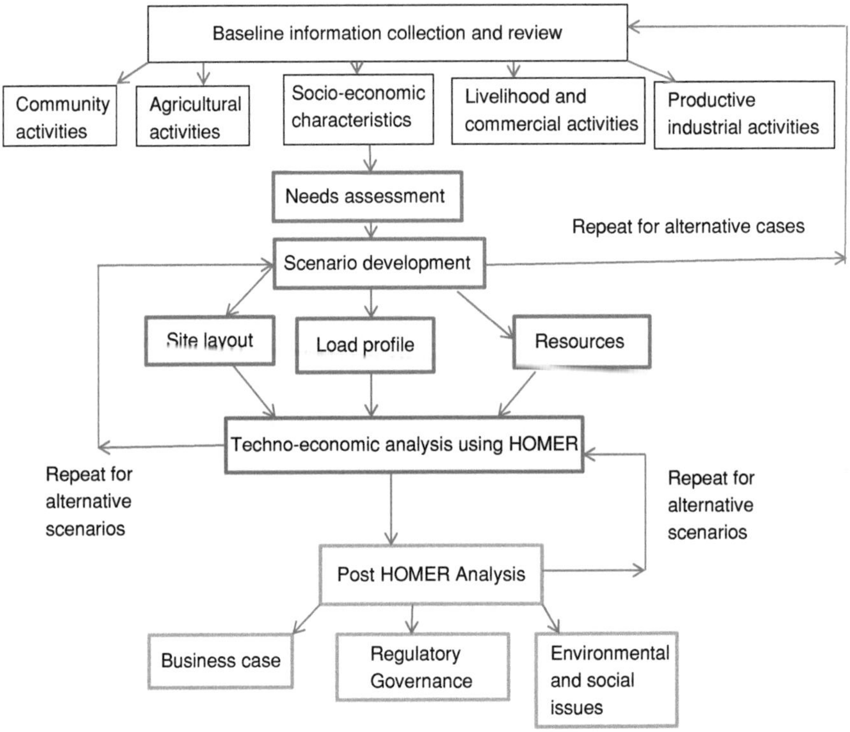

Fig. 9 Flowchart of the framework. *Source* author

The analysis starts with a detailed needs assessment which involves local information gathering to understand the socio-economic characteristics of the local population, their existing and potential livelihood, commercial and productive activities (agriculture and smart-scale industries) as well as community-related needs. Instead of developing a single point energy demand estimate, we build alternative scenarios considering different levels of energy service development (e.g. basic lighting needs, lighting and some livelihood/productive needs, service for a limited period of time, and reliable round-the-clock service, among others). It is also possible to consider multi-village systems for economies of scale.

The techno-economic analysis of appropriate electricity supply system for each scenario is then carried out using the HOMER software package developed by NREL. Each case study considers alternative resource options taking local resource availability into consideration as well as alternative scenarios for electricity needs developed in the previous step. This also leads to a further level of iteration that provides a rich set of system configurations and their life-cycle costs corresponding to alternative development paths. Information has also been used to reflect the local cost of energy system components wherever possible.

Whereas other studies end here, we take a step further to analyse the results obtained from the techno-economic analysis to consider the practical electricity supply business issues such as viability, funding, tariff and cost recovery, as well as issues related to business environment such as regulatory governance.

In the following section, the above framework is implemented using a case study.

4 Case study of a Village Electricity System in Bangladesh

4.1 Village Background

For this chapter, we have considered a non-electrified village in Netrokona district of Dhaka division. As indicated before, Netrokona has the lowest level of electrification in Dhaka division and is comparable to other poorly electrified districts of the country. Although Netrolona Palli Bidyut Samiti (PBS or village cooperative) exists and has electrified the urban areas, the villages remain non-electrified. The district is in the north of the country and its remoteness has resulted in poor level of electrification in many semi-urban and rural areas.

The chosen village, Mahishpur, comes under Atpara sub-district and is situated at 90°50′E and 24°48′N. Atpara is a remote sub-district, many parts of which are not well connected by road. The village under consideration holds 108 households with a total population of 546 people as per 2011 Census (of which 295 are male and 251 female). The average household size is 5.1 persons but the household size follows a bell-shaped curve with a minimum of two persons and a maximum of 8+ persons. This village is not electrified and does not have piped water supply. All households live in houses owned by them but more than 97 % of the houses are 'kutcha' (i.e. a house with a thatched roof). Forty seven per cent of the population is less than 14 years-old while about 5 % is above 60 years of age. Of the working-age population, women largely take care of household activities and men work in agriculture for a living. The village is connected through rural roads from Atpara and Baniajan, which are bigger villages nearby but part of it gets disconnected during the rainy season.

Being non-electrified, the local population relies on kerosene and candles for lighting purposes and fuel-wood, agricultural residues (e.g. jute sticks) and cow-dung cakes for cooking energy. The energy resources for cooking are collected or procured locally.

4.2 Needs Assessment and Scenarios

As an agricultural village, the local population is highly dependent on agricultural activities for living. The soil is fertile and generally multiple crops are produced.

The main crops are paddy, wheat, jute, mustard seed and potato. The village also supplies various fruits, namely mango, jackfruit, banana and papaya. The area receives more than 2,400 ml of rainfall during the year but the monsoon brings most of the rain, thereby causing floods in the area on a regular basis. The area, being part of the freshwater wetland ecosystem, boasts of a number of large water bodies (ponds, lakes, etc.) and fishing is also an important activity. However, due to lack of electricity no processing of food or fish takes place locally and most of the produce is sold in raw form in the nearby markets. However, natural drying of food crops, fruit and cash crops like jute takes place in the village.

To analyse the possibility of electrification through off-grid systems, we use the scenario approach to develop alternative electrification options and pathways. Given the non-electrified nature of the village, the demand is unknown but through alternative scenarios, we capture a range of demand possibilities as follows:

(a) S1: Basic residential demand—In this scenario, it is assumed that the poor households only use electricity for lighting purposes, while the middle-income and rich households use it for fans, TV and battery charging. There is no demand for productive use and the service is available for a limited period of time in the evening hours.
(b) S2: Basic residential and commercial demand during evening hours—In this scenario, some lighting load for local shops is added alongside the basic residential demand considered in S1. The supply remains limited during evening hours only.
(c) S3: Basic evening load along with daytime productive demand—This scenario extends S2 by adding demand for productive uses of electricity during any off-peak time. Such activities can include local artisanal activities and agro-based activities such as grinding, food-drying, rice milling and similar small-scale activities or even agricultural water pumping at night.
(d) S4: This scenario relaxes the limited supply constraint by providing reliable supply to all consumers. In this scenario, demand from households at any time of the day and commercial/productive demand as they arise are considered.
(e) S5: This removes the basic demand condition and replaces this with likely demand from each consumer groups but retains the restricted supply condition.
(f) S6: This is similar to S5 but removes the supply restriction to ensure reliable supply.

Although more scenarios can be created, the above cases will provide a good understanding of local demand–supply conditions and the effect on them on the techno-economic performance of the system.

In the absence of income distribution information for the specific village, we have relied on the national income distribution profile for rural areas to capture the distribution of households by income categories. We regrouped the income categories into three groups as follows: households with less than 6,000 taka/month income are classed as poor, households with income above 6,000 taka/ month but less than 15,000 taka/ month are considered as medium income and any household with income above 15,000 taka/month are considered as rich. According to the

Table 4 Electricity demand constituents by households and scenarios

Items	Poor HH	Middle income HH	Rich HH	Commercial load	Productive load
S1	2 × 10 W lighting load for 5 h in the evening (5–10 pm)	3 × 10 W lighting, 2 × 40 W fans (summer time), 1 TV 80 W for 5 h in the evening	4 × 10 W lighting, 3 × 40 W fans (summer time) and 1 TV 80 W for 5 h in the evening	Nil	Nil
S2	-do-	-do-	-do-	500 W load for 5 h in the evening	None
S3	-do-	-do-	-do-	-do-	Up to 10 kW off-peak load during day hours
S4	2 × 10 W lighting load for 5 h in the evening (5 to 10 pm) and 2 h in the morning	3 × 10 W lighting for 2 h in the morning and 6 h in the evening, 2 × 40 W fans (summer time) operating for 18 h, 1 TV 80 W for 10 h	4 × 10 W lighting for 8 h per day, 3 × 40 W fans (summer time) for 18 h and 1 TV 80 W for 10 h in the day	500 W load for 14 h a day	Up to 10 kW load at any hour
S5	Same as S1	3 × 10 W lighting, 2 × 40 W fans (summer time), 1 TV 80 W, plus additional small appliance of 80 W for 5 h in the evening	Same as S1 but an additional load of 200 W operating for 5 h	500 W load for 5 h in the evening	Up to 10 kW load during day time
S6	Same as S4	Same as S4 with an additional load of 80 W operating for 10 h	Same as S4 but an additional load of 500 W operating for 24 h	2 kW operating for 24 h	Up to 10 kW load at any time

above regrouping, 46 % of the households are poor, 15 % are rich while 39 % of the households come under the middle-income population. This results in 50 households in the poor category, 42 in the middle income category and 16 in the rich category. Based on the Household Income and Expenditure Survey 2010, it is estimated that the poor are likely to spend 468 taka/month on lighting and fuel,

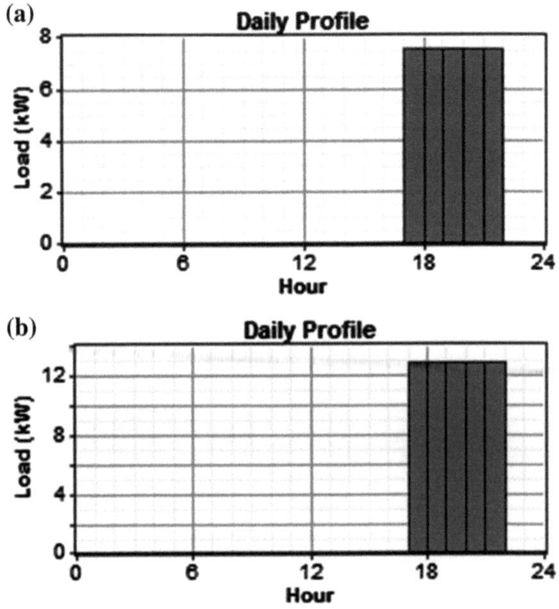

Fig. 10 **a** Daily load profile in winter for Scenario S1, **b** daily load profile in summer for Scenario S1

while the middle-income and rich households are likely to spend 572 taka and 768 taka per month respectively.

As the village is non-electrified, the consumption pattern is not available. However, past studies on Bangladesh provide a standard pattern of consumption in off-grid areas: for example Roy [6] considers that rural households use three efficient lamps of 15 W each, two or three ceiling fans of 80 W each and a television of 80–120 W. Nandi and Ghosh [5] also use the same load assumptions. Lights are operated for 6–7 h a day, fans are operated 8–10 h per day during summer and a television is operated for 5–6 h a day. Mondal and Denich [4] on the other hand consider three lights of 15 W each, two fans of 40 W each and a TV of 40 W. As our households are grouped by income class and we have considered different scenarios, the assumptions behind the needs assessment are indicated in Table 4. We have considered a number of alternative possibilities. For example, initially the load may be limited to domestic use and the supply may be limited during evening hours only. This is captured in S1. The possibility of developing limited commercial load in the evening is considered in S2, whereas some productive load during the off-peak day hours is considered in S3.

As the table suggests, the demand pattern varies by the economic condition of the households and by season (summer and winter). The overall distribution of demand is obtained by summing the demand in each category of consumer in a given period. It is possible to consider demand management options as well and more efficient technologies could reduce the demand in each case. However, for simplicity and to retain a manageable number of scenarios, we have kept the above six options.

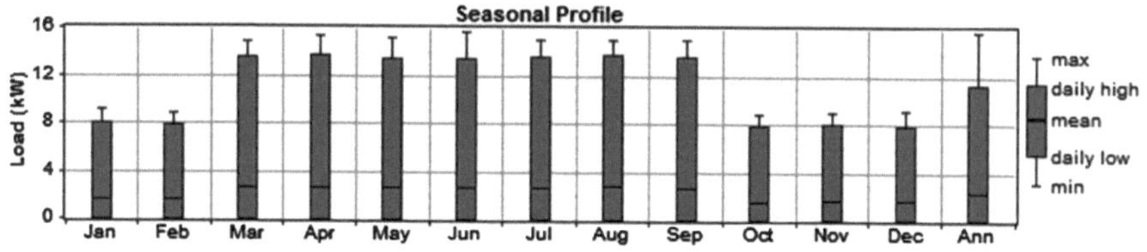

Fig. 11 Seasonal load profile for S1

Fig. 12 **a** Winter daily load profile for S2, **b** summer daily load profile for S2

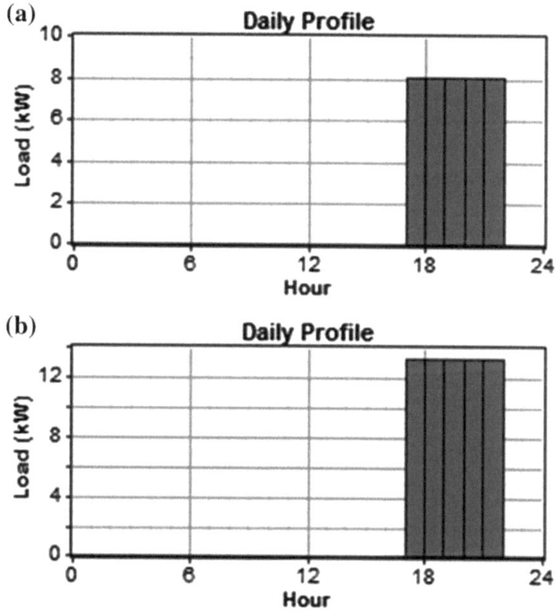

4.3 Load Profile for Different Scenarios

Given our scenarios discussed above, the load profiles are different in each case thereby allowing us to analyse a range of load situations.

(a) S1—Basic load profile
As the demand is restricted only during evening hours, following the demand logic indicated in Table 4, the daily demand profile for the basic load is shown in Fig. 10, while the seasonal profile is shown in Fig. 11. The winter load is almost half of that in summer due to absence of fan load of middle- and high-income households. A 5 % day-to-day random variation in load is assumed. The peak load of the system is 15.7 kW for 108 households and the average energy need is 53 kWh/day.

Clearly, the average to peak load is low in this case, thereby resulting in a low system load factor of 0.14.

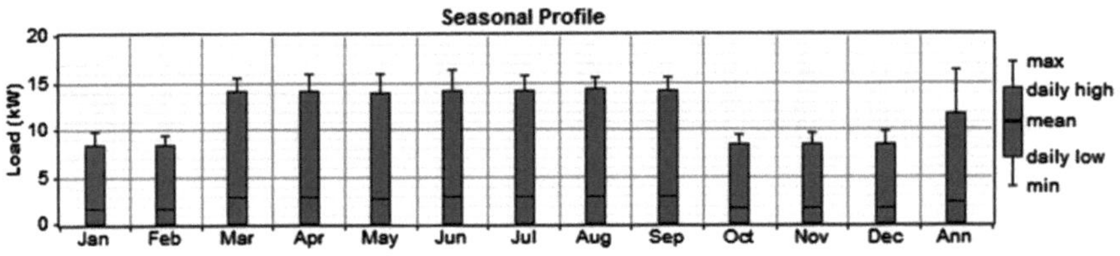

Fig. 13 Seasonal load profile of S2

Fig. 14 **a** Winter daily load profile corresponding to S3, **b** summer daily load profile corresponding to S3

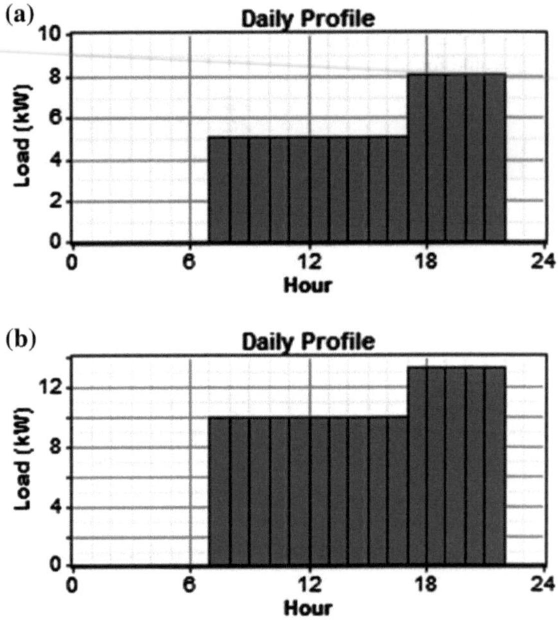

(b) Scenario S2—Basic residential and commercial load
This scenario modifies S1 by adding limited commercial demand but the restrictions on supply are maintained. As a consequence, the peak load increases slightly due to limited commercial demand considered in this case (16.3 kW peak) but the overall load factor does not change compared to S1. The daily load profiles (for summer and winter) and the seasonal distribution of load are shown in Figs. 12 and 13 respectively.
(c) S3 scenario (off-peak productive load added to S2)
In this case, it is considered that 10 kW of productive load is serviced during the daytime (between 7 am and 5 pm) during summer months while the load reduces to 5 kW during the winter months. The average load improves, resulting in a better system load factor of 0.34. The average daily energy need comes to 135 kWh. The peak load is 16.3 kW. The summer and winter daily load profiles are presented in Fig. 14, while the seasonal profile is indicated in Fig. 15.

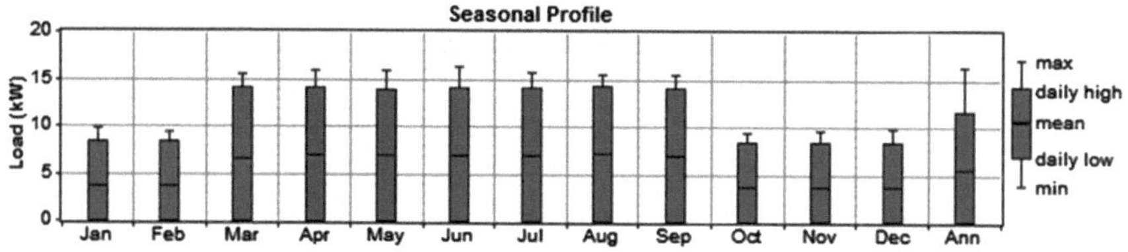

Fig. 15 Seasonal load profile corresponding to S3

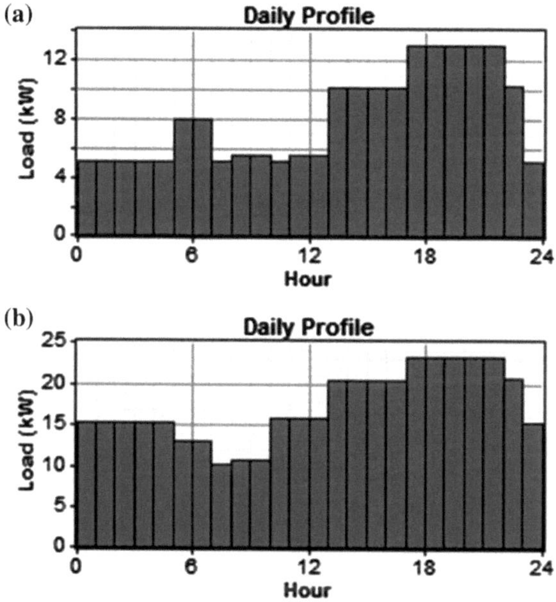

Fig. 16 **a** Winter daily load profile corresponding to S4, **b** summer daily load profile corresponding to S4

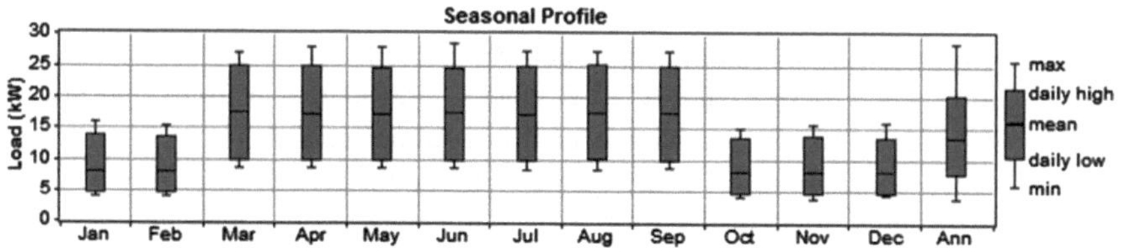

Fig. 17 Seasonal load profile

(d) Scenario S4 focuses on reliable supply in rural areas both for residential needs as well as productive needs. This allows any load to operate at any time of the day. The load profile changes considerably here as some residential load at night (mainly for fans in summer) and up to 10 kW water pumping

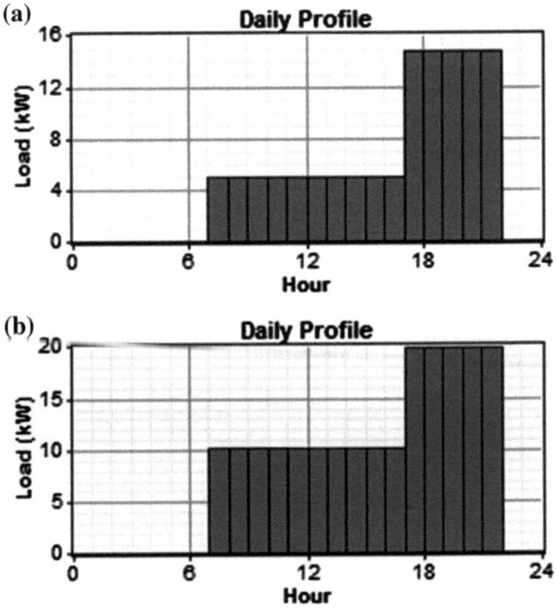

Fig. 18 **a** Daily load profile (*winter*) for S5 scenario, **b** daily load profile (*summer*) for S5 scenario

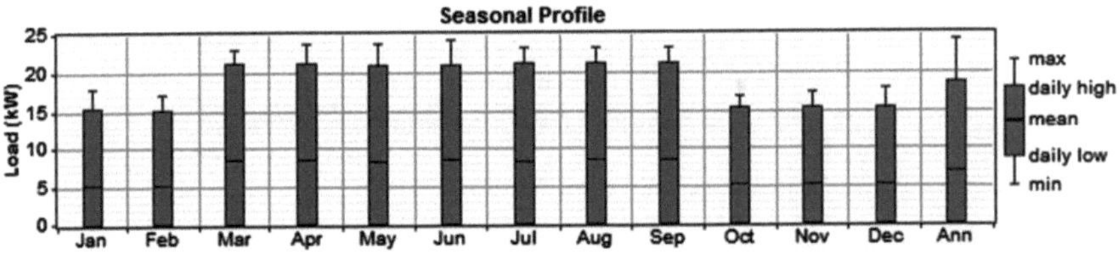

Fig. 19 Seasonal load profile for S5

load for irrigation in summer has been considered. In winter, the demand for irrigation water reduces but a 5 kW load is considered at night. During the daytime, up to 10 kW productive load in summer and up to 5 kW in winter have been considered. Here the system load factor further improves to 0.472 due to a better load distribution. The daily load profile and seasonal load profile are presented in Figs. 16 and 17 respectively. The peak load here is 28.5 kW and the daily energy demand is 323 kWh. Note that the winter base load is 5 kW whereas the summer base load is 10 kW in this case, which makes the system requirement different from the previous cases.

(e) Scenario S5 considers likely demand by removing the basic load assumptions for the domestic consumers but keeping the supply constraint in place. This provides a different load profile as some consumers are able to use more during

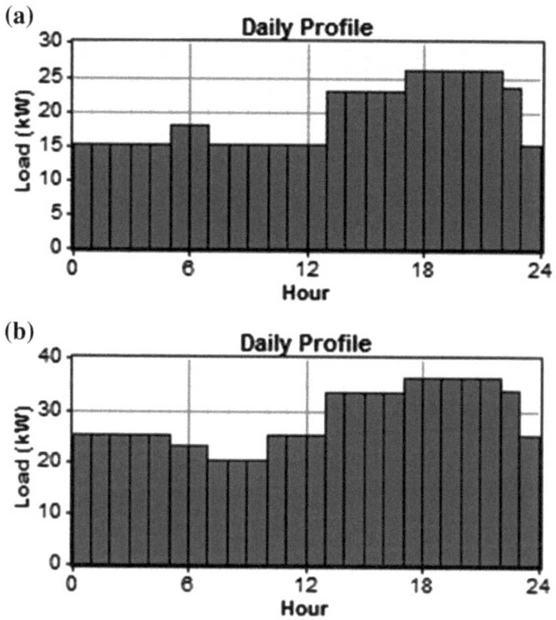

Fig. 20 **a** Daily load profile corresponding to S6 (*winter*) **b** daily load profile corresponding to S6 (*summer*)

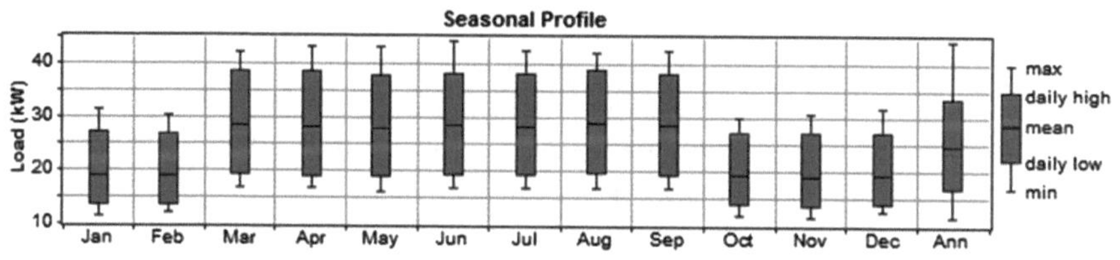

Fig. 21 Seasonal load profile

the supply period but may not reach the full potential of their demand due to limited supply duration. The supply is restricted between 7 am and 10 pm and accordingly, there is no night-time load in this case. The average energy demand comes to 167 kWh while the peak load reaches 24.3 kW (see Figs. 18 and 19). The overall load factor falls to 0.287 due to partial supply and higher peak compared to the average load.

(f) Scenario S6

The supply restriction of S5 is removed here for residential consumers and this allows full demand potential development for the rich consumers. Consequently, the peak demand increases to 44 kW here and the overall load factor improves to

Table 5 Comparison of load profiles of different scenarios

Scenarios	Peak load (kW)	Daily average energy (kWh/day)	Load factor
S1	15.7	53	0.14
S2	16.3	55.5	0.142
S3	16.3	134	0.344
S4	28.5	323	0.472
S5	24.3	167	0.287
S6	44.4	589	0.553

0.55. The average daily energy requirement increases to 589 kWh. The daily load profiles and seasonal profile are presented in Figs. 20 and 21 respectively.

Table 5 compares the load profiles of different scenarios.

4.4 Energy Resources

For the case study, three energy resources, namely solar PV, wind and diesel fuel have been considered. The site does not have any micro-hydro potential but biomass is readily available and is widely used for cooking purposes. However, as there is limited use of biomass for power generation, this option has not been considered in this study, although it may be considered in a future study.

The solar energy availability in the case study village is obtained from HOMER. Based on the latitude–longitude information for the village location, HOMER estimated the annual average solar insolation of 4.58 kWh/m^2/d. However, the insolation level increases between March and May and reduces during the monsoon season (July to September). The monthly pattern of radiation and the trend of cleanliness index are shown in Fig. 22.

The data for wind resources are not readily available for Bangladeshi villages. However, Roy [6] provides monthly average wind speed for a nearby location in the Dhaka Division. In the absence of any specific data for the village, this information has been used (see Fig. 23). It can be seen that wind blows all year round but the speed tends to be higher during the summer–monsoon months. The diurnal pattern strength of 0.0323 is used for the location and the wind speed peaks at 15 h.

For diesel generators, it has been assumed that the fuel is available from the national supply system and being incremental in nature, the village level demand will not affect the market conditions adversely. The prevailing local rates for diesel have been used, which may not reflect the true economic cost of the fuel.

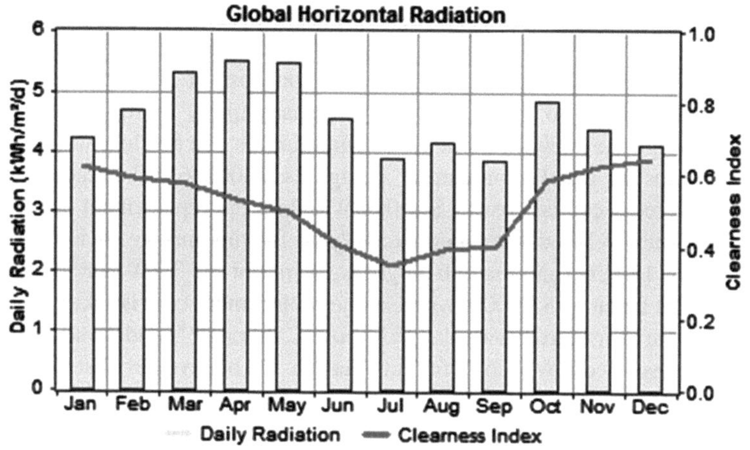

Fig. 22 Solar radiation at the case study village

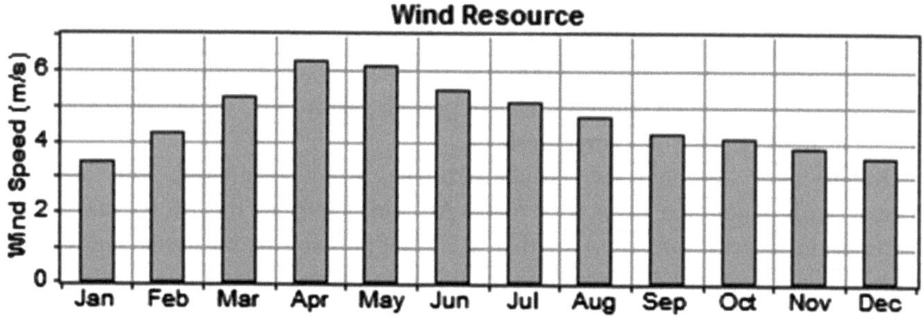

Fig. 23 Wind speed at the case study location

4.5 Component Details

Based on the above resource and demand considerations, the following components were considered: Solar PV, a generic 1 kW wind turbine, a generic 3 kW wind turbine, a diesel generator, batteries and converters. The specific details of each are provided below.

Solar PV: The unit cost of solar PV systems has declined considerably in recent time. For this study, we have considered that a 1 kW PV system requires \$2,800 and the replacement cost is \$2,000 per kW. A low operating and maintenance cost of \$10/year/kW is considered. The cost is based on ESMAP [3]. It is assumed that no tracking device is used. The life of solar panels is assumed to be 20 years. The simulation is carried out for various quantity–capacity combinations to facilitate optimal sizing of the system.

Wind turbines: We have considered two generic small wind turbines, namely of 1 kW capacity and 3 kW capacity suitable for rural application. According to ESMAP [3], the civil construction and erection cost of wind turbines can be significant compared to the equipment cost, particularly in the small capacity range. Although the capital cost of 1 kW wind turbine can be close to $2,500/kW, the overall cost of installation can be as high as $5,000–6,000. Accordingly, we have considered a capital cost of $4,000/kW, while the replacement cost is taken as $2,500. The O&M cost is taken as $50/year for this turbine. It is assumed to have a life of 15 years and the hub height is 25 m. For the 3 kW wind turbine, the capital cost is taken as $10,000, whereas the replacement cost is taken as $8,000, with $250/year considered towards O&M costs. Although Mondal and Denich [4] and others have used lower costs for Bangladesh, we believe our cost assumptions are closer to the reality.

Diesel generator: Diesel generators are widely used for electricity generation in rural areas and are widely available. There is minimal civil work involved in this case and the generator cost captures the overall investment requirement. The capital cost of 1 kW of generator is considered to be $600 and the replacement cost is considered as $500. The operational and maintenance cost of the generator is taken as $0.5/h for every 10 kW of generator size. It is assumed that the generator can be operated for 15,000 h in its lifetime and the minimum load it can take is 10 % of its rated capacity. Diesel price is taken as $0.6/l which is based on the local market price in Bangladesh.

Battery: For this analysis, Trojan L16P has been considered. This is a 6 V battery with a nominal capacity of 360 Ah and a normal life of 10 years. Four batteries in a string are used so that a 24 V bus bar can be used. The cost of batteries varies widely depending on the make and source of supply. For this study, we have taken a cost of $150 for each battery while the replacement cost is taken as $100. The O&M cost is taken as $5 per year.

Inverter: The cost of converter is taken as $200/kW and the replacement cost is taken as $150/kW. The O&M cost is taken as $25 per year. It has a normal life of 15 years and is assumed to have an efficiency of 85 %.

Other system costs: As HOMER does not include the cost of distribution network separately, we have added a capital cost of $3,000 towards the cost distribution network for 108 households and $200 per year towards fixed O&M costs.

The project life is taken as 15 years—this is done to match the project life with the debt repayment period. However, HOMER calculates any salvage value of the assets based on its remaining life and the replacement cost of the asset. A real discount rate of 5.3 % has been used in the analysis, based on the cost of capital in dollar terms.

Clearly, the economic parameters affect the overall results significantly. As mentioned earlier, some recent studies on Bangladesh have used quite different economic parameters (see Table 6 for some examples). Clearly, a lower capital and operating cost of any equipment makes it more desirable for the optimal solution and the cost of generation reduces. However, unrealistic costs reduce the relevance of the analysis and distort the optimal solution.

Table 6 Examples of cost assumption from the literature

Cost parameter	Mondal and Denich [4]	Roy [7]
Capital cost for PV	274 taka/W ($3.65/W)	$270,950 for 100 kW PV arrays
Replacement cost of PV	206 taka/W ($2.75/W)	$45,000 for 100 kW PV arrays
O&M cost	50 taka/W/year ($0.67/W)	$500/year for 500 kW
Capital cost of a 3 kW wind turbine	86,584 taka/kW ($1155/kW)	$455000 for a 300 kW turbine
Replacement cost of a 3 kW wind turbine	75,000 taka/kW ($10,00/kW)	$65,000 for a 300 kW turbine
O&M cost of a 3 kW wind turbine	1,000 taka/year/turbine (or $13)	$1,000/year for a 300 kW turbine
Capital cost of diesel generator	10,000 taka/kW (or $133/kW)	$116,883 for a 500 kW generator
Diesel price	45 taka/l (or $0.6/l)	$0.7/l
O&M cost of a diesel generator	20 taka/h for 10 kW ($0.27/h); 30 taka/h for 20 kW ($0.4/h)	$5/h

5 Results of the Techno-Economic Analysis

For the techno-economic analysis, the HOMER software package was used. The results for each scenario are presented below.

5.1 Scenario S1

Considering the demand, component cost characteristics and resource availability, a diesel generator of 20 kW emerges as the optimal architecture for this scenario. The capital cost comes to $15,000 and the annualised operating cost comes to $7,514/year (see Table 7 for cost summary). The levelised cost of electricity comes to $0.465/kWh. The diesel generator operates 1,825 h/year and consumes 8,209 l of diesel/year. As the life of the generator is limited to 15,000 h, a replacement is required after 8.2 years and an investment of $10,000 is required at this time. The generator produces 19,345 kWh in a year and there is no excess electricity production that goes waste in this case. The power generation profile is indicated in Fig. 24.

5.1.1 Scenario S2

For this scenario, the optimal solution also consists of a 20 kW diesel generator. The overall capital cost for the system comes to $15,000 and the net present value comes to $93,274. The cost of electricity is estimated at $0.453/kWh. The power generation profile is shown in Fig. 25.

Table 7 Net present cost summary of the optimal system architecture for Scenario S1

Component	Capital cost ($)	Replacement cost ($)	O&M cost ($)	Fuel cost ($)	Salvage value ($)	Net present cost ($)
Diesel	12,000	6,541	18,565	50,102	−807	86,401
Other	3,000	0	2,034	0	0	5,034
System	15,000	6,541	20,599	50,102	−807	91,436

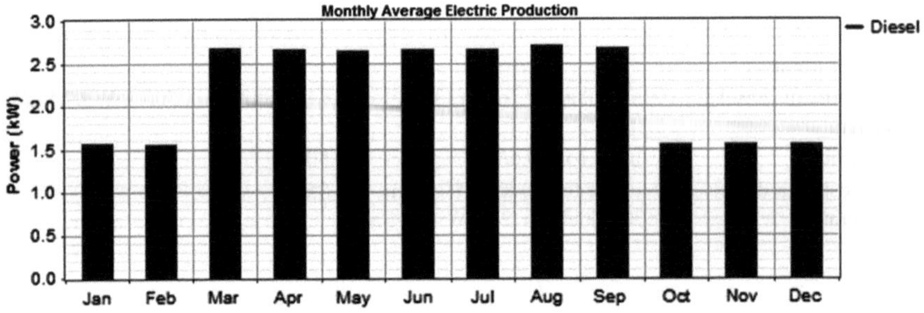

Fig. 24 Average electricity production in Scenario S1

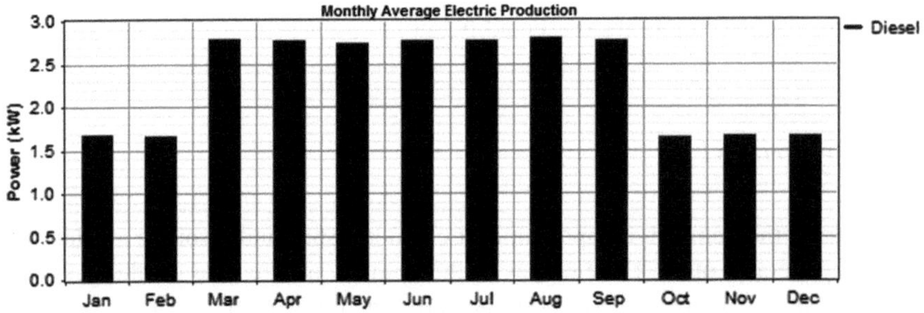

Fig. 25 Electricity output in Scenario 2

The diesel generator operates for 1,825 h and produces 20,258 kWh/year. There is no excess electricity generation in this system. It costs $15,000 in investment and consumes 8,510 l of diesel per year. The generator needs a replacement after 8.2 years at a cost of $10,000.

Given the similarities of S1 and S2, it is not surprising to see that the solutions are quite similar. In fact, the same size plant operates slightly longer to meet the demand and hence, the cost per unit reduces in the case of S2. However, the cost of electricity so produced is still much higher than the tariff charged for residential consumers. The business case of such a solution will require further investigation (as is done in Sect. 6).

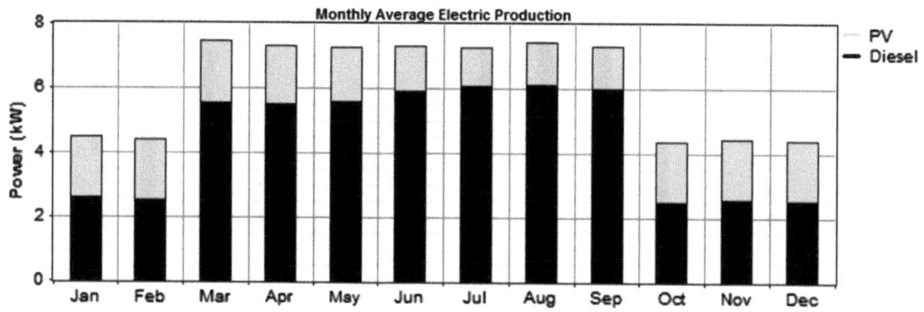

Fig. 26 Monthly average electricity generation corresponding to S3

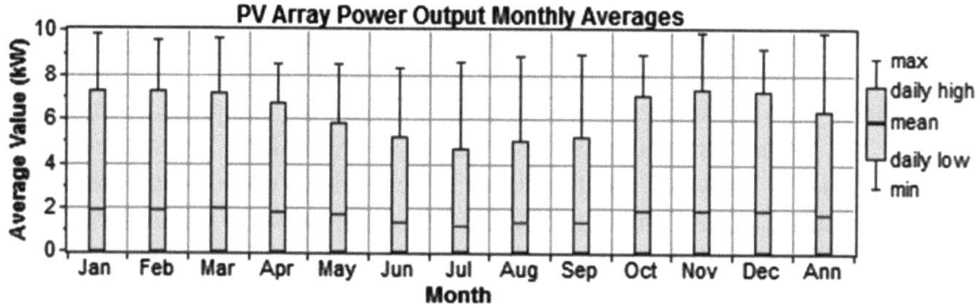

Fig. 27 PV power production in Scenario S3

5.2 Scenario S3

The optimised system architecture for this scenario consists of a 10 kW PV system and a 10 kW diesel generator alongside 72 Trojan L16P batteries, and a 10 kW inverter–rectifier. The capital cost for this system comes to $49,800 and the net present cost (NPC) comes to $184,509, with a levelised cost of $0.368/kWh of electricity generated.

Twenty seven per cent of electricity output comes from the PV system, whereas 73 % comes from the diesel generator. Practically no excess electricity is produced in the process that goes unused. The monthly generation profile is indicated in Fig. 26.

The mean output of PV system is 40.3 kWh/day and the annual electricity generation amounts to 14,719 kWh. The PV system operates for 4,378 h in a year and the levelised electricity cost from PV is $0.178/kWh. The monsoon months generally record reduced solar output (see Fig. 27). The capital cost of PV arrays comes to $28,000 and the PV system achieves a capacity utilisation of 16.8 %.

The diesel generator runs for 4,133 h per year, leading to a capacity utilisation of 44.5 %. It produces 38,970 kWh/year and consumes 14,926 l of diesel. The diesel generator costs $6,000 but adds another $16,253 for fuel costs over the project lifetime. Due to higher utilisation of the generator, its overall life reduces

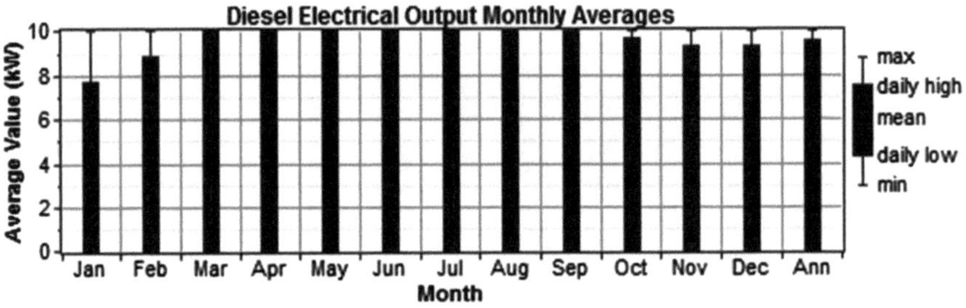

Fig. 28 Power generation profile of the diesel generator in S3

to 3.6 years and accordingly, four replacements are required during the project life. Each time, $5,000 will be required to replace the generator. The power generation profile of the diesel generator is shown in Fig. 28.

The battery system has a nominal capacity of 156 kWh and provides autonomy of 19 h. The annual throughput of the batteries is 10,261 kWh. The expected life of the batteries is 7.5 years and they have to be replaced once during the project lifetime requiring an investment of $7,200. The capital cost for the battery system comes to $10,800 and the net present value of battery-related costs comes to $19,301.

The second-best configuration consists of a 10 kW PV, 1 kW Generic wind turbine and a 10 kW diesel generator supported by a battery bank of 64 batteries and a 10 kW converter. The initial capital cost comes to $52,600, while the NPC comes to $188,049 and the levelised cost of electricity comes to $0.375/kWh. The renewable energy share improves to 30 % in this case but the reduction in diesel use is more than offset by the increased investment required for the wind turbine, thereby increasing the capital requirement and the levelised cost of supply.

A diesel-only system comes as the least capital intensive option but in terms of cost of supply, it ranks towards the bottom of the range. A 20 kW diesel generator that runs for 5,475 h and consumes 21,736 l of diesel can meet the demand effectively. But the cost of electricity increases to $0.471/kWh, making this one of the least preferred options in terms of NPC. However, the system requirement simplifies here as the generator can be operated as required. The excess electricity generation is practically non-existent in this case and the capacity utilisation improves to 28 %.

Although this scenario requires a bigger system compared to S1 and S2, a better load distribution improves the capacity utilisation of the system and hence reduces the cost of electricity generation. However, the capital cost is about 3.3 times higher than the basic system suggested in S1/S2. This requires further attention, which we consider in the business case analysis.

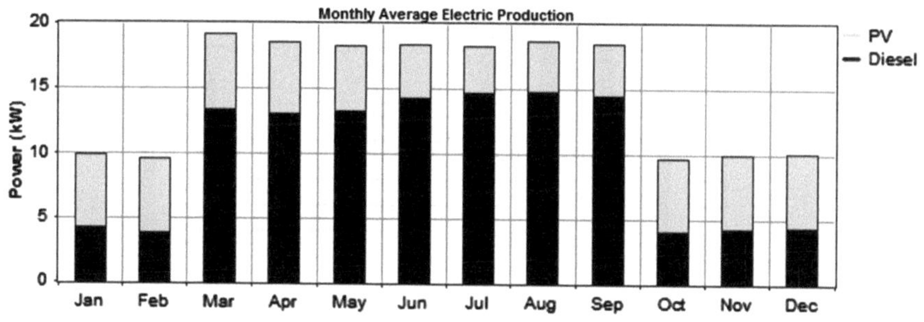

Fig. 29 Monthly electricity production profile corresponding to S4

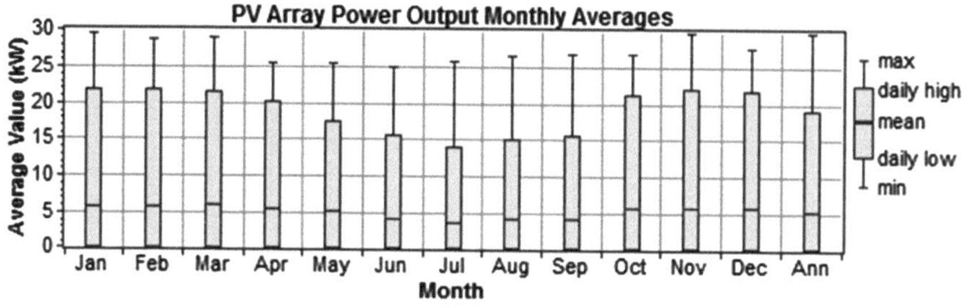

Fig. 30 PV power in Scenario S4

5.3 Scenario S4

The optimal system corresponding to this scenario requires 30 kW PV and a 20 kW diesel generator alongside 160 batteries, and a 15 kW inverter-rectifier. The capital cost comes to $126,000 while the NPC comes to $435,552. The levelised electricity generation cost comes to $0.363/kWh.

As shown in Fig. 29, PV arrays provide 34 % of the electricity output while the remaining 66 % comes from the diesel generator. The system also produces about 1 % excess electricity that remains unused. The solar PV produces 121 kWh/day and operates for 4,378 h per year producing 44,157 kWh of electricity per year (see Fig. 30). The levelised cost of solar electricity comes to $0.178/kWh and achieves a capacity factor of 16.8 %. The capital cost of PV system comes to $84,000.

The diesel generator operates for 4,939 h and produces 86,701 kWh/year. It consumes 33,550 l of diesel and achieves a capacity utilisation rate of 49.5 %. The capital cost required for the diesel generator is $12,000 but the fuel cost comes to $204,767 over the life of the project. Accordingly, the diesel system accounts for the highest share of the NPC in this scenario. The power output is shown in Fig. 31. The expected life of the generator is about 3 years and consequently, four replacements are required during the project life, requiring $10,000 each time in investment.

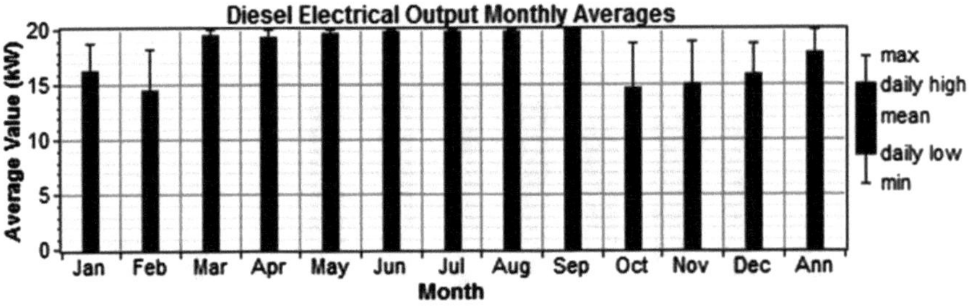

Fig. 31 Diesel power output in S4

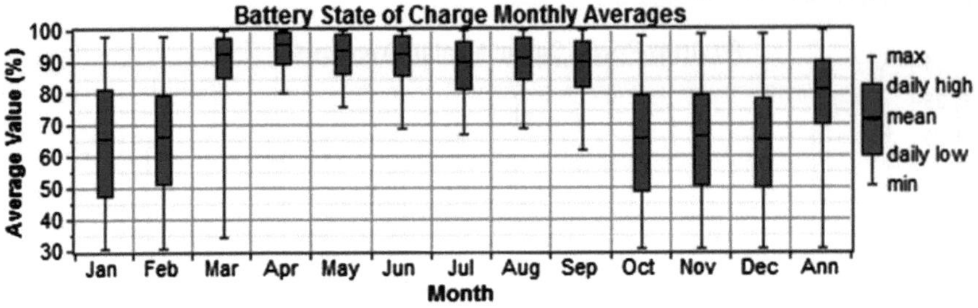

Fig. 32 Battery charging status in S4

The battery system has a nominal capacity of 346 kWh in this case and provides autonomy of 18 h. The charging status of the batteries is shown in Fig. 32. The capital cost of batteries comes to $24,000 but the expected life of batteries is 5.9 years, thus requiring two replacements during the project life.

The second-best solution comes with a 30 kW PV system alongside a 1 kW wind turbine and a 20 kW diesel generator supported by a set of 160 batteries and a 15 kW converter. The capital cost comes to $130,000 but the NPC comes to $436,792. The diesel generator requires 441 l of diesel less than the optimal case but this does not offset the capital cost of a wind turbine, making the option less attractive in terms of cost of electricity supply. However, it achieves 35 % renewable energy share compared to 34 % in the optimal case.

A 30 kW diesel generator could also meet the needs effectively and would require about $21,000 in capital investment but the operating cost makes this the least preferred solution in terms of cost of supply. The levelised cost of electricity comes to $0.463/kWh. The diesel requirement also increases to 52,045 l in this case. In terms of levelised cost, this becomes the least preferred option, despite being the least capital intensive option. However, a 20 kW diesel generator along with 56 batteries and 1 10 kW converter turns out to be a better option than a diesel generator alone, as it can serve the load at a cost of $0.379/kWh. The capital cost increases to $25,400 but the fuel requirement reduces by more than 3,600 l, thereby reducing the cost of supply substantially.

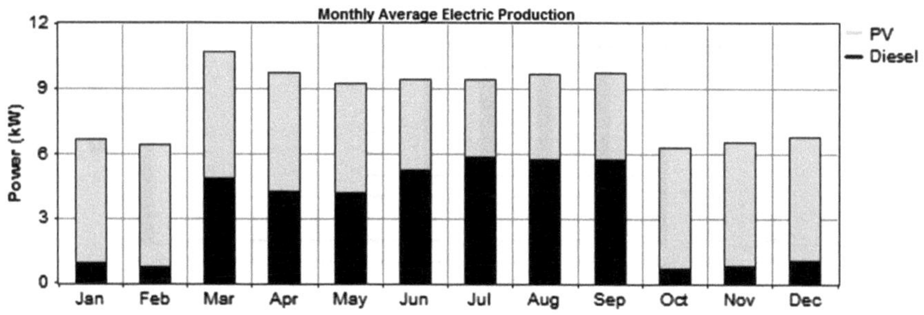

Fig. 33 Monthly electricity generation mix corresponding to S5 scenario

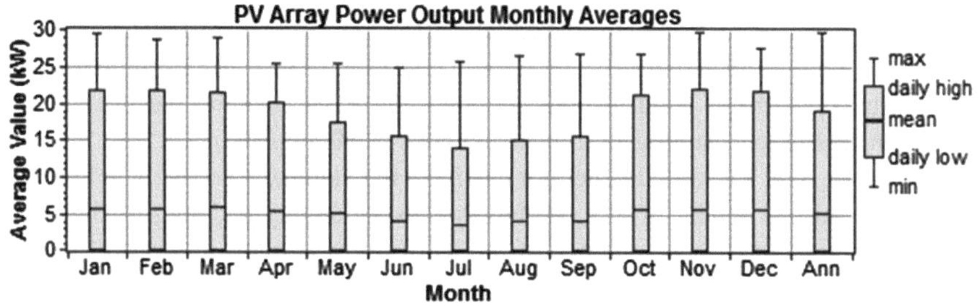

Fig. 34 PV power in Scenario S5

5.4 Scenario S5

The optimal system architecture corresponding to this scenario looks like this: 30 kW PV, a 20 kW diesel and 160 Trojan L16P batteries with a 20 kW inverter-rectifier. The capital cost of the system comes to \$127,000 while the NPC comes to \$239,709 and the levelised cost of electricity comes to \$0.387/kWh.

Sixty per cent of the generation comes from the PV system, while the remaining 40 % comes from the diesel engine. Figure 33 provides the average electricity generation mix. This system also produces about 3 % excess electricity that goes unused. PV produces 44,157 kWh of electricity per year, with an average output of 121 kWh/day. The capacity factor comes to 16.8 %. The PV power output profile is presented in Fig. 34. The capital cost of the PV system comes to \$84,000.

The diesel generator runs 1,561 h per year and consumes 11,238 l of diesel. Due to reduced use of the generator, its life increases to 9.6 years, thereby requiring one replacement during the lifetime of the project. Figure 35 provides the average monthly power output from the diesel generator.

The battery system will have a nominal capacity of 346 kWh, with battery autonomy of about 35 h. The capital cost comes to \$24,000 while the present worth of the replacement cost is \$18,980. The net present value of the battery system is the highest cost element for this configuration.

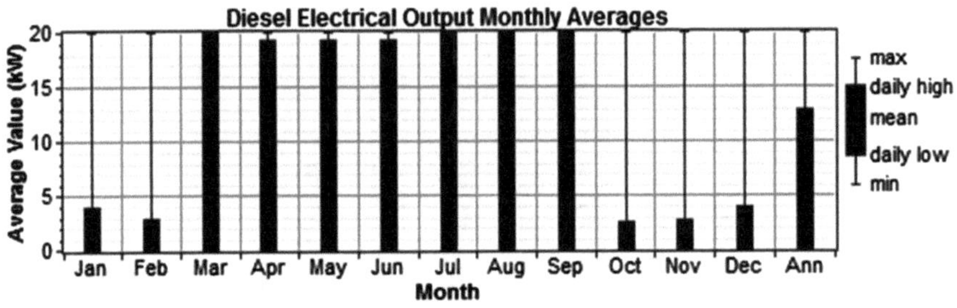

Fig. 35 Diesel generator output profile

The second-best system consists of 30 kW PV arrays, 1 kW of wind turbine and a 20 kW diesel generator, with a capital cost of $131,000. Although the diesel generator requires less fuel (10,879 l compared to 11,238 l in the optimal case), the capital cost of the wind turbine does not get offset, making the overall cost higher in this case, resulting in a levelised cost of $0.389/kWh.

As before, a 30 kW diesel generator can meet the demand effectively in this case with a capital investment of $21,100 but the cost of supply becomes $0.526/kWh and this options turns out to be one of the least preferred ones in terms of overall cost. The diesel requirement increases to 28,328 litres in this case but there is no excess electricity generation.

5.5 Scenario S6

The optimal system architecture for this scenario requires 50 kW PV, and a 30 kW diesel plant supported by 200 Trojan L16P batteries and a 25 kW inverter-rectifier. The capital cost for this system is $196,000 while the NPC comes to $752,290. The levelised cost of electricity for the system is $0.344/kWh.

The electricity generation mix for this scenario is as follows: 31 % of output comes from PV, and 69 % from the diesel plant. Thus, renewable energy penetration in the optimal system is 31 %. Like other scenarios, excess electricity amounting to about 1 % of the demand is produced which is not used. The electricity production mix is shown in Fig. 36.

The PV arrays produce 202 kWh/day and over the year produce 73,595 kWh. The monthly distribution of solar output is shown in Fig. 37. The capital cost for the PV system comes to $140,000.

The diesel generator operates 5,889 h per year and produces 160,621 kWh (see Fig. 38). It consumes 61,835 l of diesel and has an expected life of 2.55 years. Thus, although the initial capital required for the diesel generator is $18,000, the present worth of the replacement cost comes to $51,424. The present value of the fuel-related cost, $377,407, is however the most important cost element for this scenario.

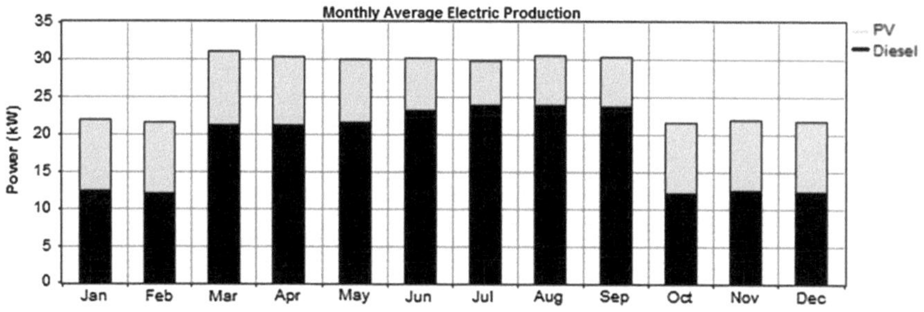

Fig. 36 Electricity mix in scenario S6

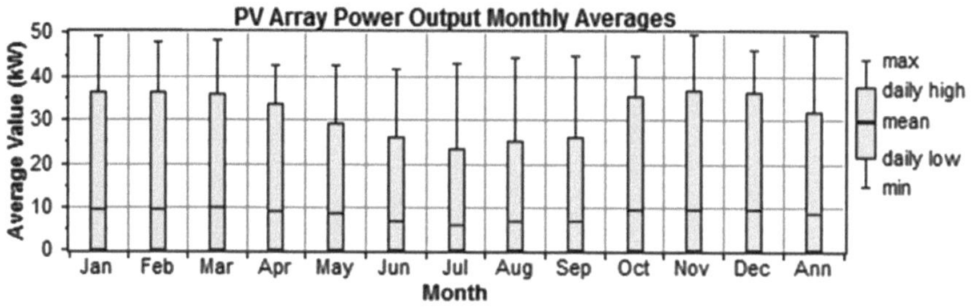

Fig. 37 PV power in Scenario S6

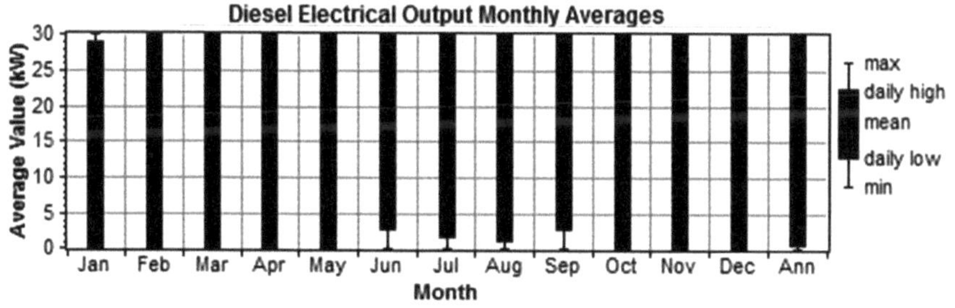

Fig. 38 Diesel power output in Scenario S6

The second-best solution consists of a 50 kW PV system, 1 kW wind turbine and a 30 kW diesel generator alongside 200 Trojan L16P batteries and a 25 kW converter. The capital cost of this system comes to $200,000 but the NPC comes to $753,517, making the levelised cost higher than the optimal solution ($0.345/kWh).

A 50 kW diesel generator can meet the demand with the least capital investment (of $33,000) but as before it emerges as a less preferred solution due to high operating cost. The diesel requirement increases to 92,845 litres and the generator operates 8,760 h per year.

Table 8 Comparison of optimal solutions

Scenarios	Architecture	Peak Load (kW)	Capital cost ($)	Capital cost per kW of peak ($/kW)	Levelised cost of electricity ($/kWh)	Installed capacity to peak load ratio	Diesel use (l)	RE share
S1	20 kW diesel generator	15.7	15,000	955	0.465	1.27	8,209	0
S2	20 kW diesel generator	16.3	15,000	920	0.453	1.23	8,510	0
S3	10 kW PV, and a 10 kW diesel generator	16.3	49,800	3,055	0.368	1.23	14,926	0.27
S4	30 kW PV, and a 20 kW diesel generator	28.5	126,000	4,421	0.363	1.77	33,550	0.34
S5	30 kW PV and 20 kW diesel	24.3	127,000	5,226	0.387	2.06	11,238	0.60
S6	50 kW PV, and a 30 kW diesel plant	44.4	196,000	4,414	0.344	1.80	61,835	0.31

The above scenarios provide alternative pathways of development of the off-grid electrification system. They also can be viewed as pathways to improve the system as the benefits of electrification lead to higher demand.

A comparison of the optimal solutions for six scenarios shows (see Table 8) the following:

- It appears that a diesel-based system is a preferable solution when the demand is limited and the supply is restricted. As the demand improves and the supply is provided round the clock, hybrid systems appear to be more appropriate.
- The initial investment cost is considerable for hybrid systems. This happens due to intermittent nature of the renewable resources that require back-up capacities. Accordingly, all hybrid systems require a significant spare capacity, thereby reducing the overall system capacity factor. The reserve capacity in all these cases is high.
- Depending on the size of excess capacity maintained in each scenario, the cost per kW of peak load serviced varies. But the initial investment cost of diesel-based systems tends to be comparatively low but the capacity replacement charges can be high for both diesel-based systems and hybrid systems. This is an important consideration for business viability analysis. While initial capital grants can help develop a system, unless there is adequate revenue generation to meet future costs, the long-term sustainability of a solution cannot be guaranteed. This aspect is hardly considered in the techno-economic analyses.
- The cost of service remains quite high for all cases and considering the size of the poor population in the area, the cost can be unaffordable to many users.

The electricity tariff approved by the Electricity Regulatory Commission for residential consumers is just $0.04/kWh for consumption up to 100 kWh. It is evident that in all scenarios the levelised cost of supply from off-grid sources is much higher. Therefore, the issue of business case for the investment needs to be considered separately, on which we focus next.

6 Business and Governance Analysis of Alternative Scenarios

The techno-economic analysis considered above is useful in analysing the optimal technology combinations for a given energy demand. However, it does not perform any financial analysis of business investment. For example, the capital requirement is different for different optimal solutions and some sub-optimal solutions in a technical sense may even make more business sense, particularly when private investment is being looked into. Moreover, as the cost of supply turned out to be high, options for reducing the supply cost become important to make supply affordable to consumers. However, any such cost reduction mechanism has financial implications for the government, or the supply business or both. Therefore, a balance has to be achieved between affordable supply to consumers and business viability from the investors' perspective. In this section we consider a number of business-related questions to see how the off-grid options considered in the previous scenarios can be delivered.

6.1 Financial Cost-benefit analysis

The analysis presented here follows the principles of financial cost–benefit analysis. It is considered that a viable investment project (from the investors' perspective) must generate positive net present benefits (i.e. the net present value of costs should be less than the net present value of benefits). In the case of our off-grid electricity supply project, the costs include initial investment, fuel-related costs, operating and maintenance-related costs, and the cost of replacing assets. The benefits on the other hand come from sale of electricity and for the financial analysis, this only considers the revenue generated from sale of electricity.

For each type of stakeholder (namely investor, consumer and the government), different aspects are considered. For example, an investor while looking for adequate return on the investment has to ensure that the debt is repaid on time and the asset is replaced on schedule so that the business can be run effectively. This requires ensuring adequate funding for debt repayment and asset replacement. Similarly, consumers of different groups pay different tariffs for grid connected supply. A similar approach is used here as well. As the consumers are likely to compare the charges for off-grid service to the tariff charged for grid-based supply, we consider this in our analysis to see if grid price parity can be achieved. The effect of grid parity tariff on other stakeholders is also considered. Further, we consider the rental charges paid for SHS as an alternative and analyse the effects of such tariffs on the business. Finally, the burden on the government finances is also analysed.

6.2 Analysis of Different Scenarios

6.2.1 Scenario S1

Here, an initial investment of \$15,000 is required, followed by an investment of \$10,000 in the ninth year. In addition, \$4,925 per year is spent on fuel and \$1,825/ year is spent on operating and maintenance costs. Accordingly, these recurring costs contribute significantly to the overall cost of electricity supply. In the S1 scenario only residential demand exists and each household, whether rich or poor, consumes less than 40 kWh per month. All consumers use electricity when it is available and hence contribute to the peak load in proportion to their demand. The regulated tariff for residential consumers using up to 100 kWh per month is set at taka 3.05 ($0.04). Is it possible to achieve grid tariff parity for the off-grid supply in this scenario?

Of the two major cost components, if the capital required for the assets is supported through a grant, the consumers would need to bear the operating costs only. Assuming that 100 % of the asset replacement costs are borne through a grant, the cost of electricity comes to $0.387/kWh. This implies that even if $15,000 is provided to the project operator as a capital grant, the cost of electricity reduces slightly and the average electricity cost remains almost 10 times higher than the grid-based electricity. If the initial capital as well as the capital required for asset replacement is provided through a grant fund, thereby reducing the entire capital-related cost, the electricity charge per unit for the operating cost recovery comes to $0.359. Thus, just capital subsidy cannot ensure grid tariff-parity for this off-grid solution—some operating cost subsidy will also be required. In fact, if grid parity pricing is charged, the net present value of revenue comes to $7,871 over the project life which will not recover even the generator operator's cost and the distribution system fixed cost. Thus, it appears that aiming for a grid parity price for the off-grid system is a non-starter from any perspective. No business case can be made for such an option.

However, a more appropriate reference point could involve a comparison with the SHS. Given that SHS are popular in Bangladesh, it is legitimate to ask whether it makes economic sense to go for a diesel-based mini-grid instead of promoting SHS in such off-grid areas. Grameen Shakti, the leading SHS provider in Bangladesh, provides the equipment costs for various system capacities. A 10 W system costs $130, a 20 W system costs $170, a 50 W system costs $380, an 80 W system costs $560 and a 135 W system costs $970.[4] In our scenario, a low-income consumer is considered to use a 20 W load, while the medium and rich consumers use 190 and 240 W respectively. As the systems are not directly comparable, we assume that the low-income groups would go for a 10 W SHS, while the medium- and high-income groups would go for 50 and 80 W systems respectively. Based on the household distribution used in our analysis, there are 50 poor households, 42 medium-income households and 16 rich families. The total system cost for SHS for all these families comes to $31,420. Even considering a 4 % discount offered for 100 % down-payment, the capital requirement comes to $30,163 (i.e. 2 times the capital requirement for the mini-grid in S1). Clearly, from the capital cost perspective, the diesel mini-grid makes economic sense. As the batteries have to be replaced at least twice over the 15-year period and the electricity output will be much less than the diesel-based system, the cost of electricity delivered from the

4 Based on Grameen Shakti cost data as reported in http://www.gshakti.org/index.php?option=com_content&view=article&id=115&Itemid=124. 75 taka = 1 US dollar is used for conversion.

Table 9 Consumer spending on electricity under different recovery considerations

Item	Unit	LI	MI	HI
Consumption in summer	kWh/month/HH	3	28.5	36
Cost at full levelised cost	$	1.395	13.2525	16.74
Consumption in winter	kWh/month/HH	3	16.5	18
Cost at full levelised cost	$	1.395	7.6725	8.37
Cost at 100 % capital subsidy for summer consumption	$	1.17	11.115	14.04

SHS would come to $0.715/kWh.[5] Thus, from the life-cycle cost perspective, the SHS investment does not make economic sense compared to the diesel-based mini-grid considered in scenario S1.

If consumers are buying SHS in Bangladesh, it is likely that consumers elsewhere will be willing to pay similar charges for electricity from a mini-grid. Grameen Shakti offers a number of financing options to SHS owners. The least demanding option requires them to pay 15 % initially and the rest 85 % in 36 equal monthly installments with a flat rate service charge of 8 %. For our three chosen system sizes of 10, 50 and 80 Wp, the initial payment comes to $20, $59 and $80, respectively, while the monthly payment comes to a flat charge of $3.3, $10 and $14.3 respectively. Can these amounts be sufficient for the off-grid service suggested in S1?

In this scenario, our households consume more in summer than in winter due to fan loads for the medium- and high-income groups but for the low-income group, the consumption pattern does not vary seasonally. Accordingly, we take summer consumption to find out their monthly expenditure at full levelised cost and with capital grant support. This is presented in Table 9.

As can be seen, the poor consumer groups would be paying about 50 % of the cost they would be paying for an SHS while the middle-income and high-income groups would pay slightly more than that for an SHS in summer months. However, it needs to be kept in mind that the SHS would not provide the same level of electricity service as they get from the diesel-based mini-grid. But if they consume less, as is shown in the case of winter months, their payment will be reduced and can be lower than that of the SHS. Similarly, with 100 % capital grant subsidy, the cost reduces but not very dramatically.

It can thus be concluded that for a limited level of supply over a fixed number of evening hours, a diesel-generator-based mini-grid option can be a suitable option that requires about one-half of the capital cost of SHS based supply for a higher installed capacity. Poorer consumers with just fixed lighting loads can be charged a fixed monthly rate, whereas other consumers can be charged based on

[5] This assumes the capital cost of $30,163, battery replacement cost of $10,800 on the 6th year and 11th year, electricity output based on a 5 h use of the system at the system peak load, and a discount factor of 5.3 % for a 15 year project life.

their consumption level. As we have considered the cost recovery based on the costs payable for an SHS, this option can be suitable for implementation by socially responsible private entities as well as by community-based organisations. Moreover, the technology in this case is widely available and can be operated using locally available skills. The option is however less environment friendly as it depends on a fossil fuel. It also faces the risk of fuel price fluctuations, but as a less capital intensive option, this offers a good starting point for building demand in off-grid areas. However, even for such a small-scale initiative, the investor has to secure more than a million taka, which may need financial and organisational support.

6.2.2 Scenario S2

This is similar to S1 in terms of configuration. Only the fuel-related costs increase due to higher plant operation but higher output reduces the levelised cost per unit. The main difference is that a small commercial load is also serviced and the regulated tariff for grid-based supply to such users is about 2 times the rate charged for the residential consumers (in the lowest block of consumption level). Although this suggests the possibility for differential charges, in our case this would not be relevant as we have considered the cost of competing supply (i.e. SHS). As the SHS system price varies by size and not by user type, our previous analysis holds here as well and is not repeated here.

6.2.3 Scenario S3

In this scenario, a daytime productive load of 10 kW has been considered in addition to the evening residential-commercial load. Thus, the productive load is serviced outside the evening peak. The system configuration changes in this case and a hybrid system emerges as the optimal choice. The capital cost required for this option is $49,800. In addition, the batteries require one replacement in the eighth year ($7,200) and the diesel generator requires three replacements in the 4, 8, 11 and 15th years. The total capital requirement for asset replacement is $27,000 but its net present value comes to $17,671.

Following the economic pricing principle, if the off-peak consumption is charged to cover the operating cost only, the tariff for productive use comes to $0.242 (or about 18 taka per kWh). Although this is about 3 times the prevailing rate for this category of consumers of grid electricity, it is cheaper than the alternative supply from a diesel generator (which comes to $0.33/kWh for operating cost coverage and $0.423 for full cost coverage). For other peak load consumers, the economic principle requires the tariff to recover full costs including capital costs. The levelised cost for full cost recovery comes to $0.368, which is lower than that for scenarios S1 and S2. Consequently, residential consumers pay less on average compared to the previous two scenarios and they can expect to

Table 10 Revenue generation using economic tariff

Item	Unit	LI	MI	HI	Commercial	Productive	Total
Summer Cons	kWh/month	3	28.5	36	75	3,000	
Winter cons	kWh/month	3	16.5	18	75	1,500	
Annual cons	kWh/year	36	282	342	900	28,500	
Tariff	$/kWh	0.368	0.368	0.368	0.368	0.242	
Revenue	$	13.248	103.776	125.856	331.2	6,897	
Av monthly expense	$/month	1.104	8.648	10.488	27.6	574.75	
Income from all consumers	$/year	662.4	4358.592	2013.696	331.2	6,897	14262.89
Revenue requirement	$/year						17676.13

reduce their spending even compared to owning an SHS. However, as shown in Table 10, the revenue so generated is not sufficient to meet the revenue requirement of the electricity supplier. Thus, the strict economic cost recovery principle cannot be applied in this case.

One option could be to allocate the balancing cost to the productive users. This can be done in a number of ways but the most common options would be either to charge a fixed per kW/ month charge in addition to the energy rate or to increase the energy rate without adding any fixed charge. The fixed charge has some merit as a part of the revenue will flow even if the user does not consume energy for any reason. Given the size of the productive load considered here, a monthly fixed charge per kW can be a logical choice.

It becomes clear that the addition of a productive load brings the average cost of supply down and improves the financial position of the supplier. This happens despite an increase in the capital requirement, although only small companies may become interested in this size of business. As the cost recovery is likely to be possible even without any government intervention, this can become a viable business opportunity in Bangladesh. However, it may be difficult to realise the full potential of productive load instantaneously. This highlights the importance of mapping local level opportunities and enlisting support of local stakeholders early in the development process. In addition, support for such ventures through some risk sharing arrangements can improve the attractiveness of the business.

Although this is a hybrid system, the diesel generator still plays an important role. Thus, this option can be viewed as an extension of the previous scenarios S1 and S2 where the operation starts with a diesel generator for a restricted period of supply and then expands to include off-peak productive load. However, the supplier is likely to continue with its diesel generator in such a case, which, as mentioned earlier, is not the least-cost option given the high fuel cost and asset replacement cost. However, such a gradual approach may make practical sense given the limited stress on initial capital requirement.

6.2.4 Scenario S4

In this scenario, the supply reliability is considered, when 24 h of service is made available, allowing consumers to use electricity at night. This changes the demand situation considerably and the system configuration changes accordingly. All consumers now contribute to the peak demand, which increases the peak capacity requirement. Accordingly, all consumers should bear the responsibility for the peak load. In such a case, a time-differentiated tariff could be appropriate but given the small volume of consumption involved, the metering cost is likely to outweigh the benefits. Accordingly, a simple pricing system with flat rates for residential and commercial consumers and a fixed charge coupled with an energy charge for the productive uses could be appropriate.

The supply system for this scenario requires more PV arrays compared to S3. The capital cost increases to $122,000 while the diesel generator requires four replacements (at a non-discounted cost of $40,000) and the batteries require two replacements (at a non-discounted cost of $32,000). The levelised cost of electricity comes to $0.363/kWh, whereas the energy-related charge comes to $0.227/kWh. However, as before, sufficient revenue will not be recovered if productive users are charged only at the energy-related charge while others are charged at the full levelised cost. Moreover, in this case, there is no justification for the preferential treatment of the productive use, particularly when part of it coincides with the peak hours. Therefore, the tariff has to be carefully designed to avoid undesirable effects. An example is provided in Table 11 where an energy-related charge of $0.31 is used for productive uses supplemented by a fixed charge of $30/kW/month. Alternative tariff schemes can be developed to suit the specific requirements but a full-scale analysis of this aspect is beyond the scope of this chapter.

Table 11 shows that the low-income consumers will still pay less than that required for owning an SHS for a comparable service. However, a comparison with the SHS cost becomes somewhat less relevant for the middle- and high-income groups as they receive round-the-clock power from the mini-grid compared to a limited supply from the SHS. Although they are likely to spend more on electricity cost for a reliable supply, the cost per unit of electricity is less. The average monthly bill between $20 and $27 for these categories is however much higher than these groups pay on fuel and electricity as per the Household Income Expenditure Survey. Moreover, the monthly bill for productive loads will be significant due to high consumption level and this can be a disincentive for promoting productive loads. A 100 % capital grant would reduce the cost to $0.261/kWh but this could still make productive activities reluctant to consume significant quantities of electricity.

As the system size increases, the capital requirement increases as well. More importantly, the cost of asset replacement becomes important. Depending on the capital structure and repayment requirement, it is possible that the supplier faces some funding mismatch. This would require access to flexible funding arrangements and short-term funding for working capital. However, unless the business is

Table 11 An example of tariff schemes for Scenario S4

Item	Unit	LI	MI	HI	Co	Prod	Total
Summer cons	kWh/month	4.2	75.9	102	210	7,200	
Winter cons	kWh/month	4.2	30.3	33.6	210	3,600	
Annual cons	kWh/year	50.4	682.8	882	2,520	68,400	
Tariff	$/kWh	0.363	0.363	0.363	0.363	0.31	
Revenue	$	18.2952	247.8564	320.166	914.76	24804	
Av monthly expense	$/month	1.5246	20.6547	26.6805	76.23	2,067	
Income from all consumers	$/year	914.76	10409.97	5122.656	914.76	24804	42166.14
Revenue requirement	$/year						41,516

not organised around a bankable contractual arrangement, securing finance from traditional sources can be a challenge.

As indicated before, this option can also be considered as an extension of the earlier scenarios, particularly S1 and S2. The advantage here is that the PV system along with the battery and converters can be appended to the diesel generator system suggested for S1 or S2. This gradual expansion of the system can work for rural areas where the demand is likely to develop once the benefits of electricity are realised by the population. Similarly, this also allows time for developing the productive load that can act as an anchor for the system.

6.2.5 Scenario S5

This scenario is similar to S3 but allows additional demand for a restricted period of supply. This in effect increases the peak demand and the residential consumers are responsible for this increase. Accordingly, they should bear the responsibility for additional system capacity addition. As the productive load comes during off-peak hours, the logic of off-peak price for them applies here as well.

However, because of increased peak load, the system capacity requirement increases, which in turn leads to higher cost of electricity compared to Scenario 3. Therefore, the residential consumers end up paying more in this case. As the demand by the poor does not change compared to other scenarios, they are not really responsible for the additional capacity requirement and should not be unduly overburdened. This leaves the burden of extra cost on the middle-income and high-income consumers. However, such a tariff policy will impose high charges on the middle- and high-income groups, making the price unattractive to them. Consequently, all consumers would end up paying more in this scenario compared to S3, making them worse off. Hence, there is limited business sense for this scenario and the supplier should avoid encouraging such skewed demand during the peak period.

6.2.6 Scenario S6

This scenario removes supply restriction and allows for full demand development. Accordingly, the high-income consumers can use electric appliances like refrigerators, while commercial consumers can use electricity at any time. Consequently, the consumption of middle-income and high-income households as well as commercial activities increases compared to Scenario S4. This scenario results in the least levelised cost of electricity of six scenarios.

As in Scenario S4, all consumer categories contribute to peak demand and accordingly are required to bear the consequences by paying appropriate charges. Although the economically efficient tariff would have to distinguish between peak and off-peak periods, the time-of-use metering cost may be difficult to justify for such small consumers. Accordingly, energy-related tariff supplemented by fixed charges may be relevant. However, as the consumption of poor households do not change

Table 12 Example of electricity bill and revenue generation at the levelised cost of electricity for Scenario S6

Item	Unit	LI	MI	HI	Co	Prod	Total
Summer Cons	kWh/month	4.2	99.9	462	1,440	7,200	
Winter cons	kWh/month	4.2	54.3	393.6	1,440	3,600	
Annual cons	kWh/year	50.4	970.8	5,202	17,280	68,400	212205.6
Tariff	$/kWh	0.344	0.344	0.344	0.344	0.344	
Revenue	$	17.3376	333.9552	1789.488	5944.32	23529.6	
Av monthly expense	$/month	1.4448	27.8296	149.124	495.36	1960.8	
Income from all consumers	$/year	866.88	14026.12	28631.81	5944.32	23529.6	72998.73
Revenue requirement	$/year						71,544

compared to S4, they may be charged at a flat rate only. As shown in Table 12, if electricity is charged at the levelised cost of energy, the required revenue can be collected but the monthly bill for average high income households, commercial users and productive consumers becomes quite big, even by developed country standards, thereby suggesting limited attractiveness of such high-level consumption for these categories of consumers. The operating cost component in the charge comes to \$0.224/kWh, which is closely related to diesel fuel use in the system. The monthly bill will not change significantly even if the charge recovers only the operating costs. This perhaps shows the limitation of a diesel-based hybrid system.

Moreover, the capital cost of this system increases to \$196,000 (or about 15 million taka), which may be attractive to medium-sized firms. As before, the capital requirement for asset replacement also increases to \$115,000 (non-discounted). Thus, financing the capital requirement becomes another constraint for this option.

Based on the above, analysis, it becomes clear that small-scale supply as indicated in scenarios S1–S4 can be developed into businesses for rural electricity delivery but as the system becomes bigger with higher demand, the monthly bill can be very high for high energy using consumers. The relatively high cost of supply may not be attractive for consumers and is unlikely to be sustainable. The capital constraint is another issue that can become difficult to overcome. Moreover, capital subsidy alone will not reduce the costs significantly as the operating costs remain high and providing capital and operating subsidy for village level supplies will not be sustainable in the long-run.

6.3 Remote Area Power Supply System

Bangladesh has set a target of providing universal electrification by 2020. The state-owned agencies like Bangladesh Power Development Board (BPDB), Rural Electricity Board (REB) and Palli Bidyut Samity (PBS) are involved in providing electricity in rural areas. In addition, Grameen Shakti, a non-profit organisation, is also actively involved in promoting renewable energy solutions, mainly the SHS. However, recognising the challenge faced by the country in reaching its target, the Government introduced a new initiative, called the Remote Area Power Supply System (RAPSS) in 2007. This allows the private sector to get involved in rural power supply and the guidelines[6] for the RAPSS indicate that:

(a) The Power Division of the government will identify the potential RAPSS areas. These areas would cover the geographical area of two or more sub-districts.
(b) The system can cover both off-grid and on-grid areas.
(c) The operator will be selected through a competitive bidding process.

6 http://www.powerdivision.gov.bd/pdf/RAPSS.pdf

Table 13 Financial support required for grid price parity

Description	Unit	S1	S2	S3	S4	S5	S6
Capital required at 100 % capital cost subsidy	$	15,000	15,000	49,800	126,000	127,000	196,000
Average tariff at 100 % capital cost support	$/kWh	0.39	0.378	0.272	0.261	0.2	0.255
Target tariff (weighted by consumption)	$/kWh	0.04	0.077	0.069	0.08	0.079	0.083
Difference	$/kWh	0.35	0.301	0.203	0.181	0.121	0.172
Amount of operating subsidy required/year	$/year	6770.75	6097.658	10002.83	21339	7375.555	36977.42

(d) The operator will operate under a licence from the Bangladesh Electricity Regulatory Commission for a period up to 20 years.
(e) A fund called RAPSS Fund will be created to support the rural electrification process and will receive funds from the government, donor agencies and other sources. The fund can be used for providing capital grant support, to provide loans of 5–10 years' duration, to subsidise connection charges and to offset duty, tax and VAT.
(f) The retail supply tariff will be set initially through the bidding process but if the tariff is significantly higher than the tariff charged by the nearest PBS, then the government may decide to provide subsidy to close the gap, depending on the funding available from the RAPSS Fund.[7]
(g) The capital cost subsidy can be given up to a maximum limit of 60 % and if the retail tariff still remains high, soft loan can be provided from the fund. The investor has to invest 20 % through equity participation.

The RAPSS Guidelines provide a framework for private sector involvement in rural electricity supply. The government has identified 30 remote sub-districts for electrification through micro-utility model. However, despite the policy framework, the micro-utility model has not progressed much and has generated limited investor interest.

From our analysis it becomes clear that even if 100 % capital cost subsidy is provided, the cost of supply will remain higher than the retail tariff approved by the regulatory commission for different categories of consumers. Table 13 shows the amount of capital subsidy required under different scenario and the operating subsidy required to reach the grid price parity in rural areas under the optimal configurations considered in this study.

Clearly, it shows that trying to reach the grid price parity will impose significant financial burden on the government, particularly for Scenarios 4, 5 and 6. They are

7 http://www.powerdivision.gov.bd/pdf/RAPSS%20Fund.pdf

unlikely to be sustainable solutions. This happens even after providing significant capital support. The first three options could still be considered as the capital subsidy requirement is not too demanding and the price parity can be restricted to poor consumers while others may be charged the levelised cost. This will reduce the operating cost subsidy.

The case of sub-district level operation can provide the required scale economy and may ensure larger systems for local grids where higher technical efficiency of operation can also be expected. This can be an area for further research where an analysis using the terms and conditions offered by RAPSS guidelines can also be considered.

7 Conclusions

This chapter has considered the village-level electrification in Bangladesh and analysed the viability and business case of a hybrid mini-grid system for a remote non-electrified village in Dhaka division. The analysis developed six alternative demand scenarios, considered local resources for electricity generation, conducted techno-economic analyses of all scenarios using HOMER and performed business analysis. The demand scenarios captured alternative development pathways—starting from basic level supply for 5 h per day to unrestricted, reliable supply consisting of residential, commercial and productive loads. The techno-economic analysis suggested optimal configurations that consisted of diesel generators for the basic level of supply and hybrid PV—diesel solutions for more elaborate services. The renewable energy share in all configurations varies between 0 % (in the basic cases) and 60 % (in S5) and the cost of electricity per kWh decreases as the system size increases. However, the hybrid systems require significant excess capacity due to intermittent nature of solar energy and consequently, the initial investment requirement increases. Moreover, during the project life some assets (such as batteries and diesel generators) need to be replaced depending on their life and extent of use. This requires significant investment at regular intervals to keep the system going.

The analysis of business case of the investments revealed that the levelised cost of electricity from the off-grid options is higher than the regulated tariff for various categories of consumers that receive grid electricity. However, the cost of off-grid supply is likely to be cheaper than the cost of owning an SHS. Low-income consumers will pay almost one-half of the cost of owning an SHS for a comparable level of energy use while the high-income users may be paying somewhat more for the restricted level of supply, although the monthly bill will not be too burdensome for low level of supplies. However, the problem arises when demand restrictions are removed allowing consumers to use high volumes of energy. Their monthly bills will be burdensome, making higher consumption unattractive. This happens due to high capacity-related costs and operating costs of the system.

It is also found that capital cost subsidy will not be sufficient to ensure grid price parity and significant amount of operating cost subsidy will be required. As the operating cost subsidy will impose a recurring burden on the government's finances, it is unlikely to be sustainable. This makes the energy access challenge significant. Our analysis suggests that the basic electricity supply provision through a mini grid is the most preferable business solution—it requires less capital, less subsidy volume and moderate monthly bills for consumers. Such a business can be organised by local entrepreneurs, private investors or local community organisations.

Bangladesh has been promoting Remote Area Power Supply System since 2007 where private investors can enter into rural electricity supply through a competitive bidding process for a maximum period of 20 years. This allows sub-district level geographical areas under the jurisdiction of the licensee. However, the objective of achieving grid-like pricing may be difficult to attain with capital subsidy and soft loans, unless the system has a low operating cost. This is an area for further investigation.

8 Appendix

Annex A.1 Details of district-level electrification rate in Bangladesh as per Census 2011

District	HH-total	Electrification (%)	Rural	Urban
Lalmonirhat	289,953	20.1	16.4	61.9
Kurigram	507,106	20.8	16.9	48.9
Bhola	371,799	25.9	21	79
Panchagarh	228,074	27	22.6	73.8
Bandarban	78,714	28.2	17.7	77.7
Barguna	214,863	28.8	23.8	78.1
Gaibandha	611,297	29.4	26.2	72.8
Netrokona	477,927	30.2	24.8	81.3
Patuakhali	345,113	31.8	26.5	84.4
Cox's Bazar	413,402	32.1	25.9	66.5
Khagrachhari	132,503	32.7	20.6	75.5
Sunamganj	438,752	34	30.1	78.4
Nilphamari	430,906	34.5	28.8	74.1
Thakurgaon	319,929	34.7	30.1	78.7
Rangpur	717,362	38.1	32.1	80
Jamalpur	562,180	39.4	33.7	71.6
Dinajpur	713,255	39.4	33.1	79.5
Noagaon	654,275	39.4	35.2	82.6
Magura	205,492	40.8	35.6	76.5
Bagerhat	350,537	40.8	36.3	84.5
Sherpur	340,769	41.2	36.5	72.4

(continued)

Annex A.1 (continued)

District	HH-total	Electrification (%)	Rural	Urban
Mymensingh	1,150,574	41.3	34.3	86.2
Satkhira	468,853	41.8	38.2	81.4
Rangamati	126,414	41.8	29	93
Pirojpur	255,059	42.9	36.6	82.8
Shariatpur	246,535	43.2	38.9	78.1
Lakshimur	364,255	43.8	39.4	81.9
Narail	162,299	45.7	41.8	75.3
Rajbari	237,352	45.8	41.5	81.8
Sirajganj	713,064	47.1	42.5	79.9
Habiganj	391,657	47.7	43.3	84.2
Chapai Nawabganj	357,246	48.6	42.7	76.6
Faridpur	418,554	48.7	43.9	85.7
Natore	422,921	48.8	44.5	78
Jhalokati	157,559	49	43.6	79
Gopalganj	248,735	49.1	45.6	82.4
Kishoreganj	623,914	49.6	43.9	80.8
Maulvi Bazar	359,715	50.8	47.1	87.3
Noakhali	590,808	51.1	47	81.2
Joypurhat	241,994	51.6	47.2	77.4
Bogra	863,600	52.6	46.3	82.6
Manikganj	323,741	52.7	49.8	85.6
Barisal	510,780	53.5	44.5	88
Chandpur	503,851	55.6	50.4	79.7
Tangail	866,578	56.2	52	82.7
Jhenaidah	421,300	58.6	54.6	81.4
Madaripur	251,581	59.3	55.5	85.5
Pabna	588,891	60.1	56.3	81.5
Chuadanga	276,910	60.6	56.6	74.3
Jessore	653,423	61.1	55.4	87.3
Meherpur	165,974	61.2	58.1	87
Rajshahi	630,331	62.1	54.1	80.2
Sylhet	589,425	62.9	53.9	93.9
Kusthia	475,989	64.1	61.6	87.3
Khulna	540,504	64.1	50.4	94.2
Brahmanbaria	537,560	71.3	67.5	93.7
Narsingdi	473,937	72.8	67.6	95.8
Feni	275,461	73.4	68.9	91.3
Comilla	1,048,984	74.9	71.8	92.7
Chittagong	1,509,717	79.1	67	95.7
Gazipur	809,761	84	78.6	95.7
Munshiganj	3,100,664	90.1	89	97.7
Narayanganj	663,088	95.3	93.8	98.2
Dhaka	2,639,630	97	92.2	98.6

Source BBS website (www.bbs.gov.bd)

References

1. BBS (2011) Report of the Household Income and Expenditure Survey, 2010, Bangladesh Bureau of Statistics, Dhaka. http://www.bbs.gov.bd/PageReportLists.aspx?PARENTKEY=66. Accessed on 10 Apr 2014
2. BBS (2012) Socio-economic and demographic report, bangladesh population and housing census 2011, Bangladesh Bureau of Statistics, Ministry of Planning, Government of the People's Republic of Bangladesh, Dhaka. http://www.bbs.gov.bd/WebTestApplication/userfiles/Image/BBS/Socio_Economic.pdf. Accessed 5 Mar 2013
3. ESMAP (2007) Technical and economic assessment of off-grid, mini-grid and grid-electrification technologies, ESMAP technical paper 121-07, The World Bank. http://siteresources.worldbank.org/EXTENERGY/Resources/336805-1157034157861/ElectrificationAssessmentRptSummaryFINAL17May07.pdf
4. Mondal AH, Denich M (2010) Hybrid systems for distributed power generation in Bangladesh. Energy Sustain Dev 14(1):48–55
5. Nandi SK, Ghosh HR (2009) A wind-PV-battery hybrid power system at Sitakunda in Bangladesh. Energy Policy 37(9):3659–3664
6. Roy RB (2012) Design and cost analysis of hybrid power system for off-grid rural areas of Bangladesh. Can J Electr Electron Eng 3(7):413–423
7. Sovacool BK, Drupady IM (2013) Energy access, poverty and development: the governance of small-scale renewable energy in developing Asia, chapter 3: Grameen Shakti in Bangladesh. Ashgate Publishing, Surrey
8. Yadoo A, Cruickshank H (2010) The value of cooperatives in rural electrification. Energy Policy 38(6):2941–2947

Application of Multi-criteria Decision Aids for Selection of Off-Grid Renewable Energy Technology Solutions for Decentralised Electrification

Dattakiran Jagu, D. Pugazenthi and V. V. N. Kishore

Abstract This chapter demonstrates a multi-stakeholder approach for selection and ranking of renewable energy technologies for decentralised electrification in India by using PROMETHEE, a multi-criteria decision aid. A graphical descriptive analysis is applied to map the various conflicts observed and to suggest possible interventions. The results show that micro-hydro is currently the best compromise solution for decentralised electrification in India, followed by bio-methanation. A substantial investment in technology standardisation of biomass technologies and associated sub-systems and significant reduction of the costs of PV-based technologies are required before they can be adopted on a wider scale. Innovative hybrids and smart mini-grids can be used in the short term for diversity in supply options.

Author Dattakiran Jagu was formerly with Department of Energy and Environment, TERI University, New Delhi, India. Author V. V. N. Kishore was former professor of Department of Energy and Environment, TERI University, New Delhi, India and presently an independent consultant in Renewable Energy.

D. Jagu (✉)
157, 26th Main, 37th Cross 9th Block, Jayanagar, Bangalore 560069, India
e-mail: jdkiran@gmail.com

D. Pugazenthi · V. V. N. Kishore
Department of Energy and Environment, TERI University, New Delhi, India
e-mail: pugazenthi.d@live.com

V. V. N. Kishore
e-mail: vvn_kishore@yahoo.com

S. C. Bhattacharyya and D. Palit (eds.), *Mini-Grids for Rural Electrification of Developing Countries*, Green Energy and Technology,
DOI: 10.1007/978-3-319-04816-1_11,

1 Introduction

The lack of access to modern energy services acts as a severe impediment to human development [19, 42] and ensuring energy access is one of the major challenges developing countries face today. The percentage of the population not using electricity or commercial energy is one of the most crucial indicators to energy access [22]. With over 1.3 billion people—primarily in sub-Saharan Africa and developing Asia—still lacking access to electricity [24] and possibly an even larger number having only intermittent access to power, the task of supplying them reliable and affordable electricity is formidable. The conventional resources that would be required to achieve this are equally enormous.

The IEA estimates that a cumulative investment of $1 trillion— an average of $48 billion per year, is required to achieve the target of universal energy access by 2030 [6]. About 90 % of this cost is estimated for electrification alone. In a similar study, Bazilian et al. [2] have estimated that till 2030, the annual cost of universal access to electricity would be between $12 and $134 billion. These are only a tiny fraction of the global investment required in energy infrastructure [6]. This kind of additional investment would be a very heavy burden for developing countries. This obligates funding agencies, government bodies and private investors to make judicious choices by adopting the right solutions that target the most energy-poor while ensuring the long-term sustainability of these solutions.

Decentralised/distributed generation systems, often using renewable energy technologies (RETs) have the flexibility in size, fuel and technology choice, capability to produce reliable power [23] and ability to induce local energy autonomy. However, RETs vary immensely in terms of the resources required, their initial and operational costs, perceived social and environmental benefits and their levels of technical maturity.

In spite of such problems in decision-making, RETs have been tried out in several developing countries for off-grid electricity generation. Micro-hydro in Nepal, solar photovoltaic (PV) home lighting systems in Bangladesh, India, Philippines and Kenya, biomass systems based on gasification, anaerobic digestion and bio-diesel in India, Sri Lanka and Cambodia and miscellaneous hybrid systems in India are some examples. Both micro-hydro and PV systems have been chosen primarily for their technology maturity and reliability. PV systems have especially been promoted with a conviction that increased productions will ultimately reduce the cost. Biomass technologies, on the other hand, were primarily selected because of reliability of local resources, lower initial costs and possibility of meeting electricity-intensive load demands such as irrigation, flour milling, and oil expelling at a lower cost. Unfortunately, experiences (both positive and negative) with hundreds of such field installations have not led to clear directions for future funding or promotional policies, investments in technology/product development, focus on applied R&D and regulation. The application of a multi-criteria approach and the use of multi-criteria decision aids (MCDA) for sustainable development decisions fosters integration of the stakeholders, makes the decision process fair and democratic and legitimises the results of decision [37].

The objective of this study is to demonstrate the identification of suitable renewable energy technologies for decentralised electrification in India using multi-criteria decision aid and also to understand the future R&D or financing needs for other promising technologies. This study has been confined to those RETs that have at least one working decentralised installation in India. This has led to shortlisting of the following technology options: micro-hydro, solar PV, biomass gasification, wind-PV hybrid and biomethanation.

It is proposed to adopt a top-down approach driven by a single problem of seeking the most suitable renewable energy technology, which is further decomposed into dimensions and objectives. The extent to which a particular RET fulfils each objective is measured with respect to one or more criteria. Next is to identify the various dimensions that need to be considered for taking a decision on a decentralised electrification project. Field visits conducted in six states spread across India and face-to-face discussions held with project developers, technical experts, non-governmental organisations and end-users gave an insight into some important factors considered by them for setting up and operation of decentralised projects. These factors fell broadly into the following categories: Environmental, Social, Economic, Resource, Technical, Operational and Regulatory.

The study has been validated by a multi-criteria decision aid PROMETHEE (Preference Ranking Organisation Method for Enrichment of Evaluations). A graphical descriptive analysis is applied to map the various conflicts and to suggest possible interventions.

2 Earlier Studies for Comparison and Selection of RETs

There is a well-developed body of literature featuring studies on performance evaluation and selection of appropriate renewable energy options. A rich diversity can be observed within these studies based on their focus of evaluation and the analytical approaches employed.

The studies can be hierarchically arranged into those that focussed on: (a) Techno-economic characteristics such as cost of energy, energy potential, life cycle cost, reliability of supply, and exergy [13, 21, 26, 31, 34–36, 40, 46]; (b) Social, Economic and Environmental impacts [3, 12, 16, 28, 41, 48]; and (c) Sustainability attributes [1, 14, 27, 38, 50]. Techno-economic evaluations, that form the bulk of these studies, can help the decision-maker in appreciating the technical and economic implications of a particular choice. The studies based on social, economic and environmental impacts alone fail to emphasise the importance of an existing local ecosystem. Studies that additionally focussed on the operational sustainability included sustainability of the technology during and beyond the lifetime of the project.

A further classification of the evaluation studies can be made based on the approach used to analyse the energy options. Evaluations based on a single criterion such as cost of energy are easier to analyse but could be inadequate for decision-making. A slightly more complex approach involves optimisation [31, 48] or a break-even analysis [34] involving two or more criteria. Variations include those that seek the optimal mix of RETs [26] when subject to input constraints. However, this approach fails to consider multiple perspectives of different stake holders, especially the end-users. A multi-stakeholder and multi-criteria approach that captures the stakeholder priorities and adopts a rational analysis method can lead to a wider consensus among the stakeholders. This, when combined with a participatory approach [14, 38] that considers the aspirations and constraints of the local communities can result in a more credible solution [5] and can build their trust in decisions and political institutions [43].

Many initial studies were based on single-criterion approach focusing primarily on the techno-economic dimension, which have been sometimes used to promote a 'favourite' technology. For example, techno-economic studies for remote village electrification in India justified the high initial investment in PV systems compared to the cost of extending the grid. However, PV systems have often provided limited electricity supply and did not meet all the current and future needs, especially irrigation in most cases. While it may be technically possible to design PV systems to meet the entire demand, the cost of doing so is likely to be very high, thereby requiring longer pay-back periods and making the cost of supply beyond reach of many consumers.

By employing analytical techniques, some studies aimed to identify the most favourable technology or a single best-optimised solution within their decision context and to suggest policy recommendations for its choice. For example, Kaya and Kahraman [28] employed a fuzzy multi-criteria decision-making technique to identify that wind energy is the most appropriate technology for Istanbul region and that Çatalca district is the best area for installing the wind turbines. Few studies went further to investigate the market mobilisation and technology transfer of the identified technology [16] or to explore newer technical opportunities [21]. A critical gap found in many studies is that the prescriptive approach adopted for solving the decision problem fails to depict the inter-dimensional and inter-perspective conflicts involved in the decision-making. Another key missing element in most studies is that they often stopped short of suggesting remedial measures or policy recommendations for the technical advancement of the non-best solutions.

3 Methodology

3.1 *Multi-criteria Decision Aids*

Multi-criteria decision analysis or multi-criteria decision aids (MCDA) evolved as a response to the inability of people to effectively analyse dissimilar information from multiple disciplines [30]. They can provide the concepts and guidelines for structuring and modelling decision problems [45], thereby aiding the decision-making process in developing suitable criteria, in gaining acceptance of stakeholders and in creating new ideas for solutions [37]. With no right solution independent of the decision process [44], the decision taken can be only as legitimate as the underlying MCDA technique.

3.2 Selecting an MCDA Technique

Numerous MCDA techniques and their classifications have been comprehensively studied in various literatures [4, 17, 20, 47]. Despite the large number of MCDA techniques, none is perfect [33] and the success or failure of a particular technique depends primarily on the context in which it is being applied. A comparison of the MCDA methods revealed that the most crucial quality criteria are the method's ability to deal with complexity, possibility to consider non-substitutability, ability to invoke stakeholder participation and ability to provide information to stakeholders to make better decisions [15]. For RET selection, Polatidis et al. [39] recommend that an analyst should necessarily consider the technique's treatment of the sustainability issue, modelling of the decision matrix preferences, technical features, treatment of uncertainty and consideration of practical aspects such as ease of use, ability to handle multiple criteria, qualitative inputs and support for multiple decision matrices. In our context, we chose to apply PROMETHEE—an outranking method, along with a multi-stakeholder approach in identifying the criteria for evaluation, prioritising amongst them and then evaluating the RETs.

There are at least four reasons for choosing PROMETHEE for this study. First, it is flexible in accepting poorly shaped stakeholder inputs such as environmental, economic and social impacts of the RETs. Second, both qualitative and quantitative data can be dealt with simultaneously, each in its own units. Third, it can provide two types of rankings—with and without incomparability amongst the RETs, which helps in appreciating the relative strengths and weaknesses of each RET. Finally, by using the GAIA (Geometrical Analysis for Interactive Aid) tool, it permits a visual depiction of the decision problem which aids in a better understanding of the inter-dimensional and inter-stakeholder synergies and conflicts thereby ensuring debate and consensus building among the stakeholders.

3.3 PROMETHEE and GAIA

PROMETHEE [7, 8, 10, 11] is an out-ranking technique typical of the European school of MCDA. It is based on the principle of pair-wise comparison of the alternatives. After forming an evaluation matrix of performance of alternatives on all the criteria, the PROMETHEE process involves:

Table 1 Types of preference functions

Preference function	Remarks
Usual preference function	Simple, does not include any threshold, best suited for criteria with a few very evaluation criteria
U-shape preference function	Introduces the idea of an indifference threshold
V-shape preference function	Special case of the Linear preference function where the *Q* indifference threshold is equal to 0, well suited to quantitative criteria when even small deviations should be accounted for
Level preference function	Suited to qualitative criteria when the decision-maker wants to modulate the preference degree according to the deviation between evaluation levels
Linear preference function	Best suited for quantitative criteria when a *Q* indifference threshold is wished
Gaussian preference function	An alternative to the Linear, has smoother shape but it is more difficult to set up because it relies to a single *S* threshold that is between the *Q* and *P* thresholds and has a less obvious interpretation

3.3.1 Assigning a Preference Function to Each Criterion

The preference functions translate the difference of performance of alternatives on a given criterion in terms of a preference degree measured between 0 (no preference) and 1 (absolute preference). Six possible types of preference functions (Table 1) [11] can be assigned by the decision-maker, each representing a different perception of measurement scales for criteria.

3.3.2 Assigning Weights to the Criteria

No specific guidelines exist in PROMETHEE for determining weights to the criteria. A simple method of calculating weights by pair-wise comparison of criteria used in Kohli et al. [29] has been adopted in this chapter. When weights calculated for each of the stakeholders are averaged out, this method prevents any one individual's judgment to dominate the process.

3.3.3 Estimating the Outranking Degree of Options

Using the criteria weights and preference functions, a multi-criteria preference index π (a, b) is computed as the weighted average of the preference functions P_j (a, b).

$$\pi(a,b) = \frac{\sum_{j=1}^{J} w_j P_j(a,b)}{\sum_{j=1}^{J} w_j}$$

where, $w_j > 0$ is the normalised weight allocated to the *j*th criterion (the more important the criterion, the larger w_j), and $P_j(a, b)$ is the value of the preference function for criterion *j*th criterion when action *a* is compared to action *b*.

$\pi\ (a, b)$ is a number between 0 and 1. It expresses how much '*a*' is preferred to '*b*' taking into account all the criteria and their weights.

The positive flow (strength) Φ+ expresses how much an alternative is dominating the other alternatives and the negative flow Φ− expresses how much it is dominated (weakness) by the others. A higher value of Φ+ or a lower value of Φ− indicates better performance. Positive and negative flows usually induce somewhat different rankings of the alternatives. The net flow Φ, which is the balance between the positive and negative flows, defines the net outranking.

$$\Phi + (a) = \frac{1}{n-1} \sum_{b \neq a} \pi(a, b)$$

where: alternative '*a*' is dominating the other n-1 alternatives.

$$\Phi - (a) = \frac{1}{n-1} \sum_{b \neq a} \pi(b, a)$$

where: $n - 1$ alternatives are dominating alternative '*a*'.

$$\Phi(a) = \Phi + (a) - \Phi - (a)$$

Two main PROMETHEE methods have been used to rank the RETs:

- PROMETHEE I—provides a partial ranking based on Φ+ and Φ− and permits incomparability between the alternatives.
- PROMETHEE II—provides a complete ranking based on Φ assuming comparability amongst all the alternatives.

The prescriptive (ranking) approach of PROMETHEE is complemented by a descriptive (visual) approach called Geometrical Analysis for Interactive Aid (GAIA). GAIA involves computation of uni-criterion net flows by normalisation and projecting them onto a plane for visual analysis. The two-dimensional representation of multi-criteria data and of the technology profiles helps to identify conflicts among criteria, fix the priorities and seek possible compromise solutions. A detailed discussion regarding the PROMETHEE methods and GAIA analysis can be found in [9, 32].

3.4 Visual PROMETHEE Software

Visual PROMETHEE is a up-to-date software implementation of the PROMETHEE and GAIA multi-criteria decision aid (MCDA) methods. It is developed by Professor Bertrand Mareschal from the Solvay Brussels School of Economics and Management of the Université Libre de Bruxelles (ULB). An academic edition of Visual PROMETHEE software has been used for this study.

3.5 Method of Ranking

The first step is to identify the key stakeholders who would be involved in decentralised energy planning. Franco and Montibeller [18] defined key stakeholders as those individuals, or groups, who have the power to affect the decision under consideration; or those groups that are affected, or perceived to be affected, by the decision. The following types of key stakeholders in developing countries are initially identified: project developers, technology experts, policy makers in government bodies, international donor agencies, private sponsors and the local citizens. Inclusion of project developers and private sponsors with prior experience in RET installations increased the legitimacy of the decision, and inclusion of technical experts increased its technical competency. Inclusion of international donor agencies brought in their expertise in similar contexts. Policy makers contributed to the political and regulatory perspective of the decision. In addition, discussions with the local citizens are helpful in capturing the aspirations of local communities and therefore to develop suitable criteria that measure the ability of RETs to fulfil these aspirations.

The methodology given in Fig. 1 was used for our purpose to rank the RETs. The sequential process makes it simple and easy to follow each step and the recursive steps improve the quality of decision.

Minimum inputs in the form of available alternatives, evaluation criteria and their relative importance are required from the experts. Subjectivity is eliminated with the formation of a decision matrix which captures all the inputs, thus making the process objective and transparent. Iterations during performance evaluation help in judging the most important alternatives, while eliminating the clearly inferior ones. This leads to a compact decision matrix and lesser information for the decision-maker to comprehend. To account for uncertainty in the experts' opinions and to handle the evolving nature of the technology development and variations in costs, a sensitivity analysis was proposed for the results obtained.

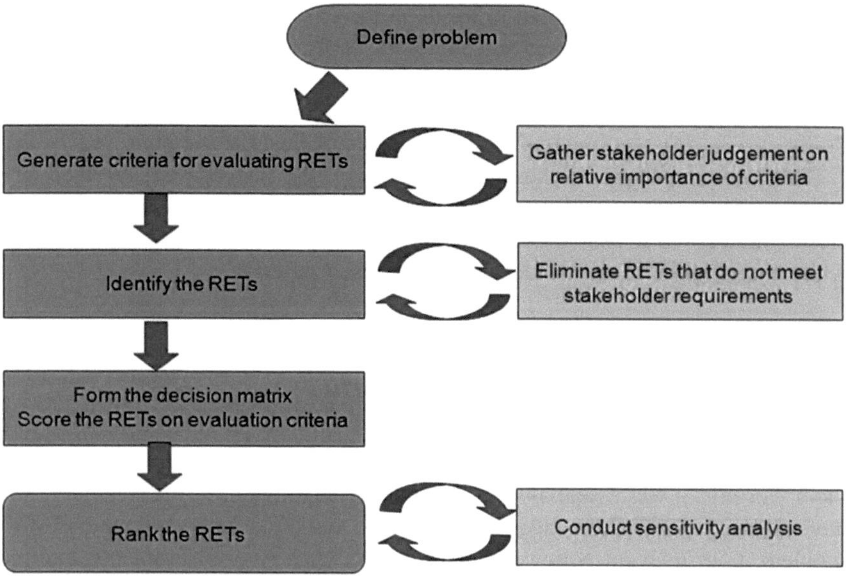

Fig. 1 Ranking methodology

3.6 Generation of Evaluation Criteria

In the review of MCDA techniques, Wang et al. [49] identified at least 25 criteria, broadly categorised into Technical, Economic, Environmental and Social indicators. For studies on rural electrification, Ilskog [25] proposed 39 indicators, grouped under five dimensions: Technical, Economic, Social/ethical, Environmental, and Institutional sustainability. Karger and Hennings [27] compiled 86 sustainability criteria, arranged in the form of a value tree to evaluate different explorative scenarios of decentralised electricity generation.

From the reviews, efficiency and reliability of service and technology maturity are clearly the most widely used technical indicators. Among the economic indicators, capital cost, operational and maintenance (O&M) cost and cost of energy (CoE); and among the social indicators, contribution to local employment and social acceptance are the ones often used. Reduction of greenhouse gas (GHG) emissions and land use were the oft-used environmental indicators.[1] Other critical indicators included the availability of the renewable energy resources and availability of human resource for operating and servicing the technology.

Discussions with local citizens in remote unelectrified and partly electrified villages in three underdeveloped states of India further revealed that factors such

[1] However, this does not necessarily cover all environmental damages. For example, toxic waste water from biomass gasification system is a problem but the environmental indicator has not covered this aspect.

as the number of jobs provided to locals in the construction and operation of the project, involvement of women and equity in supply are among the most important social benefits perceived by them. Few citizens also opined that the quality of electricity supply needed to run productive loads or appliances that reduce their drudgery is at least as important as the number of hours of supply.

The evaluation criteria chosen for our study, the factors used for their construction/measurement and the scales of measurement are given in Table 2. The criteria are chosen so as to incorporate all the important dimensions of the decision while ensuring the relative independence of each criterion.

3.7 Prioritisation of Evaluation Criteria

Once the evaluation criteria are identified, the next step was to prioritise amongst them. For this, it was essential to invoke the participation of key stakeholders who have an insight into all the facets of the criteria and the different viewpoints of the problem. Four relevant policy makers in central and state governments, one representative of an international donor agency, three project developers and four private sponsors of existing projects were identified from multiple geographic locations in India to conduct the interviews.

3.7.1 Interviews with Key Stakeholders

The 12 key stakeholders identified were interviewed either face-to-face or over the phone with a written survey questionnaire (Table A.2) to elicit their opinion on the relative importance of the criteria. Semi-structured interviews were conducted in a conversational format. Care was taken to ensure that there was a fair representation from rural areas and that the stakeholders were randomly identified from different states in India. Due to time and resource constraints, a full-scale survey was not possible. The prioritisation process in our study can only provide an overview of varied stakeholder priorities. Therefore, further sensitivity analysis of stakeholder priorities was done to analyse the stability of the rankings obtained.

The goal was to compare all the alternatives pair-wise and assign a weight coefficient to each of them. To enable easier comparison, the questionnaire was formed such that the stakeholder could enter the priorities on a scale of 1 (highest priority) to 10 (least priority) or could categorise the criteria into five categories of importance—Very high, High, Medium, Low, Very low.

3.7.2 Analysis of Survey Results for Prioritisation of Criteria

It was observed that the different stakeholders displayed a great deal of variation on some criteria, while some criteria are consistently given similar priority by all

Table 2 Evaluation criteria

Criteria	Scale	Factors assessed
Environmental benefits	Qualitative	Reduction in CO_2 and reduction in environmental degradation
Social benefits	Qualitative	Jobs to local people in construction and operation, increased income/productivity, improvement in health, safety and education, involvement of women in operation
Resource availability and variability	Qualitative	Availability of energy resource in the specific region, its variability during the day and throughout the year
Initial capital cost (ICC)	Numerical	Cost of Equipment, civil works, battery storage, wiring and installation cost, T&D lines
Operation and maintenance cost (O&M)	Numerical	Costs for fuel, labour, servicing and battery replacement
Technology maturity	Qualitative	Number of technology providers in the country, availability of standards, status of R&D, ease of use
Technology performance	Qualitative	Ability to handle high-powered loads, plant load factor (PLF)
Supply chain availability	Qualitative	Availability of local manufacturing, local availability and cost of spares and service
Operational sustainability	Qualitative	Availability of stable quality and quantity of fuel/feed stock throughout the lifetime of the project, availability of human resource for operation, scalability of technology
Policies and Regulations	Qualitative	Policies favouring decentralised electrification (subsidies available for different technologies, different schemes implemented in the past), environmental regulations, policies for promoting private enterprise

the stakeholders. This high variation in stakeholder priorities (Table 3)—particularly of criteria such as social benefits, environmental benefits and policies and regulations clearly depicts the conflicting views among the various groups of stakeholders and the subjective nature of their opinions.

After each survey, weights were calculated for each criterion using a method adopted from Kohli et al. [29]. A weight is obtained for each criterion by first establishing a rank order for the criteria and then converting this into quantitative weights by pair-wise comparison in an $N \times N$ matrix. This method requires relatively simpler calculations and permits inconsistency in priorities between varied types of stakeholders. An example for calculation of weights is provided in Table A.1. To handle the variations in the priority rankings and to arrive at a single weighting of the criteria, the weights obtained from all the stakeholders' surveys were averaged.

Table 3 Differences in stakeholder priorities

Priority level	Criteria	Mean priority	Standard deviation
High	Resource availability and variability	3.83	1.79
	Initial capital cost	4.00	2.73
	Technology performance	4.5	1.78
	Operation and maintenance cost	5.08	2.19
	Operational sustainability	5.5	2.15
	Technology maturity	5.58	2.35
	Environmental benefits	5.33	4.12
	Social benefits	5.67	3.42
	Supply chain availability	6.67	2.42
Low	Policies and Regulations	7.5	3.09

3.8 Identification of RETs

The following RETs have been tried out for off-grid electricity generation in South Asia: micro-hydro, solar PV, biomass gasification and biomethanation. Combinations of a few of the RETs were also tried in some instances. For limiting our study to a finite number of RETs, we chose to include only one hybrid: Small wind turbine—Solar PV hybrid in our study. The RETs identified for our study are mentioned in Fig. 2.

Solar home systems and small wind home systems, though ideal for individual homes and battery charging stations, cannot be considered for mini-grids. Due to their inherent characteristic, they are not capable of generating enough power to cater to the demands of an entire village/hamlet. Hence, these two alternatives are dropped from further analysis.

3.9 Formation of a Decision Matrix and Scoring of RETs

Once the alternatives are determined, the next step is to evaluate the alternatives for the criteria defined. Data is obtained by field visits to many working installations for decentralised electrification in India and one installation in Nepal (Table 4).[2] Various stakeholders such as technology experts, project developers and NGOs who were responsible for the installation and operation are asked to rate the particular technology in which they have experience/expertise. Where criteria could not be quantitatively measured, suitable qualitative scales with nine levels of measurement are constructed.

[2] The visited sites are not necessarily rural electrification projects although they relate to decentralised electrification.

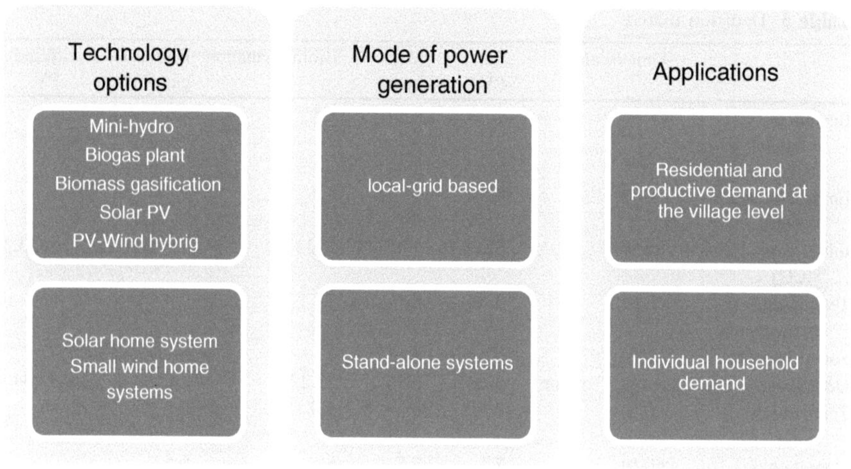

Fig. 2 RET options identified

Table 4 List of decentralised electrification sites visited

SI. No.	Site location	Technology used	Capacity of the plant (kW)
1	Mandya, Karnataka, India	Wind-PV	5
2	Radhapur, Madhya Pradesh, India	Biomass gasification	10
3	Jabalpur, Madhya Pradesh, India	Biomethanation	10
4	Kasdol, Chattisgarh, India	Solar PV	3
5	Sanawadia,Madhya Pradesh, India	Wind-PV	2
6	Bangalore, Karnataka, India	Solar PV	100
7	Kabbigere, Karnataka, India	Biomass gasification	500
8	Satna, Madhya Pradesh, India	Biomethanation	50
9	Bhavanipatna, Orissa, India	Micro-hydro	12
10	MalekhuKhola II, Mahadevsthan VDC, Dhading, Nepal	Micro-hydro	26

At least two installations for each RET identified for our study was visited. The capital costs obtained excluded any subsidies, because subsides vary for different projects. As subsidies distort the decision-making process, the subsidy component has been removed from the analysis. To remove discrepancies in costs due to variation in plant capacity, the capital costs and the monthly operational expenses are calculated per unit size of the plant. Further, since the site conditions are not always uniform, the two sets of scores obtained for each RET are averaged to obtain one score per criterion. The RETs to be evaluated and the criteria on which they are evaluated are arranged in rows and columns to form a decision matrix. The decision matrix thus constructed is given in Table 5.

Table 5 Decision matrix

	Units/scale	Solar PV	Micro-hydro	Biomethanation	Biomass gasification	Wind-PV
Resource availability and variability	(1–9)	7	8.5	8.5	4.5	7
Environmental benefits	(1–9)	8.5	9	8.5	9	9
Initial capital cost (ICC)	\$/kW	5277.78	3787.44	3222.22	1194.44	5833.33
Technology performance	(1–9)	3.5	8	7	7	6
Social benefits	(1–9)	6	9	7	8	7
O&M costs	$\$\ kW^{-1}\ month^{-1}$	30.93	6.56	22.22	18.18	41.11
Technology maturity	(1–9)	7.5	8	3.5	6	8
Operational sustainability	(1–9)	7.5	8.5	8	6.5	7.5
Supply chain availability	(1–9)	6	6	6.5	4	5.5
Policies and regulations	(1–9)	7.5	7	2	3.5	5

Source Field visits (*Note* A conversion of 1 USD = 45 INR)

The decision matrix consists of qualitative and numerical parameters. The numerical parameters such as initial capital cost and O&M costs have been obtained from the field visits, while all other parameters are qualitative in nature.

4 Results and Discussion

4.1 Ranking of the RETs

With the decision matrix and the criteria weights as inputs, PROMETHEE analysis is performed using Visual PROMETHEE software. A 'preference function' is then assigned to each criterion in order to model the way the decision-maker perceives the measurement scale of the criterion. The purpose of the preference function is to translate the deviation in scores to a preference degree between 0 and 1: 0 means no preference at all and 1 means an absolute preference.

Preference functions with multiple levels of preference are assigned to each criterion. Indifference and preference thresholds are defined for a few criteria such as costs, technology maturity and supply chain availability, so as to factor in their ever-evolving nature and also the subjective nature of the stakeholder judgements.

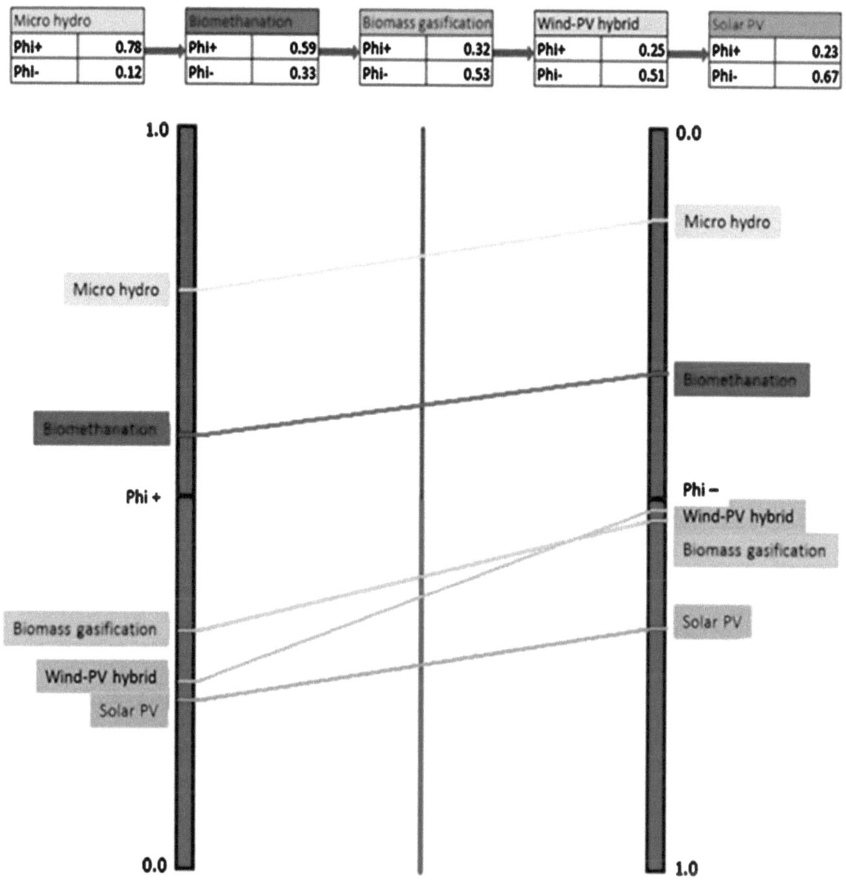

Fig. 3 PROMETHEE I ranking

In the PROMETHEE I Partial Ranking (Fig. 3), the leftmost bar shows the ranking of the RETs according to Phi+: Micro-hydro is on top, followed by Biomethanation, Biomass gasification, Wind-PV hybrid and Solar PV. The rightmost bar shows the ranking according to Phi−: Micro-hydro is still on top, followed by Biomethanation. But it is now followed by Wind-PV hybrid, Biomass gasification and Solar PV. PROMETHEE I provides a partial ranking based on the positive ($\Phi+$) and negative ($\Phi-$) flows of the alternatives. PROMETHEE II (Fig. 4) provides a complete ranking based on the net flow (Φ).

Visual PROMETHEE software allows simultaneous viewing of both rankings using a two-dimensional diamond-shaped representation (Fig. 5). Inside the PROMETHEE Diamond, the RETs are placed based on their relative strengths ($\Phi+$) and weaknesses ($\Phi-$). The two axes are angled such that each RET is a cone. The intersection of the two flows gives the partial ranking and the projection

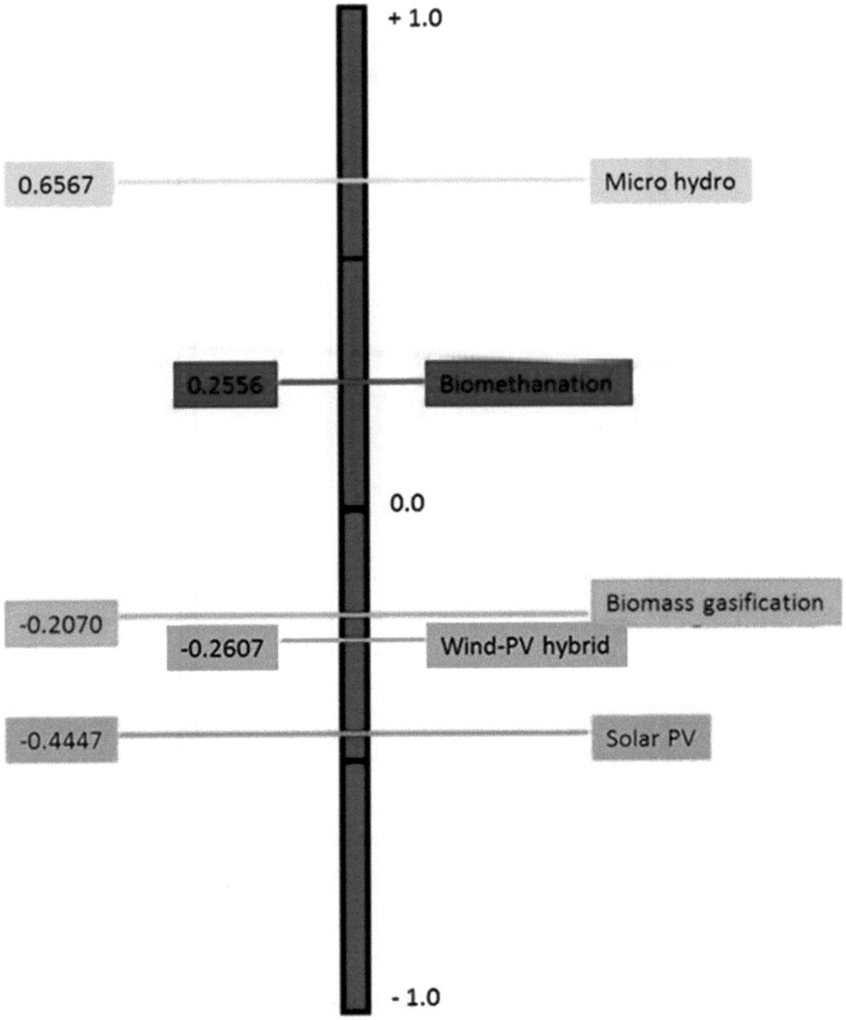

Fig. 4 PROMETHEE II complete ranking

of the meeting on a vertical axis gives the Net flow (Φ) (i.e., complete ranking). Overlapping of cones indicates incomparability.

Micro-hydro clearly emerges as the best-suited technology for India, followed by biomethanation in both the rankings. Biomass gasification is incomparable with Wind-PV in PROMETHEE I because biomass gasification performs better on few criteria while the other performs better on the other criteria. However, PROMETHEE II ranking places Biomass gasification slightly ahead of Wind-PV. Solar PV is ranked the lowest in the rankings. Its low Phi+ scores could possibly be due to its good performance on factors such policy and regulations and technology

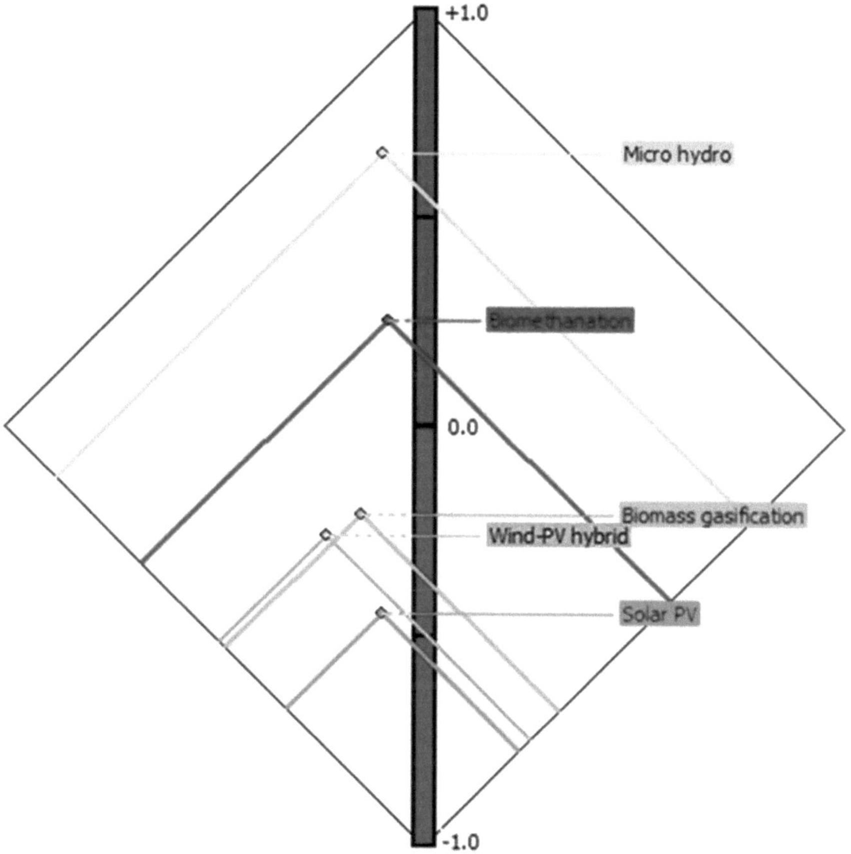

Fig. 5 PROMETHEE diamond-shaped rankings

maturity—factors that had low weightage. The reason for high Phi—scores could be due to its bad performance on factors such as initial capital cost, technology performance and social benefits. As this simulation indicates, RETs selection is not dominated by costs alone but is also greatly affected by technology maturity, technology performance, resource availability and social benefits.

The results suggest that in geographical regions with a sufficient resource, micro-hydro technology should be the most preferred solution for decentralised electrification. The two biomass technologies, despite their lower technical maturity and lack of policy support, are still worthy alternatives that cannot be overlooked, especially in countries such as India with abundant biomass availability.

Further sensitivity analysis is done to verify the stability of the rankings while changing the weightage of these criteria (see Figs. 7 and 8 for more details).

4.2 Visual Analysis of Results

GAIA is the descriptive companion method to PROMETHEE. GAIA starts from a multidimensional representation of the decision problem with as many dimensions as the number of criteria (ten in this study). A mathematical method called Principal Components Analysis is used to reduce the number of dimensions while minimising the loss of information. Using Visual PROMETHEE software, the multidimensional problem can be reduced to three dimensions—*U*, *V* and *W*. The GAIA plane (Fig. 6) is the two-dimensional representation on *U* and *V*.

On this *U*-*V* plane, the RETs are represented by points, and the criteria are represented by axes. The position of each RET is related to its evaluation on the ten criteria such that RETs with similar profiles are closer to each other. For example, Solar PV and Wind-PV hybrid are very close to each other. Each criterion is represented by an axis drawn from the centre of the GAIA plane.

Criteria expressing similar preferences have axes that are close to each other, while conflicting criteria have axes that are pointing in opposite directions. For example technology performance and O&M costs are relatively close to each other. This suggests that it is possible to find solutions (RETs) that are good on both criteria simultaneously. Initial capital cost is pointing in almost opposite direction to supply chain availability. This suggests that it might not be possible to find an RET that has low initial capital cost and simultaneously has a good supply chain.

The relative positions of actions and criteria also reveal interesting information. The projection of an RET in the direction of an axis represents how good the RET performs on that criterion. This information is of course highly dependent on the quality of the GAIA plane represented by delta (δ). A 'δ' value of greater than 70 % indicates that most of the information could be represented on the GAIA plane.

The Decision Axis (the thicker red axis) is the projection of the Decision Stick (i.e. the axis representing the weights of the criteria in the multidimensional space) onto the two-dimensional GAIA Plane. The orientation of the decision axis indicates which criteria are in agreement with the PROMETHEE rankings. The decision axis points to the direction of the best compromise solution (if one exists).

We can observe the axes are oriented in multiple directions, reflecting the varied dimensions to the decision problem. The technologies too are spread across in multiple directions, indicating their relative strengths and weaknesses. The biomass technologies, for instance, are located away from technology maturity thereby indicating their current low maturity levels (these are also located away from supply chain and policy axes, which is not surprising). Other than micro-hydro option, no other RET lies in the direction of the decision axis, thereby indicating their poor performance on one or more criteria. This suggests that in the current decision context, micro-hydro could be the only compromise solution in India.

Fig. 6 Global visual analysis

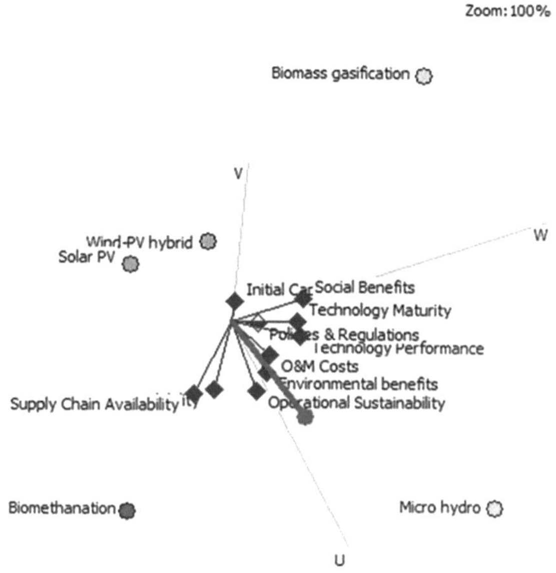

4.3 Sensitivity Analysis

The inconsistency in stakeholder priorities could be attributed to the differences in expectations from each stakeholder group. For instance, while emissions reduction (environmental benefits) is a high priority for international donor agencies, other criteria such as the operational cost of the RET and employment generation for the local community are far more important priorities for private sponsors. The subjective nature of the value judgments, the conflicts observed in priorities for different stakeholder groups and the uncertainty in investment priorities warranted sensitivity analyses to be performed on the rankings. Weights of one or more criteria are increased while proportionately decreasing the weights of the rest. The change in the ranking is viewed in real time. Two types of sensitivity analyses are thus performed: (i) by varying stakeholder priorities and (ii) by varying investment priorities.

In accordance with the varied perspectives of different stakeholder groups, three different scenarios are created for the criteria in which maximum conflict was observed during criteria prioritization and the result is shown in Fig. 7:

Economic scenario

- Caters to the primary concerns of private sponsors.
- Initial capital cost and O&M cost criteria get highest priority.
- ICC and O&M together get 50 % weight.

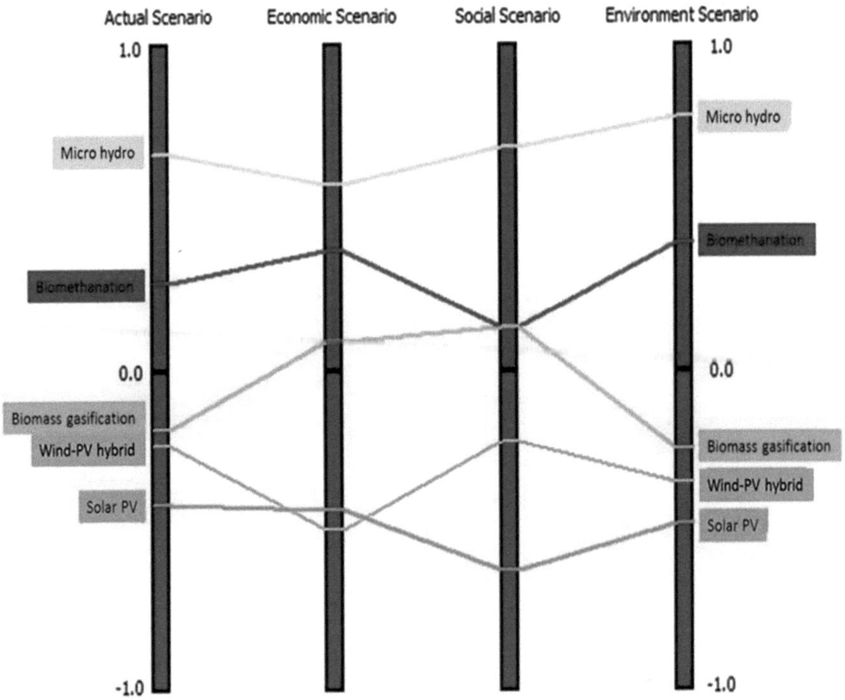

Fig. 7 PROMETHEE II ranking for varying stakeholder priorities

Social scenario

- Caters to the primary concerns of government and local citizens.
- Social benefits get 50 % weight.

Environmental scenario

- Caters to an increased awareness in protecting the environment.
- Environmental benefits get 50 % weight.

Sensitivity analyses by modifying the stakeholder priorities displayed very little change in the overall ranking (PROMETHEE II), which in turn revealed the stability of the initial ranking. Incomparability between biomass gasification and wind-PV hybrid was noticed in the environmental scenario. Also, biomass gasification was ranked close to the biomethanation technology in the social scenario. An interesting observation for the economic scenario was that the biomethanation technology was ranked very close to the dominant alternative. This suggests that it could be the best compromise solution in a 'low cost' scenario and in the absence of a sufficient hydro resource.

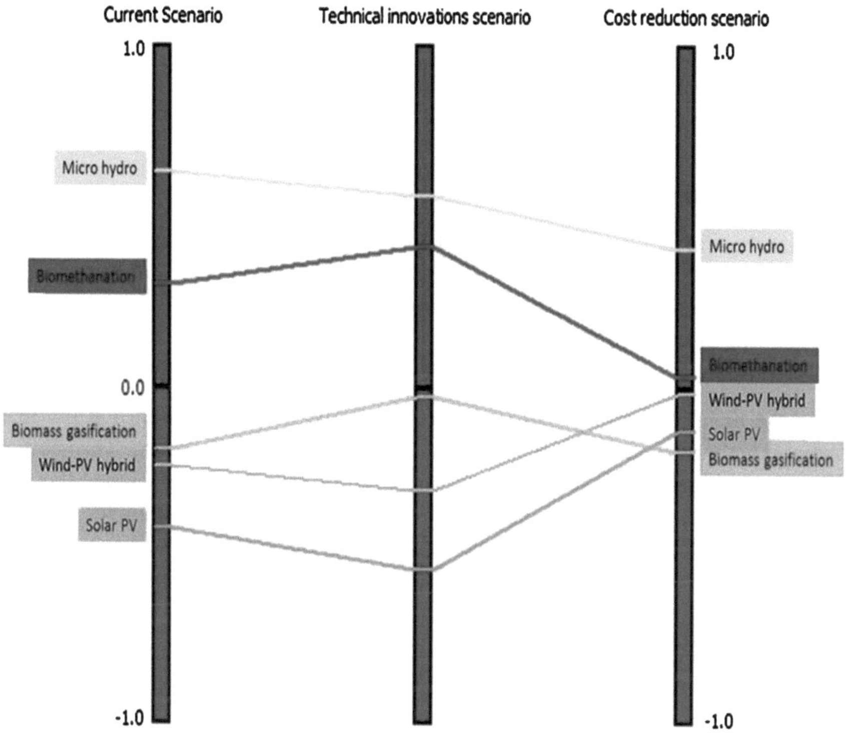

Fig. 8 PROMETHEE II ranking for varying investment priorities

Two exploratory scenarios are created to appreciate the probable areas in which investment would be necessary to promote the non-optimal technologies and to identify the technologies that could potentially develop into the best compromise solutions. The scenarios are developed in view of the scope for reduction in the prices of solar photovoltaic panels and in acknowledgement of the scope for technical improvement of the biomass technologies in developing countries. In either scenario, the performances of the technologies that received an impetus are modified to reflect the effect of the interventions. The result is shown in Fig. 8.

Technical innovations scenario

- Maximum investment in applied research to promote the least matured RETs;
- Extensive focus on technical innovations (For example, a high efficiency, standardised engine for biogas, simple anaerobic systems for specific substrates such as poultry waste, simple and efficient pelletizing machine for producing standardised fuel, among others) and supply chain improvement for biomethanation and biomass gasification (for example, involvement of forest officials in supplying sized biomass, employing women's groups in cutting the fuelwood, pelletization, and so on).

- Technical maturity of both the biomass technologies was increased to equal the dominant RET.
- Their supply chain performance was changed to 'good'.
- The modified PROMETHEE rankings thus obtained are shown in Fig. 8.

Cost reduction scenario

- Extensive focus on cost reduction of the costliest RETs—Solar PV and Wind-PV hybrids.
- Initial capital cost of both reduced two fold.
- Their O&M cost reduced by half.

In the technical innovations scenario, with interventions to improve the least matured technologies, biomethanation ranked alongside micro-hydro. The score of biomass gasification too showed a significant improvement and is ranked the next best. It reveals that technical innovation alone is not a critical parameter for solar PV and Wind-PV hybrid for ranking the RETs. In the cost reduction scenario, the scores of both Solar PV and Wind-PV increased noticeably. However, neither of them scored as well as the dominant RET—micro-hydro. Biomass gasification is the least preferred option in this case.

5 Conclusions

This chapter demonstrates the use of MCDA for selection and comparison of RETs for decentralised electrification in the context of a developing country. It also presents a comprehensive picture of the decision-making problems and assists the stake holders in India and South Asia to take a well-informed decision.

Analysis using PROMETHEE and GAIA revealed that micro-hydro is currently the best compromise renewable energy technology for decentralised electrification in India. The two biomass technologies—biomethanation and biomass gasification are ranked next. Low initial capital costs favour their choice but their low technical maturity level and poor supply chain availability are the major constraints affecting their adoption. Currently, biomethanation appears to be the next best compromise solution in geographical regions where sufficient hydro resource is unavailable. A dedicated investment in applied research and an extensive focus on localised innovation strategy can help boost the maturity of the biomass technologies and promote their accelerated diffusion in developing countries. Wind-PV hybrid and Solar PV are ranked the least. To improve these non-optimal technologies, there is a need for rapid decline in prices and the evolution of suitable financing and business models.

In the interim, new hybrid solutions and smart mini-grids can be adopted to effectively utilise the core strengths of each of the RETs in addition to maintaining diversity in supply options for decentralised electrification.

However, it merits mentioning the possible limitations of the study. The results of the study cannot be taken as sacrosanct, as the results largely correlate with input data obtained from field visits in India; the local needs, and resource availability. These constraints should be considered before interpreting the results to similar contexts in other developing countries. Therefore, it is difficult to generalise the findings of the study. A further generalisation requires information on a wider set of projects.

Appendix

Table A.1 Estimation of weights for criteria—an example

Criteria	1	2	3	4	5	6	7	8	9	10	Total	Weight
1. Initial capital costs	1	1	2	2	3	3	4	4	5	5	30	0.2206
2. Technology maturity		1	2	2	3	3	4	4	5	5	29	0.2132
3. O&M costs			1	1	2	2	3	3	4	4	20	0.1470
4. Technology performance				1	2	2	3	3	4	4	19	0.1397
5. Operational sustainability					1	1	2	2	3	3	12	0.0882
6. Social benefits						1	2	2	3	3	11	0.0808
7. Supply chain availability							1	1	2	2	6	0.0441
8. Resource availability and variability								1	2	2	5	0.0368
9. Policy and regulations									1	2	3	0.0220
10. Environmental benefits										1	1	0.0074
Total											136	

Note The method of calculating weights was adopted from a technical paper by Kohli et al. [29]

After placing the criteria into five categories of importance, they are compared pair-wise and values are given for each comparison as below:

- 1 if a criterion is being compared with itself.
- 1 if two criteria being compared belong to same category.
- 2 if first criterion is placed one category higher than second criterion.
- 3 if first criterion is placed two categories higher.
- 4 if first criterion is placed three categories higher.
- 5 if first criterion is placed four categories higher.

The weight coefficient for a particular criterion is obtained by dividing the individual score of that criterion by the total score for all the criteria.

Table A.2 Survey for evaluating a technology on the identified criteria

Technology evaluated

Mode of interview Location Date

Instructions:
Please assign a score to the technology for each of the ten criteria below
A score reflects how good/bad this technology performs for that particular criterion as compared to all other technology options available
For cost criteria, please enter the actual costs noted at your site
For criteria requiring qualitative scores, please circle the option that matches the most with the field conditions

Selection criteria	Factors being assessed	Score
Initial capital costs	Cost of equipment, civil works, battery storage, wiring and installation	______ (Lakhs per kW)
O&M costs	Costs for fuel, labour, servicing and battery replacement (calculated pro rate)	______ (Rs. per kW per month)
Resource availability and variability	Availability of energy resource in the area/region	No resource availability
		Very low resource availability
	Variability of the resource throughout the year	Low resource, variable
		Low resource, always available
	Availability of land for installation and water for running the plant	Medium resource, variable
		Medium resource, always available
		High resource, variable
		High resource, always available
		Very high resource availability, always available
Technology maturity	Number of technology providers in the country	Technology barely use
		Very low maturity
	Availability of standards	Low maturity
		Low-to-moderate maturity
	Status of R&D	Moderately mature
		Moderate-to-high maturity
	Ease of use	Highly mature but some areas for improvement
		Highly mature with very minor concerns
		Extremely mature technology

(continued)

Table A.2 (continued)

Selection criteria	Factors being assessed	Score
Technology performance	Ability to handle high-powered loads	Technology barely runs
		Very low performance
	Plant load factor	Low performance
		Low-to-moderate performance
		Moderate performance
		Moderate-to-high performance
		High performance, moderate downtime
		High performance, low downtime
		Excellent performance
Supply chain availability	Availability of local manufacturing	No supply chain exists
		Very low supply chain for few components
	Local availability of spares and services	Low supply chain for many components
		Low-to-moderate supply chain
		Moderate supply chain for key components
		Moderate supply chain for many components
		Good supply chain for most components
		Good supply chain for all components
		A reliable mature supply chain
Operational sustainability	Availability of stable quality and quantity of fuel/feedstock throughout the lifetime of the project	Unsustainable
		Very low sustainability
	Availability of human resource for operation of the plant	Low sustainability on all criteria
		Low sustainability on few criteria
	Scalability of the technology	Moderately sustainable
		High sustainability on few criteria
		High sustainability on all criteria
		Very high sustainability on all criteria
		Most sustainable

(continued)

Table A.2 (continued)

Selection criteria	Factors being assessed	Score
Policies and regulations	Policies favouring use of technology for decentralised electrification (includes subsides available, scheme implemented in the past)	Detrimental policies
		Benign policies
	Environmental regulations	Negligibly favourable
		Slightly favourable
	Policies promoting private enterprise	Moderately favourable
		Favourable but many areas for improvement
		Mostly favourable
		Highly favourable
		Extremely favourable policies
Environmental benefits	Reduction of CO_2 emissions (as compared to coal based power)	No benefits to the environment
		Very negligible benefits
	Reduction in environmental degradation at the site	Low benefits on most factors
		Low benefits on some factors
		Moderate benefits on some factors
		Moderate benefits on most factors
		High benefits
		Very high benefits
		Extremely beneficial to environment
Social benefits	Jobs to the local people in construction and operation of the plant	No benefits to the society
		Very negligible benefits
	Increased income due to productive uses such as milling, de-husking	Low benefits on most factors considered
		Low benefits on some factors
	Involvement of women in the operation	Moderate benefits on some factors considered
		Moderate benefits on most factors
	Improvement in health, safety and education	High benefits
		Very high benefits
		Extremely beneficial to society

References

1. Amer M, Daim TU (2011) Selection of renewable energy technologies for a developing county: a case of Pakistan. Energy Sustain Dev 15:420–435
2. Bazilian M, Nussbaumer P, Haites E, Levi M, Howells M, Yumkella KK (2010) Understanding the scale of investment for universal energy access. Geopolitics Energy 32:19–40
3. Beccali M, Cellura M, Mistretta M (2003) Decision-making in energy planning. Application of the Electre method at regional level for the diffusion of renewable energy. Renewable Energy 28:2063–2087
4. Belton V, Stewart TJ (2003) Multiple criteria decision analysis: an integrated approach, 2nd edn. Kluwer Academic Publishers, London. ISBN 0-7923-7505-X
5. Bhattacharyya SC (2007) Review of alternative methodologies for analysing off-grid electricity supply. Renew Sustain Energy Rev 16:677–694
6. Birol F (2011) Energy for all: financing access for the poor. International Energy Agency, Paris
7. Brans JP (1982) Engineering decision: development of instruments to support the decision method PROMETHEE. In: Nadeau R, Landry M (eds) The decision support: nature, instruments and future prospects. Press Laval University, Quebec, pp 183–213
8. Brans JP, Mareschal B (1994) PROMCALC & GAIA: a new decision support system for multicriteria decision aid. Decis Support Syst 12:297–310
9. Brans JP, Mareschal B (2000) How to decide with PROMETHEE. Systems Science, pp 1–5
10. Brans JP, Vincke P (1985) A preference ranking organization method: the PROMETHEE method for MCDA. Manage Sci 31:647–656
11. Brans JP, Mareschal B, Vincke P (1986) How to select and how to rank projects: the PROMETHEE method for MCDA. Eur J Oper Res 24:228–238
12. Catalina T, Virgone J, Blanco E (2011) Multi-source energy systems analysis using a multi-criteria decision aid methodology. Renewable Energy 36:2245–2252
13. Chen F, Lu S, Tseng T, Lee S, Wang E (2010) Assessment of renewable energy reserves in Taiwan. Renew Sustain Energy Rev 14:2511–2528
14. Cherni JA, Dyner I, Henano F, Jaramillo P, Smith R, Font RO (2007) Energy supply for sustainable rural livelihoods. A multi-criteria decision-support system. Energy Policy 35:1493–1504
15. De Montis, A., De Toro, P., Franke, D., Omann, I., & Stagl, S. (2005). Criteria for quality assessment of MCDA methods. Getzner, M., Spash, C., & Stagl, S. (Eds). Alternatives for Environmental Valuation. 1st ed. London: Routledge Publishers, ISBN 0-415-31012-1, p 181-193
16. Doukas H, Marinakis V, Karakosta C, Psarras J (2012) Promoting renewables in the energy sector of Tajikistan. Renewable Energy 39:411–418
17. Figueira J, Greco S, Ehrgott M (2005) Multiple criteria decision analysis: state of the art surveys. Springer Verlag, London. ISBN 0-387-23067-X
18. Franco LA, Montebeller G (2009) Problem structuring for multi-criteria decision interventions. The London School of Economics and Political Science, London, ISSN 2041-4668
19. GNESD (2007) Reaching the millennium development goals and beyond: access to modern forms of energy as a prerequisite. Global Network for Energy for Sustainable Development, Roskilde
20. Guitouni A, Martel J (1998) Tentative guidelines to help choosing an appropriate MCDA method. Eur J Oper Res 109:501–521
21. Hepbasli A (2008) A key review on exergetic analysis and assessment of renewable energy resources for a sustainable future. Renew Sustain Energy Rev 12:593–661
22. IAEA (2005) Energy indicators for sustainable development: guidelines and methodologies. International Atomic Energy Agency, Vienna

23. IEA (2002) Distributed generation in liberalised electricity markets. International Energy Agency, Paris
24. IEA (2010) World Energy Outlook 2010. International Energy Agency, Paris
25. Ilskog E (2008) Indicators for assessment of rural electrification—an approach for the comparison of apples and pears. Energy Policy 36:2665–2673
26. Kanase-Patil AB, Saini RP, Sharma MP (2010) Integrated renewable energy systems for off grid rural electrification of remote area. Renewable Energy 35:1342–1349
27. Karger CR, Hennings W (2009) Sustainability evaluation of decentralized electricity generation. Renew Sustain Energy Rev 13:583–593
28. Kaya T, Kahraman C (2010) Multicriteria renewable energy planning using an integrated fuzzy VIKOR & AHP methodology: the case of Istanbul. Energy 35:2517–2527
29. Kohli S, Raman P, Kishore VVN (1988) Evaluation of various fuels for gasification. In: Energy options for the 90's. Tata McGraw Hill, Solar Energy Society of India. New Delhi, pp 508–514
30. Linkov I, Varghese A, Jamil S, Saeger TP, Kiker G, Bridges T (2004) Multi-criteria decision analysis: a framework for structuring remedial decisions at contaminated sites. Linkov I, Ramadan AB (eds) Comparative risk assessment and environmental decision making. Kluwer Academic Publishers, Dordrecht, pp 15–49, ISBN 1-4020-1895-9,
31. Mahmoud MM, Ibrik IH (2006) Techno-economic feasibility of energy supply of remote villages in Palestine by PV-systems, diesel generators and electric grid. Renew Sustain Energy Rev 10:128–138
32. Mareschal B, Smet YV (2012) New developments of the PROMETHEE & GAIA MCDA methods
33. Martel JM (1999) Multicriterion decision aid: methods and applications. CORS-SCRO annual conference, Windsor, Ontario, 7–9 June 1999
34. Munuswamy S, Nakamura K, Katta A (2011) Comparing the cost of electricity sourced from a fuel cell-based renewable energy system and the national grid to electrify a rural health centre in India: a case study. Renewable Energy 36:2978–2983
35. Nguyen KQ (2007) Alternatives to grid extension for rural electrification: decentralized renewable energy technologies in Vietnam. Energy Policy 35:2579–2589
36. Nouni NR, Mullick SC, Kandpal TC (2009) Providing electricity access to remote areas in India: niche areas for decentralized electricity supply. Renewable Energy 34:430–434
37. Omann I (2012) Multi-criteria decision aid as an approach for sustainable development analysis and implementation. Ph.D. Thesis, University of Graz
38. Polatidis H, Haralambopoulos D (2004) Local renewable energy planning: a participatory multi-criteria approach. Energy Sources 26:1253–1264
39. Polatidis H, Haralambopoulos DA, Munda G, Vreeker R (2006) Selecting an appropriate multi-criteria decision analysis technique for renewable energy planning. Energy Sources 1:181–193
40. Purohit P (2007) Financial evaluation of renewable energy technologies for irrigation water pumping in India. Energy Policy 35:3134–3144
41. Ramanathan R, Ganesh LS (1994) A multi-objective evaluation of decentralized electricity generation options available to urban households. Energy Conserv Manage 35:661–670
42. Ramani KV (2004) Energy for sustainable development: challenges for Asia and the Pacific and lessons from UNDP projects in the region. United Nations Development Programme, New York
43. Rowe G, Frewer LJ (2000) Public participation methods: a framework for evaluation. Sci Technol Human Values 25:3–29
44. Roy B (1996) Multicriteria methodology for decision aiding. Kluwer Academic Publishers, Dordrecht. ISBN 0-7923-4166-X
45. Roy B, Vanderpooten D (1996) The European School of MCDA: emergence, basic features and current works. J Multi-Criteria Decis Anal 5:22–38
46. Thiam D (2010) Renewable decentralized in developing countries: Appraisal from microgrids project in Senegal. Renewable Energy 35:1615–1623

47. Triantaphyllou E (2000) Multi-criteria decision making methods: a comparative study. In: Pardalos PM, Hearn D (eds) Applied optimization. Kluwer Academic Publishers, London, ISBN 978-1-4419-4838-0
48. Varun, Prakash R, Bhat IV (2010) A figure of merit for evaluating sustainability of renewable energy systems. Renew Sustain Energy Rev 14:1640–1643
49. Wang JJ, Jing YY, Zhang CF, Zhao JH (2009) Review on multi-criteria decision analysis aid in sustainable energy decision-making. Renew Sustain Energy Rev 13:2263–2278
50. Yi S, Sin H, Heo E (2011) Selecting sustainable renewable energy source for energy assistance to North Korea. Renew Sustain Energy Rev 15:554–563

Energising Rural India Using Distributed Generation: The Case of Solar Mini-Grids in Chhattisgarh State, India

Debajit Palit, Gopal K. Sarangi and P. R. Krithika

Abstract Conventional grid extension has been the predominant mode of electrification in India. However, solar photovoltaic technology has also been used for providing electricity access in remote, forested habitations and islands. Under the Remote Village Electrification Programme by the Government of India, around 12,000 villages and hamlets have been electrified using renewable energy. The state of Chhattisgarh in Central India has alone been able to electrify around 1,400 remote and forested villages through solar mini-grids. This chapter attempts to examine the development and operation of the solar mini-grid model for enhancing electricity access in India, with special focus on the state of Chhattisgarh. The work, based on extensive literature review, interview with key stakeholders and field visits to selected remote forested villages in the state of Chhattisgarh, shares the experiences and lessons of the solar mini-grid programme for rural electrification in the state by comprehensively analysing multiple dimensions of the programme such as coverage and trend, technical designs, institutional arrangements, financial mechanism and operation and maintenance aspects, which were key to the success of the solar mini-grids. We observe that robust institutional arrangement, strong policy support and an effective maintenance and an oversight mechanism have been the key contributing factors for the success of this initiative.

D. Palit (✉) · P. R. Krithika
Social Transformation Division, The Energy and Resources Institute,
Lodhi Road, New Delhi 110003, India
e-mail: debajitp@teri.res.in

P. R. Krithika
e-mail: krithika@teri.res.in

G. K. Sarangi
Department of Policy Studies, TERI University, Institutional Area,
Vasant Kunj, New Delhi 110070, India
e-mail: gopal.sarangi@teri.res.in

S. C. Bhattacharyya and D. Palit (eds.), *Mini-Grids for Rural Electrification of Developing Countries*, Green Energy and Technology,
DOI: 10.1007/978-3-319-04816-1_12,

1 Introduction

Having the largest rural population in the world, India confronts a huge challenge of rural electrification, especially electrifying the remote, forested and tribal habitations located in far off places. Despite conscious efforts undertaken by the federal and provincial governments since 1951, the level of household electrification and power availability in India remains far below the global average. According to the Census of India 2011, electricity as a source of lighting energy in India (both rural and urban) is limited to 67 % of total households, whereas the world average stands at approximately 81.5 % [7]. In rural areas, the electrification coverage scenario is further bleak where around 59 %, constituting about 72.4 million rural households, still rely on kerosene or other traditional fuel for lighting. Despite various policy and programme specific thrusts (refer to [19] for a detailed discussion of various rural electrification programmes), energy access challenges, especially in the rural areas, continue to derail the overall development process of the country.

While the centralised grid-based electrification has been the most common approach,[1] decentralised renewable energy technologies, especially solar PV (photovoltaic) systems, have also been adopted and being increasingly considered as a cost-effective mode of electrification, especially for areas where it is not techno-economically feasible to extend the electricity grid or in areas where electricity supply from the grid is inadequate to meet the demand [16, 18]. These off-grid communities, often characterised by scattered settlements, consist of small, low-income households. Hence, they are economically unattractive for electricity distribution companies (discoms) to extend the grid. While extending the grid to such areas might be economically unattractive for the discoms, they have not attempted to cover the off-grid areas with decentralised distributed generation systems either, though they are the licensees to provide electricity services in all areas. The vacuum has been largely fulfilled in many cases by NGOs, through pilot projects by raising funding support from donor agencies and support received through funds from corporate social responsibility initiatives, and by the state renewable energy development agencies, established in different states by the state governments, which work under the aegis of the Ministry of New and Renewable Energy (MNRE). Of late, private entrepreneurs have also ventured into the field, foreseeing the business prospect of the sector.

Statistics from the MNRE indicate that more than 12,000 villages in India have been covered through renewable energy-based mini-grids and solar home systems (SHSs), especially under the Remote Village Electrification Programme (RVEP). Solar PV projects (>1 kWp capacity) that include solar mini-grids with a capacity ranging from 1 to 500 kWp have been installed in the country with a cumulative capacity of 96.61 MWp (as of August 31, 2012). Though the off-grid village electrification coverage may be only around 1.5 % of the total village

[1] Currently about 94 % of the inhabited areas in India are covered through grid electrification [24].

electrification coverage in the country [18], its benefit and impact has been immense in bringing about social and economic upliftment of communities in India, through incremental livelihood opportunities and better facility for health and education [4, 8, 21, 30]. At the same time, literature also shares that many of the renewable energy-based projects in India have met with limited success because of technical, institutional, financial and regulatory issues [11, 22, 25, 32].

While there exists ample literature analysing the solar PV programmes for rural electrification at the country level and also the impact of solar-based electrification projects [8, 10, 12, 18, 23, 28, 30, 31], there is limited literature that has comprehensively examined the solar mini-grids delivery model(s) as a means to enhance rural electricity access. The solar mini-grid model, however, has been receiving lot of attention both in South and South East Asia as well as in Africa to achieve the energy access objectives of the 'Sustainable Energy for All' by the year 2030. This chapter makes an attempt to examine the development and operation of the solar mini-grid model(s) for enhancing electricity access in India, with a special focus on Chhattisgarh, a province in the central region of India.

India has been implementing solar mini-grids since the 1990s as part of the technology demonstration programme by MNRE (called Ministry of Non-conventional Energy Sources before 2004). The state of Chhattisgarh, which is the focus of the study, has reportedly implemented the highest number of solar mini-grids in India, providing electricity to more than 57,000 households, spread across 1,439 villages and hamlets in the state [9]. Despite the state having such a large number of operational mini-grids in India and possibly also globally, it has not received adequate attention in the literature, barring a select few [1, 13, 14]. Millinger et al. [14] carried out a socio-technical bottom-up assessment to study two important dimensions of project, i.e. technical and maintenance factor and impact on beneficiary, more specifically on the women groups while [13] briefly provided the status and implementation of the model. Buragohain [1] carried out an impact evaluation study of the remote village electrification programme in six states that also covered the Chhattisgarh state.

This chapter attempts to examine the nuances of solar mini-grids implementation in India and then attempts a comprehensive assessment of the solar-based mini-grids deployed in Chhattisgarh by examining multiple dimensions such as such as technical design, delivery model, financing aspects, policy and regulatory architecture, impacts, etc.

1.1 Study Methodology

This study applies a triangulation of research methods to comprehensively assess the solar mini-grid village projects in Chhattisgarh. Review of secondary literature has been supplemented by visit to selected field sites and interview with key stakeholders to gather relevant information for the study. While secondary information has also been collected from Chhattisgarh State Renewable Energy

Development Agency (CREDA), the nodal agency implementing the mini-grids in the state, primary data were obtained by visiting eight mini-grid project sites across two different districts—Raipur and Korba—in the state. The field visits were limited to villages in two districts because of logistical constraints and security concerns.[2] However, these districts and villages were selected because they are a representative sample of solar mini-grids of the entire state. The secondary data on status of implementation of the mini-grids, technical design and cost parameters for more than 600 mini-grids were collected from CREDA and detailed information on solar power plant-wise number of connections, month-wise and annual energy generation and consumption, operation and maintenance details was obtained from the CREDA regional offices at Raipur and Korba districts. The field visits consisted of focus group discussions (FGDs) and semi-structured interviews with key stakeholders such as mini-grid consumers, village energy committee (VEC) members, local health centre staff, tribal hostels, schoolteachers, mini-grid plant technicians, plant operators and cluster managers. In addition, some quantitative information was also obtained during the FGDs with mini-grid consumers and VECs to assess the impact of the programme. This was also supplemented by interviews with technology suppliers and CREDA officials. Further, some of the private sector players in other states as well as West Bengal Renewable Energy Development Agency (WBREDA) officials were also interviewed to draw a comparison of mini-grid deployment in Chhattisgarh and other parts of India for cross-learning. Based on the interactions and field surveys, a critical analysis of the mini-grid model and its performance was carried out to understand what worked, what did not and why.

The chapter starts with Sect. 2, which attempts to capture the development of mini-grid model(s) and the current status of implementation including the institutional model, financial aspects and policy architecture instrumental in their promotion. Thereafter, Sect. 3 comprehensively captures the dissemination of mini-grids in the state and analyses the programme based on various parameters. Section 4 discusses key aspects of mini-grid development that have contributed to the success of the model in Chhattisgarh. Finally, Sect. 5 summarises the study.

2 Solar PV Mini-Grids in India

The concept of solar PV mini-grids in India was pioneered in the 1990s in the Sunderban delta region in the state of West Bengal, and in the forested region of Chhattisgarh state (then part of Madhya Pradesh state). A solar PV power plant of 25 kWp capacity installed in 1995–1996 by the WBREDA in Kamalpur Village of Sagar Island, continues to provide electricity to its consumers until today.

[2] The districts affected by severe left wing extremism were not considered for field study due to security concerns.

In Chhattisgarh, the first solar power plant was installed at a village called Lamni in Bilaspur district, which is still reportedly operational. Thereafter, solar mini-grids, stand-alone or in hybrid mode, have been implemented in various states, notably, Andaman and Nicobar Islands, Bihar, Chhattisgarh, Lakshadweep, Madhya Pradesh, Meghalaya, Odisha, Uttar Pradesh and West Bengal.

Solar mini-grids in India have evolved with time and changing priorities of societies. The evolution has not only been limited to technical dimensions of project development, but also to other dimensions such as new and innovative delivery and financial models. Putting it in a chronological order, the first phase of deployment of mini-grids that happened during early 1990s till early 2000 focused on developing pilots, technology demonstration and testing of the institutional models. The second phase focused on deployment of these models as an effective vehicle of electrification of remote and far off villages primarily under the government sponsored remote village electrification programme, which was initiated in 2001. This phase also experimented in deploying larger capacity mini-grids as well as hybrid mini-grids. This was also the phase when the Electricity Act [29] was enacted and mini-grids as decentralised distributed generation was included as a means for providing rural electricity supply. The current phase of mini-grid development, since the last few years, is experiencing the entry of private sector developers thereby also bringing in technical and institutional innovations and also the development of smart mini-grids for better supply and demand side management.

2.1 Solar Mini-Grid Coverage in India

The solar mini-grids in India have been predominantly deployed under the RVEP or lately as part of Jawaharlal Nehru National Solar Mission (JNNSM) by the Ministry of New and Renewable Energy. Specifically, WBREDA has set up more than 20 solar mini-grids with an aggregated capacity of around 1 MWp, thereby benefitting around 10,000 households. CREDA, another key proponent of the model, has electrified 57,968 households with electricity supply from low capacity (2–6 kWp) solar mini-grids [9]. Figure 1 presents a comparative position of states in deploying off-grid solar power plants in India.

Apart from government led initiatives, some NGOs such as TERI, WWF India and private companies have also been implementing solar mini-grids. SCATEC Solar, a Norwegian solar manufacturer has electrified 30 villages in Uttar Pradesh, Jharkhand and Madhya Pradesh. Mera Gao Micro-grid Power, a start-up company is setting up solar DC micro-grids in Sitapur and Barabanki districts of Uttar Pradesh to provide lighting services by using energy efficient LEDs. Husk Power Systems, which is more famous for its biomass gasifier-based electricity supply systems, has also reportedly ventured into solar DC micro-grid space and is connecting un-electrified households in their existing operational areas in the state of Bihar. Other private sector companies which have recently initiated extending

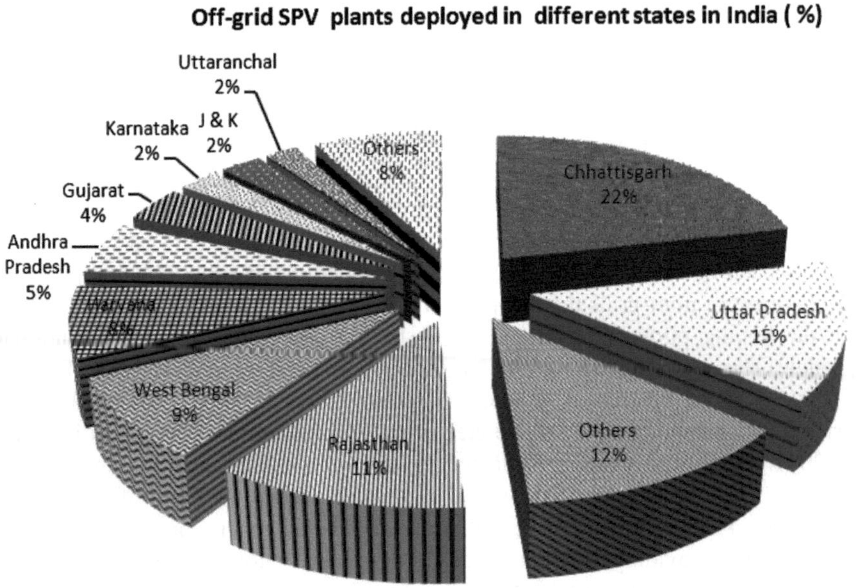

Fig. 1 Off-grid solar power plants deployed in different states in India (*Source* Ministry of Statistics and Program Implementation 2012)

electricity services to poorly electrified villages either through solar AC or DC mini-grids are Sun Edison, Kuvam Energy, Minda NextGen technologies, Gram Power Gram Oorja, among others [26].

2.2 Technical Features

The solar mini-grids are designed to generate electricity centrally and distribute the same for various applications to households and small businesses spread within a particular area. They usually supply 220 V 50 Hz three-phase or single-phase AC electricity (depending on the installed capacity) through distribution network. They consist of (i) Solar PV array for generating electricity (ii) a battery bank for storage of electricity (iii) power conditioning unit consisting of charge controllers, inverters, AC/DC distribution boards and necessary cabling, etc. and (iv) local low-tension power distribution network [23].

These mini-grids vary in capacity and size, typically in the range of 1–200 kWp, with different states adopting different sizing and models depending on their local requirements. While mini-grids in the state of Chhattisgarh are mostly small in capacity (<6 kWp capacity), the ones in Sunderban region and Lakshadweep are of higher capacities (more than 100 kWp solar PV). These solar mini-grids have also been using state-of-the-art inverters and storage systems, available at the time of installation, to ensure long life and reliable field performance. Innovations have also been brought in depending on technological development and communities' changing needs over time. Till 2000, solar mini-grids in the capacity range of 25–26 kWp were mostly implemented by WBREDA. Larger capacity schemes were not commissioned, as the acceptance of the concept and technology was not yet proven. However, observing the strong growth in interest and demand from community, WBREDA also started building power plants with higher capacities (>100 kWp) and in some locations also installed additional generation units using other forms of renewable energy such as small wind-generators and biomass gasifiers to meet the incremental power requirements [23].

The solar DC micro-grids, promoted mainly by private sector companies, generate DC electricity using one or more solar panels and are distributed over a short distance from the battery bank to the cluster of households or shops within the village. They usually supply at 12 or 24 V DC for providing lighting services for 5–7 h using LED lamps of 2–3 W per households. These micro-grids typically cover around 20 to 100 households, providing lighting and mobile charging facilities, generating energy using 100–500 Wp solar panels/array. While the DC micro-grids installed/operated by TERI and Mera Gao Power has a central storage system, Husk Power Systems on the other hand, are reportedly implementing micro-grids using decentralised storage battery in the consumers' households connected smartly from a centralised solar PV systems. Additionally, smart technologies have also been used to conduct remote monitoring and prepaid payment to keep a track of the daily performance of the plant [23].

2.3 Service Delivery Model

Most of the solar mini-grids that have been implemented in India are structured around the community, i.e. they follow the Village Energy Committee (VEC) approach for management of local generation and supply. In such a model, the VEC, formed by members from the community, plays the role of power producer, distributor and supplier of electricity, though they may not have any legal status. They collect payments from users for the electricity and also resolve disputes in the case of a disruption in power supply [18].

In addition to the above community led mini-grid models, the projects implemented by private sector are based on a commercial approach. For instance, Mera Gao Power is implementing DC micro-grids using a micro-utility approach, whereas Husk Power System has evolved a franchise-based business model for deployment of mini-grids. These models by private operators are purely service driven, however, to some extent; they also involve local stakeholders to tackle issues like mobilising people at the local scale, dealing with social problems, and realising better community responses.

2.4 Financing Approach

A major component of capital costs in case of mini-grids implemented under the RVEP came as subsidy from MNRE. For remote areas, the subsidy covers up to 90 % of the project cost and the balance 10 % cost of projects could be financed through sources such as state government funds, contributions from local Member of Parliament or Legislators and corporate sector as part of their social responsibilities. The individual consumers own the internal wiring and appliances and pay for the services they use. However, in case of households below the national poverty line benchmark, internal wirings and service connections are also taken care of from the subsidy funds. JNNSM, which is now the apex scheme for dissemination of solar devices, provides capital subsidy for off-grid solar products and mini-grids either to meet unmet community demand for electricity or in un-electrified rural areas. On the other hand, the DDG program of RGGVY considers technology with the lowest marginal cost for a given area and extends subsidy of 90 % of the project cost and some operational subsidies. The subsidy is released on annuitized basis based on performance of the system for 5 years.

In case of private sector initiatives, a major part of the project cost is borne through financing from venture capital investors and equity contribution by the company. Donor funding and corporate support under their social responsibility initiatives have also been instrumental in supporting mini-grid projects in some instances. Wherever subsidy can be availed, these companies also avail such funds subject to the norms of the government. The entire cost is recovered through retail tariff, resulting in high tariffs, as compared to the negotiated tariffs for projects implemented by state renewable energy development agencies or electricity retail tariff set by regulators in case of grid electrified villages. Moreover, the projects by private investors are usually in 'not so remote' areas where paying capacity of consumers is high and in the absence of any off-grid regulation, the tariff is negotiated between the private service provider and the consumers.

Tariff structures for consumers in different solar mini-grid projects also do not follow a uniform pattern. The tariff is usually based on a flat rate such as INR 30–200 per connection per month. Since, the number of light points and time of supply in the solar mini-grids are fixed and the socio-economic profiles are usually similar in remote villages, the fixed tariff was found to be much easier to administer as compared to metered tariff for such low consumption. However, a disadvantage of the system has been the inability to have control on the over drawl by some households thereby putting extra pressure on the system.

2.5 Policy and Regulatory Landscape

The initial renewable energy-based mini-grids in India, especially based on solar PV, were setup as part of the technology demonstration programme of MNRE during the 1990s. The launch of RVEP by MNRE in 2001 and Rural Electricity

Supply Technology (REST) Mission in 2002 saw a programmatic approach for deploying mini-grids. Under the REST Mission, the renewable energy-based decentralised generation technologies including mini-grids were considered for the first time under the mainstream rural electrification efforts. The Mission, designed to ensure an integrated approach, attempted to change the legal and institutional framework by promoting, financing and facilitating alternative approaches in rural electrification. During the same time, the first focused attempt by the Government of India to look into issues related to decentralised generation, particularly in the context of off-grid electrification also happened through the Gokak Committee.[3]

Thereafter, the Electricity Act 2003 [29],[4] enacted with the overall objective of developing the electricity industry and provide electricity access to all areas, envisaged a two pronged approach for improving rural electricity access—a national policy for rural electrification to extend the reach of grid connected supply together with enlistment of local initiatives in bulk purchase and distribution of electricity in rural areas and a national electricity policy to encourage additional generation and distribution of electricity through renewable sources of energy including mini-grids [29]. It also opened the door to off-grid generation to a much greater extent than what existed before. Under Section 4, it made the Central Government responsible to prepare and notify a national policy, permitting standalone systems[5] (including those based on renewable sources of energy) as a mode for rural electrification. Section 14 also exempts a person intending to 'generate and distribute electricity' in a rural area, notified by the State Government, from obtaining any license from the regulator.[6] However, Section 53 of the same Act also mandates that such persons shall have to conform to the provisions relating to the measures which may be specified by the Central Electricity Authority from

---

[3] The Committee recommended decentralised generation (especially mini-grid mode) may be considered for remote area electrification and the decisions between grid connection and decentralised generation should be made on the basis of technical, managerial and economic issues, viz., distance from existing grid; load density, system losses and load management.

[4] The EA 2003 [29] made the government (both state and central) obligated to supply electricity to rural areas including villages and hamlets. Section 6 of the act mandates the hitherto implied Universal Service Obligation by stating that the government shall endeavour to supply electricity to all areas including villages and hamlets.

[5] Section 2(63) under Electricity Act 2003 [29] defines stand- alone system as the electricity system set up to generate power and distribute electricity in a specified area without connection to the grid;

[6] While there is no requirement to obtain a license to generate and distribute electricity in rural areas, this also implicitly means that off-grid operators do not get the benefit of cross-subsidisation that are normally extended to rural electricity consumers. Absence of regulatory interventions though have helped in setting up of off-grid projects by different proponents, at the same time, most of these projects set up in remote areas with low paying capacity of consumers become operationally unviable after some months of commissioning [20]. Further, in the absence of off-grid regulation, many projects are also set up without following necessary electrical safety standards (such as using bamboo poles for distribution line) to keep their installation cost low. And in some cases, the tariff is negotiated at much higher price than the prevailing rural electricity tariff.

time to time. The National Electricity Policy and Rural Electrification Policy also state that wherever grid-based electrification is not feasible, decentralised generation together with local distribution network would be provided. The policy development made inclusion of decentralised generation also a part of the RGGVY, which was a great step forward in mainstreaming off-grid technologies within the ambit of the national rural electrification strategy.

3 Solar Mini-Grids in Chhattisgarh

3.1 Demographic and Socio-Economic Profile of the State

Before elaborating the nuances of rural electrification process in Chhattisgarh, it merits highlighting some of the key demographic and socio-economic characteristics of the state in order to better comprehend the context within which remote rural electrification has been carried out in Chhattisgarh. About 44 % of the total area in the state is covered with forests, more than double the national average, and this has important bearings on the rural electrification access in the state. Many of the remote villages are located within the forested areas, out of reach of the centralised grid, therefore, requires alternative energy supply systems such as mini-grids or solar home systems (SHSs) to electrify these far flung areas. Social stratification statistics reveal that out of 25.55 million total population of the state [2], scheduled caste (SC) and scheduled tribe (ST) population constitute around 43.4 % compared to the national average of 25 %. The latest poverty estimates in India indicate that about 48 % of total population in Chhattisgarh lives below the national poverty line. The Multi-dimensional poverty index of Alkire et al. ranks Chhattisgarh as the fourth poorest state in India with 70 % of the population being multi-dimensionally poor [14]. Moreover, poverty levels among SCs and STs are also higher than other social groups with more than two-third of SCs and STs in the state are below poverty line (Govt. of India, 2012). In addition, presence of social problems such as left wing extremism in as many as 16 districts in the state (out of 27 districts in the state) also poses a severe challenge to extend electrification efforts in the state [15].

3.2 Status of Electrification in Chhattisgarh

Creation of the state of Chhattisgarh in the year 2000 led to a major institutional and organisational overhaul especially for electricity provisioning. Two new institutions were created, i.e. Chhattisgarh State Electricity Board (CSEB)[7]

[7] CSEB which was functioning as a vertically integrated power utility was unbundled into five companies in with effect from January 1, 2009.

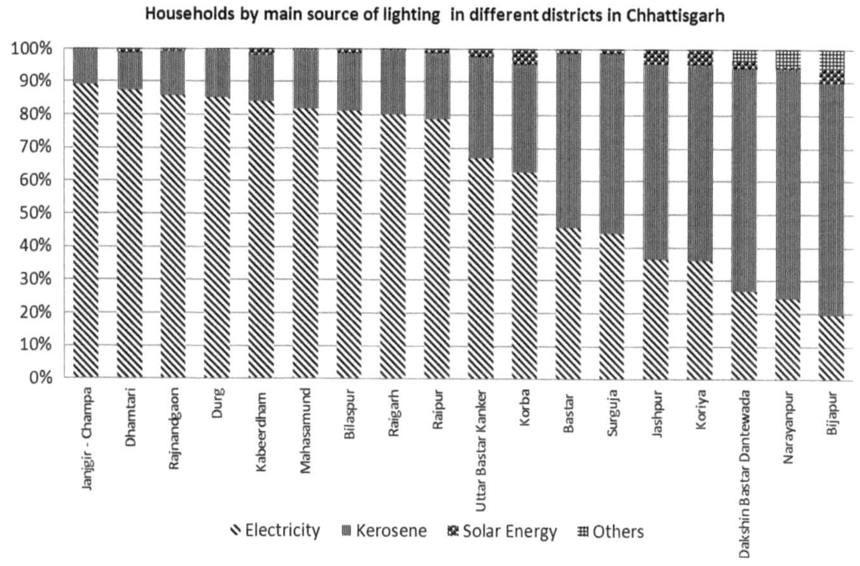

Fig. 2 Households by main source of lighting in Chhattisgarh (*Source* Census of India [2])

and Chhattisgarh State Renewable Energy Development Agency (CREDA). While CSEB was entrusted to provide electricity through grid electrification, CREDA was made responsible for promoting renewable energy sources including off-grid/ decentralised energy systems. Though this state is one of the few states in India having electrification coverage better than the national average and also has surplus power supply, a large share of rural people continues to be outside the centralised grid electricity supply as they inhabit in difficult terrains within the forested areas.

In order to understand the macro picture of the electrification status in the State, it merits dwelling on some recent electrification statistics. The overall percentage of households connected to the grid electricity in the state is around 75.3 % whereas the village level electrification stands at 97.1 % [3]. Around 563 villages (out of the total 19,744 inhabited villages) in the state remain to be electrified, with most of these un-electrified villages being in the tribal dominated districts of Sukma, Dantewada, Narayanpur and Bijapur.[8] Some interesting insights about the use of various source of energy for lighting across districts[9] in Chhattisgarh could be drawn from the Indian Census 2011 figures. In case of lighting, huge disparities exist among districts and social classes in the use of modern sources of energy for lighting (Figs. 2 and 3). The districts with poor electrification status are also the same districts that have predominantly poor population and are also affected by

[8] The electrification of all villages in the state is planned to be completed by end of 2014.

[9] At the time of 2011 Census survey in Chhattisgarh, there were only 18 districts, which increased to 27 in 2012.

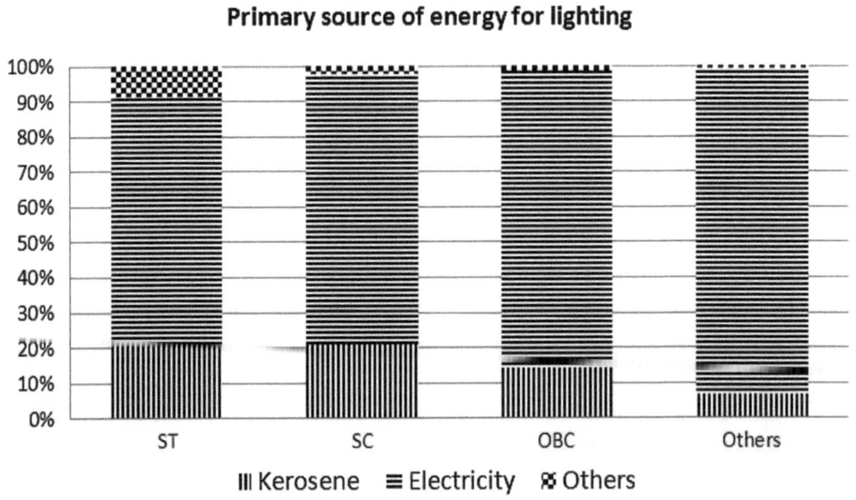

Fig. 3 Primary source of energy for lighting among social classes in Chhattisgarh (*Source* NSSO, 66 Round Survey, 2009–2010)

left wing extremism [15]. However, there is also a positive trend that some of these districts such as Koriya, Jashpur, Korba, Bijapur are utilising solar energy to electrify their households. Almost 1 % of the total households in the state are using solar energy as a source for lighting, which is second highest (after West Bengal at 1.2 %) among the larger states in India and more than double of the national average (~0.4 %).

While Chhattisgarh is abundantly blessed with sunshine, it was not the sole reason for the promotion of solar mini-grids in the state. CREDA had also implemented biomass gasification technology and bio-fuel generator sets under RVEP and VESP. However, these projects were reportedly not successful due to technological mismanagement and fuel collection issues. The lack of operation and maintenance of biomass-based plants and unavailability of qualified man-power were found to be the key factors responsible for the limited success of biomass-based electrification. In addition, the availability of only standard capacities of biomass projects irrespective of load demand has also been a constraint for their use. All such projects were eventually converted into solar energy-based projects by CREDA. The solar home systems deployed in the state have their own set of problems. While CREDA implemented solar home systems as early as 2003, the effort did not result in visible success primarily because of two major reasons. First, the beneficiaries did not realise the value of these systems, as it was heavily subsidised thereby resulting in mortgaging of the system at a low price by the owner of the system whenever they faced any financial crunch. Second, social problems like large-scale prevalence of theft negated the very purpose of the deployment [13]. Realising this, CREDA started exploring the option of installing solar mini-grids and the first mini-grid, after the creation of the

Table 1 Profile of the villages surveyed in Chhattisgarh

District	Village	Year of commissioning	Capacity of the plant (kWp)	Total number of households
Raipur	Kouhabehra	2003	3	50
Raipur	Rawan	2004	7	72
Raipur	Mohda	2004	4	72
Raipur	Latadadar	2004	3	45
Raipur	Murumdeeh	2003	5	46
Korba	Surkha	2007	4	50
Korba	Sapalwa	2005	5	83
Korba	Raha	2007	6	75

Source Authors' Field Survey 2012

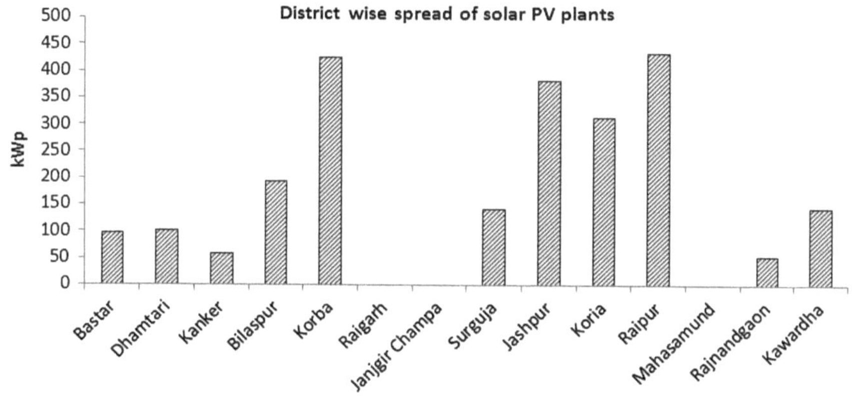

Fig. 4 District-wise spread of solar PV plants (*Source* CREDA 2012)

state, was commissioned in 2004. However, solar home systems as an option have not been discarded altogether. Rather, solar-based electrification in remote areas is carried out in two different ways. Larger villages with concentrated settlements are electrified through solar mini-grids whereas hamlets and villages with scattered households are provided with stand-alone solar home systems.

The following section provides a detailed analysis of the CREDA delivery model including the details of projects deployed, technical contours of the model, institutional arrangement, financial mechanism, functionality and performance, operation and management aspects, system of monitoring and oversight and the observed impacts of off-grid electrification carried out using solar PV technology in the state. As mentioned in Sect. 1.1, TERI team carried out field surveys in eight villages of Raipur and Korba districts to understand the multiple dimensions of project impact on communities. Table 1 shows the profile of the villages surveyed. The focus has been on these two districts as they have the highest capacity of solar mini-grids (Fig. 4). These districts also represent different physiographical and socio-economic profiles. While in Raipur, both urban and rural population are almost equal as the state capital is in the Raipur district; the major population in Korba lives in rural areas. Korba is heavily forested and is designated as tribal district whereas Raipur represents plain region in the state [6].

3.3 Coverage and Trend of Solar Mini-Grids

Chhattisgarh has a cumulative capacity of around 3.5 MWp of solar power plants covering 1,439 villages (as on June, 2012) spread across different districts in the state [9]. Figure 4 provides the district-wise installed capacity and Fig. 5 provides the year-wise installation and dismantlement by CREDA. Most of the power plants have been installed in Raipur, Korba, Jashpur and Koria districts, which account for around two-thirds of the total installed capacity. These districts are under dense forest cover, especially districts such as Korba, Koria and parts of Raipur and thus traditionally have higher number of un-electrified villages.[10]

Of the total installed capacity, around 45 kWp capacity was also found to be dismantled by CREDA mainly during the last 5 years. While in few cases, the power plants have to be dismantled due to panel theft and poor management by villagers, in most other cases, dismantling was deliberate owing the grid extension by the CSEB. While CSEB and CREDA are well coordinated while taking up of a village for solar-based electrification, with CREDA taking up the off-grid route only after CSEB's concurrence, often political compulsions lead to grid extension to village where solar mini-grid also exists. However, the dismantled systems have also been utilised mainly in two ways—(1) those villages which are facing an increasing demand for solar power and need capacity augmentation were provided with new capacities using the dismantled systems (2) the systems are retrofitted in those power plants which have some mal-functioning components such as faulty battery, panels, etc. Further, the annual installed capacity figure also portrays a reverse U shape, indicating that annual incremental increase in the installed capacity up to the year 2007 and declining thereafter (Fig. 5). This gradual decline in annual installation capacities is primarily due to increasing annual electrification rates thereby reducing the leftover villages for electrification.

3.4 Technical Contours of the Model

As mentioned in Sect. 2.2, the solar mini-grid essentially consists of a centralised solar power plant and a power distribution network that distributes the electricity to the households. It was found that in Chhattisgarh, capacity of the installed

[10] Grid extension through forests is not permitted in India as they may require the uprooting of forest trees for which necessary permission is required to be taken from the central government [17].

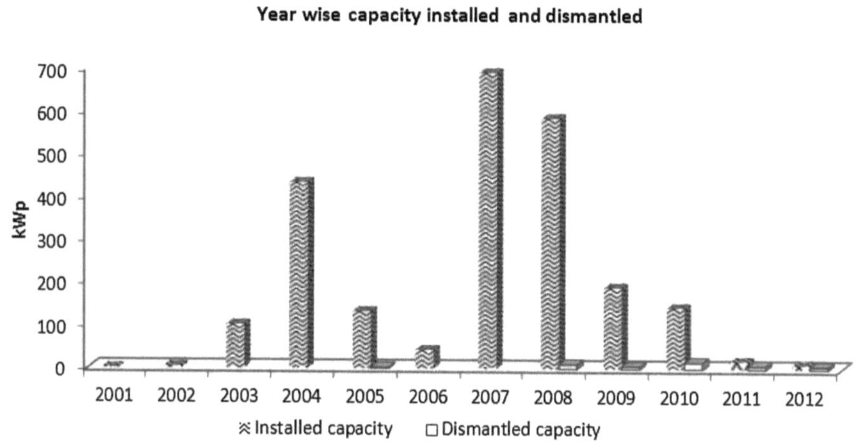

Fig. 5 Year wise solar PV plants installation and dismantlement in Chhattisgarh (*Source* Authors compilation 2013)

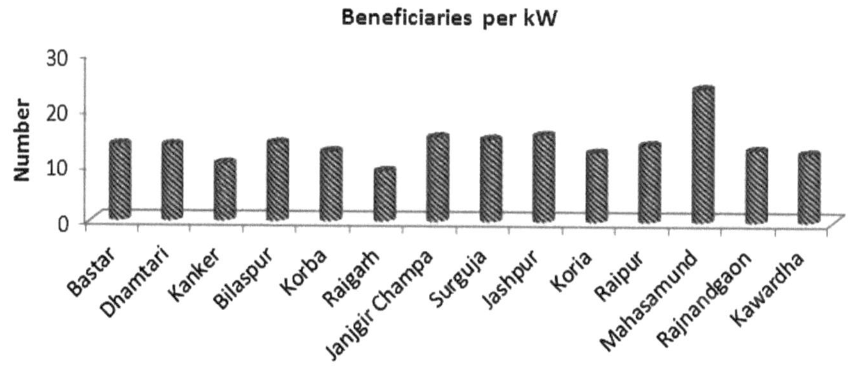

Fig. 6 District wise average number of beneficiaries connected per kWp (*Source* Authors compilation 2013)

systems ranges between 2 and 6 kWp and supply electricity in single phase. The lower capacity of installed systems is because of the smaller habitation size in terms of population (mostly 50–75 households per village or hamlets) and also because the projects were implemented mainly to provide lighting access either for residential needs or community load such as street lighting, lighting for schools, health centres and community hall, and do not include any productive load. Typically, 4 kWp plant capacity was found to be the most preferred across the different districts. It was also observed that an average of 12–14 beneficiaries are connected to solar power plant per kWp capacity across the different districts in Chhattisgarh, with average load (in terms of per household) being 73 W (Fig. 6).

Every village is provided with household as well as streetlight connections. A household connection consists of two 11 W CFLs and the supply is given for 6 h per day in most of the villages—2 h in the morning from 4 to 6 am and 4 h in the evening from 6 to 10 pm. The streetlights are mandatorily provided in all the villages mainly for ensuring security at night and operate for the same time as the domestic connections. Since the solar mini-grids are designed to cater only to the lighting requirements, duration of supply is thus designed for the optimal use of the system. However, the systems have been designed such that they generate 10–20 % incremental power to meet possible future demand. At present, the power plant is not automated through load limiter and is switched on and off by the operator who stays in the village. In fact, the primarily responsibility assigned to the operator is to do this job. There is, however, a battery protection panel, which is a switching device meant for isolating the battery bank.

A key feature of the CREDA model is the technical standardisation that contributed to the ease of operation and maintenance and minimization of cost. While the systems with installed capacity of 1–3 kWp have a battery bank of 48 V and inverter rating of 3 kVA, the systems with 4 to 6 kWp installed capacity have 96 V battery bank and inverter rating of 5 kVA, respectively. While in the beginning of the programme, solar panels were mounted on the ground and fencing provided around the same, for all later installations the civil structure has also been standardised with panel mounted in the rooftop and inverter and battery kept in the room. The height of the room is designed at a little more than usual, which reportedly helped in prevention of theft of the panels. Also, roof mounting reduced the cost of maintenance which otherwise is incurred for ground mounted panel on expenses such as fencing, theft, damage during monsoon, etc.

The battery bank in most power plants was found to have last for around 8 years, much higher than the usual battery life of 4–5 years prevalent in most others states such as in the Sunderban region where solar mini-grids have been implemented extensively [30]. The longer battery life has been ensured by CREDA by keeping a tab on the quality of battery purchased and by controlling the damage through misuse or overloading of battery capacity through routine monitoring (mandatory installation of ampere-hour meter to check battery health) and maintenance, using the three-tier model of operation and maintenance (refer to Sect. 3.8).

3.5 Institutional Arrangements

In Chhattisgarh, off-grid electrification through solar mini-grids is mainly a state-led programme by CREDA in partnership with the private sector. While this is essentially a top-down approach, there are several actors involved in the process including state, private as well as local actors imbibing the essence of a pro-poor public–private partnership model. The provincial and federal governments provide policy support and financial assistance to set up and operate the power plants, while the private sector is responsible for installation, operation and maintenance services.

As a first step in mini-grid installation, CREDA, which can be regarded as the off-grid electricity utility in the state, carries out an extensive survey in the village and builds consensus among villagers about setting up of the plant. One of the important initial steps for selection of a village is to get consent from CSEB regarding the future target of CSEB for grid electrification of that village. Once CSEB gives a green signal that the village is not going to be electrified through centralised grid system, then CREDA takes up the village to electrify through the off-grid mode. The next step is to carry out surveys, which are aimed at estimating the demand, decide on the potential capacity of the plant, keeping in mind the short-term to medium-term energy demand in the villages. Consensus is built by convincing village people about the utility of solar lights and the benefits it brings to rural people. Limitations of the proposed plant such as limited application of electrical appliances, and limited hours of supply are also informed to the potential beneficiaries of the plant. As a symbolic gesture and to build in sense of ownership, the villagers are also asked to hand over a piece of community land identified for setting up the plant. A village energy committee is also formed to facilitate the project implementation and management. Often, these activities are carried out through direct supervision of CREDA. Survey is followed by cost estimation for installation of the plant and power distribution network (PDN). After the completion of the estimation exercise, it is then sent to the state government and MNRE for necessary approval and release of fund for actual installation.

CREDA then floats tenders inviting technical and financial proposals from solar PV companies and private service providers known as 'system integrators', for setting up the power plant and also for maintaining it. A critical requirement for participation in the tender is that the solar companies or the system integrators are having their offices in the state. This is done to ensure that fly-by-night operators and non-serious bidders cannot take part in the bidding process. One or several companies are then chosen to implement the projects, under the supervision of CREDA. In addition, CREDA also enter into an annual maintenance contract (AMC) with the selected contractor for operation and maintenance of the plant. In some cases, both AMC contractor and system integrator are the same entity; however, in most cases they are different (Fig. 7).

At the last rung of the institutional hierarchy, lies the village energy committee (VEC), which acts as an interface between consumers, CREDA and private service providers and also acts as a local monitoring body. A local youth, selected by the VEC, is provided the responsibility by CREDA to operate the power plants and to take care of minor maintenance (details on roles and responsibilities are highlighted in Sect. 3.8). The entire process of implementing this model, selection of service provider and monitoring and supervision of service providers post implementation, communicating with VEC for feedback, lies with CREDA, which spearheads off-grid rural electrification process in the state.

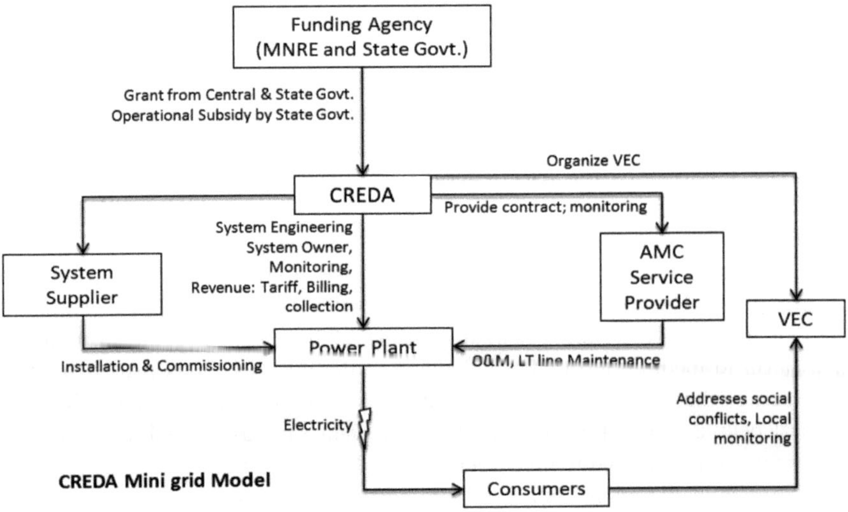

Fig. 7 The CREDA mini-grid model (*Source* Authors' compilation)

3.6 Financing Mechanism

Since a majority of the mini-grids have been deployed under the RVEP, installation costs were taken care of through the capital subsidies by Government of India supplemented by support from the State Government. On an average, the cost incurred to set up a solar power plant in the remote areas in the state has been INR 350,000 to INR 400,000 per kWp (1 US dollar is equivalent to 60 INR at present). In addition, civil construction costs were INR 150,000 to INR 400,000 depending on the capacity of the power plants and type of civil construction. Similarly, INR 150,000 is the average cost of power distribution network per kWp of solar power plant including household connectivity. Typically, the cost of connecting a household through solar mini-grids is INR 25,000/- [13]. Initially, connection charges were not imposed and were provided free of cost. However, lately CREDA has started charging for electricity connection. CREDA records show that the cost for implementing the power plant, civil construction, PDN and electricity connection to households was shared between the MNRE and state government, with MNRE's share varying between 38 and 48 % during various years (2002–2003, 2003–2004 and 2004–2005) [5]. The variation in MNRE's share is due to the fact that MNRE's share is dependent on their benchmark cost while the actual cost may be different for different power plants based on remoteness of the site and distance covered by the distribution line in the village.

Tariff structures have been devised keeping the socio-economic profiles of the beneficiaries under considerations. The tariff is INR 30/- per household per month per connection. Out of which INR 25/- is from the state government as operational subsidy and the balance is collected from the beneficiaries. After meeting the

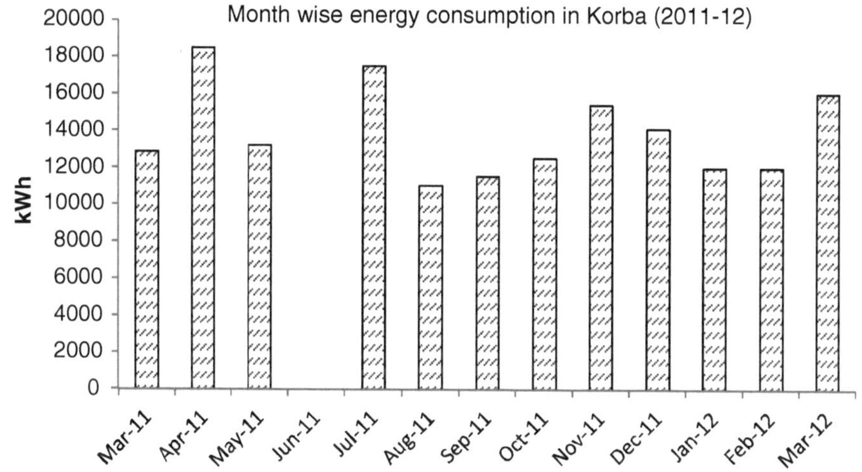

Fig. 8 Monthly energy consumption in Korba in 2011–2012 (*Source* Authors' compilation, 2013)

expenses towards operators and cluster technicians, the balance amount is also utilised to buy spares and other incidental maintenance activities. It must be mentioned here that contribution per beneficiary has been decided in line with the existing subsidy schemes under grid electricity supply for the BPL households. Under this scheme, known as '*ekalbatti yojana*', a tariff subsidy is provided for single point connection for below poverty line (BPL) households.

3.7 Aspects of Functionality and Performance

During survey in the eight villages, it was observed that almost all the plants were functioning and more than 90 % of the home light connections (HLC) were in working order. However, for street light connections (SLC) the functional status was found lower at around 70 %. The faults were mainly related to CFL blackening or loose connections. The power plants mainly serve the domestic load of the village and cater to the lighting needs of the community (Figs. 8, 9, 10 and 11). Thus, in terms of energy consumption pattern, the consumption per month for Raipur was found at 3.34 kWh/household while that for Korba it was found at 2.57 kWh/household. Based on these figures for energy consumption, it can be said that the power plants are operational for around 4–5 h as against the designed 6 h of operation. Further, the statistics for March 2012 reveals that in Raipur, a cumulative capacity of 429 kWp was serving about 10,170 home light connections and 1,120 street light connections. While in Korba, a similar cumulative capacity (426 kWp) was serving 5,350 home light connections and 874 streetlights.

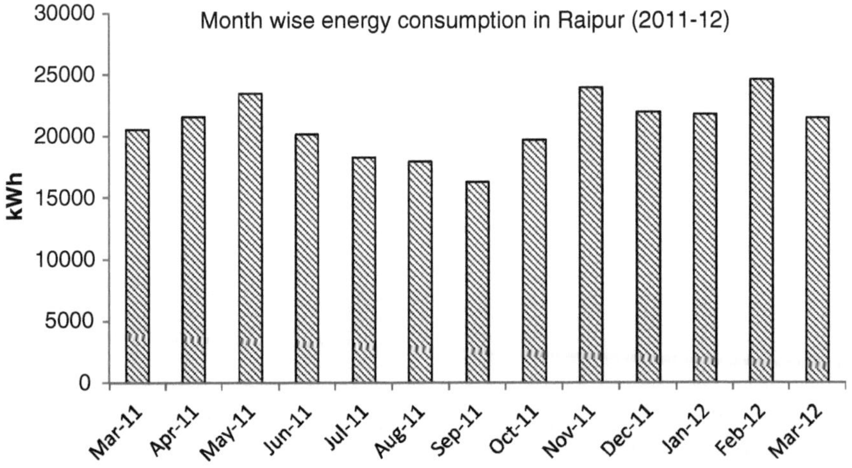

Fig. 9 Monthly energy consumption in Raipur in 2011–2012 (*Source* Authors' compilation, 2013)

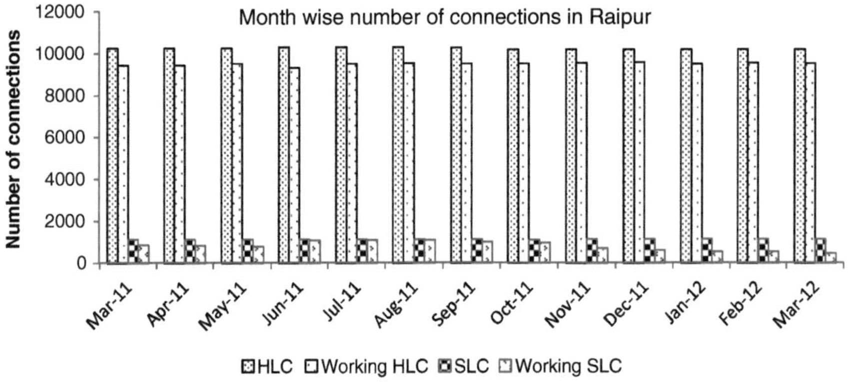

Fig. 10 Month-wise no. of connections served by solar mini-grids in Raipur (*Source* Authors' compilation, 2013)

3.8 Operation and Maintenance Aspects

In most of the states in India, where mini-grids have been implemented either by state renewable energy development agencies or NGOs, the system operator under the supervision of the VEC is always made responsible for O&M of the mini-grids. However, in Chhattisgarh, CREDA has developed its own innovative model by standardising all aspects of O&M. The model is called 'Cluster based service delivery model' or as 'Group the partners, Organize their skills, Allocate load in

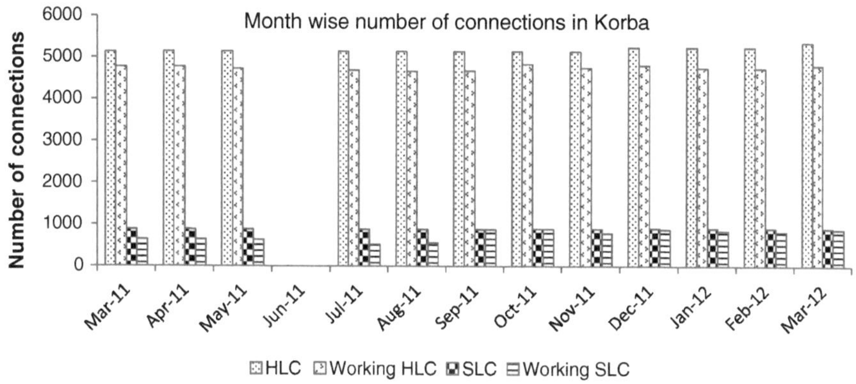

Fig. 11 Month-wise no. of connections served by solar mini-grids in Korba (*Source* Authors' compilation, 2013)

villages, Deliver service' (GOLD), where the installation is steered by CREDA and operation and maintenance of the plants is undertaken through a three-tier setup. CREDA signs an annual maintenance contract with the system integrators/AMC contractors for operation and maintenance of the solar power plants. O&M at the village/individual plant level is carried out by the operator, selected from the village and paid by the system integrators. The operator is responsible for cleaning the solar panels, switching the plant ON–OFF, daily reading of the energy metre and recording and checking the battery top-up. At the next level, a "cluster technician"—who handles cluster of villages—performs more advanced maintenance deals. The cluster technicians, engaged by the system integrators, are assigned the responsibility to supervise the operators, make weekly visits to the villages for preventive maintenance, check wiring and power distribution network, attend to any break down calls/repairs, and duly fill up and send the monthly monitoring report as per the format provided by CREDA. Every technician is provided with a motorbike by the service provider, so facilitating weekly visits to villages. The supervision of the activities at multiple clusters or a block level is done by the supervisor who monitors the activities of technicians, makes monthly visits to villages and collect reports for submission to CREDA. The appointment of the operator, technicians and supervisors is the responsibility of outsourced companies and the salaries and incentives of staff are paid by the AMC contractor or outsourced company. However, regular CREDA staff is engaged in parallel with AMC staff to ensure that assigned tasks like routine maintenance are being properly carried out.

A third-tier is managed by CREDA, which monitors all of its installations through monthly reports and replaces damaged equipment. This record based on continuous power output, has not only minimised cases of stealing and selling solar modules, it has also fuelled commercial demand for solar systems in the region. CREDA's maintenance framework also ensures that the mini-grids provide an uninterrupted power supply and an adequate supply of replacement lamps is kept in stock with each technician in case a light burns out.

Continuing with the innovative mechanism, CREDA has a structured mechanism in place to address the fault management. Especially during the monsoon season, when heavy rain can make roads inaccessible, spare inverters for each cluster is supplied by the authorities. Thus, if failures occur, inverters can easily be replaced thus reducing the external dependency on the manufacturer's technician to come and repair it. Since only two rating of inverters are used with the power plant, the replacement is easily done as spare inverters of such rating only need to be kept at the regional office of CREDA. In addition, CREDA have also ensured that extra inverter circuit boards of the two selected ratings are provided by system integrators during the setting up of any power plant, so that faulty circuit boards can easily be replaced, from the spares supplied, by the cluster technicians. Furthermore, the drivers of the vehicles, who transport technicians and supplies, are also provided basic training on replacement of lamps and topping up of batteries in cases of emergency.

Technicians and operators are trained by CREDA and are also provided periodic refresher training courses every 6 months. All the expenses for the trainings are borne by CREDA. CREDA has provided training to more than 1,400 operators for carrying out responsive maintenance of the mini-grid systems [13]. Further, 75 technicians and some 60 supervisors repair inverters and other electronics. CREDA has also trained more than 500 people to install and maintain various solar power systems that have been implemented in the state. The training session for the technicians is, in general, around 1 week with 6 h of training per day teaching them about electricity, electronic components, batteries and solar power and maintenance [14].

Every cluster has around 15–25 villages with around 50–75 households in each village. Given the operational subsidy provided by the state government, the revenue from each cluster is to the tune of INR 22,500 per month, while the expenses inclusive of operator, technician and other staff salaries are around INR 16,500 per month ensuring the operational viability of the model (Table 2). During the field visits, the model was found to be effectively working with regular payments ensured to the staff. In addition, some additional fund is also provided by the state government to meet the expenses like replacement of battery and inverter and for conducting the trainings. This fund is generated by keeping 10 % of the project cost at the beginning in bank fixed deposits.

3.9 System of Monitoring and Oversight

Strong monitoring is ensured through the presence of well-structured hierarchical monitoring system developed by CREDA, depending on the strength of each stakeholder. Starting from the village level up to the top of the hierarchy of monitoring at CREDA head quarter, the presence of a strong and routine

Table 2 Income and expenses for cluster based service delivery of solar mini-grids (in INR)

Revenue/month/cluster			
No. of villages/cluster			15
No. of customers/village			50
Total No. of customers/cluster			750
Collection per customer			30
Through CREDA from government			25
Direct from beneficiary			5
Total collection/cluster/month			22500
Expenses/month/cluster	No.	Rate	Total
Technician	1	4000	4000
Helper	1	2000	2000
Conveyance	1	2000	2000
Others	1	1000	1000
Operators	15	500	7500
Expenses/month/cluster			16500

Source CREDA 2012

monitoring system is playing a key role in the better functionality of the plants. Importantly, since operation and maintenance activity is outsourced to private companies, CREDA officials are employed at each level of hierarchy for oversight on the activities of the private companies.

The overall monitoring and oversight is the responsibility of CREDA, however, at the village level VEC is also involved for local monitoring of systems. The VEC keeps an eye on the performance of the operator and receives first level of complaint from users. Unlike in mini-grid projects implemented under VESP or RVEP in other states, here the VEC is not involved in technical operation of the power plant. Instead, they act as a lubricant for the social engineering required for smooth operation and management of the plant. VECs are also entrusted to resolve social conflicts and control theft and other management aspects of plants. Though their role has been kept limited, their importance in the whole institutional hierarchy cannot be ignored.

3.10 Project Impacts

While the focus of the study was more to understand the institutional and financial aspects and nuts and bolts of operation and maintenance, the study also attempted to qualitatively assess some of the direct impacts of the mini-grids. Based on the results from the focus group discussion, it can be said that the solar mini-grid systems in the state have been able to contribute to the socio-economic development of the community in many ways. During the visit it was observed that some households in the villages were also using mobile phones, fans and entertainment facilities such as TVs and radios. This has led to increase in flow of information resulting in higher level of awareness among beneficiaries. In specific terms, it was found that educational benefits are accruing through provision of lighting during the evening hours. Environmental benefits are being accrued through reduction in use of kerosene use. Information obtained through focus group discussion with community reveals that before these interventions were made, a household on an average used to spend about INR 100 per month on kerosene. However, with the provision of solar lighting systems, kerosene use has been drastically reduced and average monthly expenditure on kerosene has come down to less than INR 50. The study by Millinger et al. [14] also observes that the average saving on kerosene for households with mini-grid connection in Chhattisgarh is INR 30 per month. It should be mentioned here that since the lighting facility is for limited hours in the evening, some households still rely on kerosene, to some extent. In addition, electrification process has also led to income generation and employment creation. For instance, in the surveyed villages of Surkha, Saplowa and Raha, villagers reported working till late in the night for dona pata making (sal leaf cup-plate), thereby generating some incremental income for their families. In addition, at the system level, installation of mini-grids and other solar devices has been able to create a vast pool of skilled and semi-skilled human resources in the state, thereby meeting the employment needs and contributing to the socio-economic development of the state. CREDA also organises regular training courses for developing solar technicians in the state. Such trained technicians are then engaged by the service providers maintaining the solar mini-grids and other solar devices in the state.

4 Discussion

While CREDA solar mini-grid model has been comprehensively discussed in the previous sections, this section highlights the key aspects contributing to the success of the solar mini-grids in the state.

4.1 Robust Institutional Framework

One of the key ingredients of mini-grid's success could be linked to the strong and robust institutional structure created, nurtured and enlivened by CREDA. This is evident in multiple dimensions of the operational and management of mini-grids starting from the policy makers and planners in the state down to the very level in the hierarchy like operator of a single plant. Policy level support has been an important part of this institutional structure, which has been acting as a major impetus for the success of the model. In addition, it is also noteworthy to mention the public–private partnership structure engaged in the management and operation of the plant. The CREDA model also contradicts the notion that bottom-up approach with community involvement is the key to success of any decentralised project. While many researchers and practitioners have argued that decentralised energy systems should be implemented with a decentralised approach, with full planning and implementation by the community, the CREDA model clearly indicates that top-down approach of using standard designs, implementation and maintenance, with community's involvement for only local social engineering work, which they can do best, is a better (if not best) model to implement and operate the decentralised electricity infrastructure projects in remote communities. The outsourcing of maintenance to private management also brought in better efficiency of service into the project operation and management, which is difficult to ensure in a publicly devised and operated model.

Further, tariff collection is done by the AMC technician and supported by the operators in his effort. It is important to mention that there has been a clear-cut fragmentation of the amount collected from tariffs and salary of operators. Irrespective of the tariffs collected in a village, the operator gets a fixed monthly remuneration from the AMC contractor. This has led to successful operation of the plant as operator is assured to get their monthly payment. This is also corroborated from the study by Palit et al. [22] on Village Energy Security Program where one of the key reasons for failure was that the project management was completely entrusted with the VEC without ascertaining whether they are adept at managing the project or not. Further, the same operator was made responsible for system operation and tariff collection and in-case of non-payment by consumers, the operators' salary was delayed by the VEC, which acted as a disincentive for the operator. The solar mini-grid experience from the Sunderban region also shows that decentralised projects are more successful when implemented and managed in an organised way with clear cut roles and responsibilities of different stakeholders [30]. Shrank [27] also observes, that the community management system does not always create incentives for maximising profit at each power plant, thus creating problems for the coverage of costs of the power supply.

4.2 *Social Engineering is Pivotal*

The study also indicates that one of the key reasons to the success of solar mini-grids in Chhattisgarh is social engineering through regular and continuous engagement and making the community adopt the system. While, there have been places where social conflicts have led to dismantling of the installed system, overall the VEC has played their role in shaping the social engineering. This was also made possible as the VEC could focus on this aspect only, as the technical operation and maintenance was outside their purview. Extensive capacity building and awareness generation of local stakeholders also helped as the social engineering is executed through regular training programmes, both formal and informal, to different stakeholders.

4.3 Electricity Supply and Creation of Productive Activities

While provision of electricity has been primarily for lighting and no productive loads have been targeted as part of the initiative, there have been positive impacts on the local livelihood. Direct impacts are in terms of doing additional hours of work in the evening. For instance, during the *tendu* leaves collection season, extended hours of work could be possible as lighting is available during the evening. Further, in some cases, appliance uses such as use of radios, televisions, etc. was also found in some of the households in the villages visited. Indirect impacts are in the form of growing awareness by use of mobiles, radios, televisions, etc. Further research on this subject may be required for Chhattisgarh to quantitatively and qualitatively investigate the impacts of solar electrification on local economies in the state. It may be relevant to do a social cost benefit analysis of installing smaller capacity systems, as have been done in the state, to see if benefit outweighs costs. Though productive applications has its own importance in increasing local income, it was also important for the Chhattisgarh government to facilitate electricity access for lighting for such remote population and provide necessary exposure, where they can initiate or get involved in some income generation activities to pay for the services availed. The critical aspect here is that since the solar mini-grids have been primarily used to provide electricity services, any new demand for load enhancement can now be met using the existing grid infrastructure by increasing the solar panel and using additional inverter (as productive demand may be during daytime, additional storage will not be required) at a marginal cost which otherwise may not be possible with individual solar home systems.

4.4 Enabling Policy Environment

Enabling policy regime for promotion of solar mini-grids has been instrumental in accelerating the growth of off-grid electrification in the state. Realising the importance of off-grid electrification for the state, the government of Chhattisgarh has been very proactive in providing supports to mainstream off-grid electrification options for geographically difficult locations of the state. While the solar mini-grids have been implemented under the REVP, the state government has also equally been generous of sanctioning fund to as high as 55 % of the project cost to install mini-grids at remote locations. Further, the tariff subsidy mechanism provided to grid electricity consumers was also extended to subsidise households getting electricity through off-grid generation. While it can be debated whether we can call the CREDA model a success or not, as operational subsidy is also provided by the government in addition to capital subsidy, it is to be kept in mind that the provision of electricity is for villages with extremely poor economic conditions. Thus, envisaging a business sense through energy intervention becomes difficult. Moreover, the state is following an equitable approach whereby the off-grid areas are also getting similar support provided in the grid connected areas, which otherwise is absent in most of the other states (where only grid connected consumers are provided tariff subsidy or cross-subsidy). Provision of electricity in these remote and far off localities could at best be considered as merit goods as also envisaged in Integrated Energy Policy of the Government of India. Lately, however, there is an on-going thinking within CREDA that now the time has come for a strategic shift in the provision of electricity from subsidised service to a paid one in at least few select sites. The idea is to spilt up the entire beneficiaries into two different groups. Households will be provided at a marginal fee whereas business units such as rural banks, tailoring shops and grocery shops will be provided on the basis of full recovery of operation and maintenance cost.

5 Concluding Remarks

This chapter has attempted to present an analysis of the development and operation of the solar mini-grid model for enhancing electricity access in India, with special focus on the state of Chhattisgarh. It is observed from the study that the rate of success of mini-grids is directly dependent on the government's commitment to create an enabling environment, which includes having a clear-cut policy framework and milestones, systems for defining and enforcing appropriate technical standards, financial support mechanisms both towards installation and operation, and support for capacity building. Chhattisgarh has developed a robust institutional framework not only for implementation, but also for ensuring responsive after-sales service and maintenance of the solar mini-grids in the state paving the way for success of the programme.

While there may be a debate on what constitutes a success as the model followed in Chhattisgarh is highly subsidised and the power plants are designed to take care of only lighting load, there is no denial of the fact the provisioning of basic minimum electricity access to the population at the extreme base of the pyramid is also important and requires innovative approach for success. While many mini-grid projects fail in such remote areas because of lack of strong institutional framework and maintenance services, the implementing agency here has been successfully operating and maintaining the power plants by utilising the fund made available by the state government towards the operational subsidy.

The mini-grids have proved to be a reliable solution for such remote areas in comparison to solar home systems. Technically, mini-grids are preferred over other modes like solar home-lighting systems and solar lamps, as mini-grids can provide electricity services for lighting as well as to run various appliances, whereas solar home-lighting systems and solar lamps typically provide only lighting services. Organisationally, Chhattisgarh experience shows that managing mini-grids may be easier compared to individual systems due to their centralised operation in a village through a proper institutional structure. Further, the design also ensured that any future demand, in the form of new household connection or power for community load, could be catered to. The standardisation of design and operation and maintenance model by local-level service providers also ensured that the solution is cost-effective at the local scale. The system also does not lock-in the community to a particular development path as the mini-grid capacity can easily be enhanced by addition of modular capacity in case of enhanced demand for any productive loads in future, which otherwise is a constraint in case of individual solar home systems. In fact, the study clearly brings out this fact of capacity enhancement, which has been carried out in case of some villages both through installing new power plants and also using dismantled systems from other project villages where grid has reached over time.

Another key lesson from the mini-grid experience reveals that appropriate support systems should be a mixture of both 'participatory approach' and 'top-down approach'. While issues of a local nature could be better addressed through a participatory governance structure, technical, policy and financing matters can be dealt with at the appropriate intermediary and/or higher level. It is important to a design support system, so as to ensure that plans and policies match the needs of all stakeholders—consumers, owners and technology suppliers. Further, divided ownership models, where operation and revenue collection are done by separate verticals and or different individuals, seem to bring better focus on generation and service delivery.

Lastly, for the renewable energy-based, rural electrification sector to reach significant scale, implementation agencies need to work on overcoming the challenges of supply, demand and scalability and at the same time adopt standard processes and metrics, which will also help them to attract the necessary level of investment from financial institutions in support of 'energy access' programmes.

Acknowledgements We are grateful to Mr Shailendra Kumar Shukla, Director, CREDA, Mr Rajiv Gyani, Executive Engineer, CREDA and the CREDA district level officials from Raipur and Korba districts for providing secondary information on the various projects and sharing their insights with the TERI team.

References

1. Buragohain T (2012) Impact of solar energy in rural development in India. Int J Environ Sci Dev 3(4):334–338
2. Census of India (2011) Office of the Registrar General and Census Commissioner, India
3. Central Electricity Authority (2013) Progress report of village electrification as on 31-05-2013 http://www.cea.nic.in/reports/monthly/dpd_div_rep/village_electrification.pdf
4. Chakrabarti S, Chakrabarti S (2002) Rural electrification programme with solar energy in remote region—a case study in an island. Energy Policy 30:33–42
5. CREDA (2013) Chhattisgarh State Renewable Energy Development Agency, 2013. http://new.credacg.org/
6. FSI (2011) Forest and Tree Resources in States and Union Territories. Forest Survey of India, Dehradun

7. IEA (2012) World Energy Outlook, 2012 Electricity Access Database, International Energy Agency, Paris
8. Imai K, Palit D (2013) Impacts of electrification with renewable energies on local economies. In: The case of india's rural areas; working paper series vol 2013–12, The International Centre for the Study of East Asian Development, Kitakyushu, Japan, pp 20
9. Jain S (2012) Talk delivered at the round table on solar energy policy—challenges and way forward for energy access and large-scale deployment of solar energy in India. Organized by Centre for Science and Environment, New Delhi
10. Komatsu S, Kaneko S, Shrestha RM, Ghosh PP (2011) Non income factors behind the purchase decisions of solar home systems in rural Bangladesh. Energy Sustain Dev 15(3):284–292
11. Kumar A, Mohanty P, Palit D, Chaurey A (2009) Approach for standardization of off-grid electrification projects. Renew Sustain Energy Rev 13:1946–1956
12. Mainali B, Silveira S (2011) Financing off-grid rural electrification: country case Nepal. Energy 35:257–265
13. Malviya J (2011) Rural electrification in Chhattisgarh. Sun Wind Energy 7:180–182
14. Millinger M, Marlind T, Ahlgren EO (2012) Evaluation of Indian solar electrification: a case study in Chhattisgarh. Energy Sustain Dev 16(4):486–492
15. Ministry of Home Affairs (2013) Government of India. http://mha.nic.in/pdfs/NM-SRE-Scheme.pdf. Last accessed in 30 Dec 2013
16. Mishra A, Sarangi GK (2011) Off-grid energy development in India: an approach towards sustainability. In: OASYS Working Paper 12, 2011. http://oasyssouthasia.dmu.ac.uk/docs/oasyssouthasia-wp12-dec2011.pdf
17. MoEF (2011) Guidelines for linear infrastructure intrusions in natural areas: roads and power lines. National Board for Wildlife, Ministry of Environment and Forests, Government of India 2011
18. Palit D, Chaurey A (2011) Off-grid rural electrification experiences from South Asia: status and best practices. Energy Sustain Dev 15:266–276
19. Palit D, Chaurey A (2013) Off-grid rural electrification experiences from South Asia. In: Bhattacharyya S (ed) Rural electrification through decentralised off-grid systems in developing countries. Springer, London, pp 75–104
20. Palit D, Malhotra R, Kumar A (2011) Sustainable model for financial viability of decentralized biomass gasifier based power projects. Energy Policy 39:4893–4901
21. Palit D, Singh J (2011) Lighting a billion lives—empowering the rural poor. Boiling Point 59:42–45
22. Palit D, Sovacool BK, Cooper C, Zoppo D, Eidsness J, Crafton M, Johnson K, Clarke S (2013) The trials and tribulations of the Village Energy Security Programme (VESP) in India. Energy Policy 57:407–413
23. Palit D (2013) Solar energy programs for rural electrification: experiences and lessons from South Asia. Energy Sustain Dev 17(3):270–279
24. Rajiv Gandhi Grameen Vidyutikaran Yojana- at a glance 2013. http://rggvy.gov.in/
25. Raman P, Murali J, Sakthivadivel D, Vigneswaran VS (2012) Opportunities and challenges in setting up solar photo voltaic based micro grids for electrification in rural areas of India. Renew Sustain Energy Rev 16:3320–3325
26. Sanyal S (2013) Micro-grids in Uttar Pradesh and Bihar—a NVI briefing note; new ventures India 2013. http://nvindia.biz/documents/NVI_BriefingNote_Micro-Grid-UP-Bihar_Apr2013.pdf
27. Shrank S (2008) Another look at renewables on India's Sagar Island. In: Working Paper #77. Stanford Program on Energy and Sustainable Development 2008
28. Sovacool BK, Drupady IM (2011) Summoning earth and fire: the energy development implications of Grameen Shakti in Bangladesh. Energy 36(7):4445–4459
29. The Electricity Act (2003) The Gazette of India, Ministry of Law and Justice, Government of India, 2003

30. Ulsrud K, Winther T, Palit D, Rohracher H, Sandgren J (2011) The solar transitions research on solar mini-grids in India: Learning from local cases of innovative socio-technical systems. Energy Sustain Dev 15:293–303
31. Wijayatunga PDC, Attalage RA (2005) Socio-economic impact of solar home systems in rural Sri Lanka: a case-study. Energy Sustain Dev 9(2):5–9
32. Wong S (2010) Overcoming obstacles against effective solar lighting interventions in South Asia. Energy Policy 40:110–120

Poverty Amidst Plenty: Renewable Energy-Based Mini-Grid Electrification in Nepal

Gopal K. Sarangi, D. Pugazenthi, Arabinda Mishra, Debajit Palit, V. V. N. Kishore and Subhes C. Bhattacharyya

Abstract Providing access to electricity to a large section of rural population in Nepal has traditionally been a challenging exercise. This has been exacerbated by difficult geography, poor-socio-economic profile of rural Nepal and moreover by the on-going energy crisis. This chapter conducts an objective assessment of the renewable energy-based off-grid electricity sector in Nepal, with specific focus on micro-hydro-based mini-grid systems by applying a mixed method research design built on both qualitative and quantitative research techniques. While the country's experiences of developing micro-hydro- and solar energy-based off-grid interventions are captured by qualitative analysis, a standard techno-economic analysis of a micro-hydro mini-grid project is conducted to explore the possibility of introducing additional productive loads and to examine the cost efficacy of generating energy from micro-hydro vis-à-vis solar. Assessment of off-grid electrification options reveals that despite visible progresses, there still exist multiple

---

Author V. V. N. Kishore is presently an independent renewable energy consultant and former Professor and Head of Department of Energy and Environment, TERI University, New Delhi.

G. K. Sarangi (✉) · D. Pugazenthi · A. Mishra · D. Palit
TERI University, New Delhi, India
e-mail: gopal.sarangi@teri.res.in

D. Pugazenthi
e-mail: pugazenthi.d@live.com

A. Mishra
e-mail: amishra@teri.res.in

D. Palit
e-mail: debajitp@teri.res.in

V. V. N. Kishore
Department of Energy and Environment, TERI University, New Delhi, India

S. C. Bhattacharyya
De Montfort University, Leicester, UK
e-mail: subhesb@dmu.ac.uk

S. C. Bhattacharyya and D. Palit (eds.), *Mini-Grids for Rural Electrification of Developing Countries*, Green Energy and Technology,
DOI: 10.1007/978-3-319-04816-1_13,

roadblocks to scale up. Absence of clearly spelt out policy goals, weak institutional designs, low load factors, and lack of adequate finance and overall regulatory concerns stand as major obstacles for off-grid electricity sector development in the country. In addition, project-specific analysis reveals that solar loses out as a cost-effective option compared to micro-hydro. But optimal use of energy generated from micro-hydro-based mini-grids requires creation of productive applications at the local scale on a sustainable basis.

1 Introduction

Nepal, a land locked mountainous country of South Asia, is located between India and China. The country occupies a total land area of about 147,180 km^2 with a population of 30.5 million in 2011. About 40 % of the total land area of the country is covered by forests and shrubs. Country's socio-economic profile is not very encouraging as 44 % of country's population lives below poverty line [30]. The economy of Nepal is ranked as one of the slow growing economies by the World Bank [31], with per capita income of $742 and growth rate of 4.5 % during 2011–2012 [9]. About 82 % of country's population lives in rural areas. A large section (i.e. about 70 %) of population in Nepal is dependent on primary economic activities such as forests and agriculture, whereas the rest are engaged in secondary and tertiary sector activities. On human development front the performance has also not been encouraging. With a Human Development Indicator (HDI) of 0.463, Nepal is outranked by 156 countries [30]. Presence of political uncertainty and disruptive political activities are believed to have bred the debilitating and inefficient governance system and culture pervaded in all spheres of socio-economic life, including the energy sector in the country [15].

Providing access to energy in Nepal has traditionally been a challenging exercise. While at the global scale, about 1.2 billion people do not have access to electricity and about 2.6 billion people do not have access to clean cooking facilities [5], Nepal, being located in one of the least electrified regions of the World, i.e. in South Asia, has not escaped from this hard reality. About 7 million people (about 27 % of the total population) do not have access to any form of modern lighting energy [5]. Energy access challenges get exacerbated by geographical variations, poor transportability; scattered settlements, illusive energy development strategies, and lack of adequate capital [22]. Two pertinent aspects need further qualification. First, physiographic features of the country, as it has bearings on the state of energy system in the country and second, depth and intensity of on-going energy crisis, an outcome of multitude of factors like long persisting political instability, lack of adequate resources, and poorly crafted policy and regulatory framework.

Physiographic features of the country are characterised by rough physical terrain conjugated with a low, scattered and sparse population density. It is recognised at the policy sphere that providing grid electricity to all areas in Nepal seems to be a huge task in the country in the foreseeable future [12]. Grid-based centralised electrification system is considered to be relatively expensive and time-consuming to electrify scattered settlements located in difficult geographical terrains of the country [2, 3, 18]. Studies indicate that about 10 million people (which constitute about 33 % of the total population of the country) live in such remote locations requiring 5–18 days of walk to reach [34]. Marginal cost of grid expansion in Nepal is very high due to physical isolation, low electricity loads, and scattered low income consumers [8].

It is pertinent to highlight the persistence and deepening energy shortages the country has been experiencing since long, which has culminated into a 'great energy crisis'. Economic Survey of Nepal 2011–2012 acknowledges this on-going energy crisis in the country and states that "energy crisis has been the largest obstacle for country's economic development" [9]. The crisis gets embodied in multiple dimensions of energy supply, energy production, and energy consumption. It is posited that the energy crisis in Nepal gets accentuated by rapid urbanisation and growth of industries [16]. Load shedding of about 12–14 hours per day for almost all the on-grid households (about 2.4 million households are connected to the grid) is a clear manifestation of the magnitude of such a crisis. This high load shedding is often attributed to the poor power planning in the country resulting from inadequate capacity additions and growing electricity consumption in the country [27].

In the above backdrop, this chapter focuses on off-grid electrification development in Nepal, with a specific thrust on micro-hydro-based mini-grid systems. Though recently Nepal has decided to go for solar mini-grids for electrification, it has not yet been physically taken up; therefore we have not considered solar mini-grids in our analysis. The analysis dwells on the following set of research objectives.

- Assesses the state of renewable energy-based off-grid electrification in Nepal.
- Presents a critical evaluation of policy and institutional landscape governing the renewable energy-based off-grid electrification in Nepal.
- Conducts a techno-economic analysis of a micro-hydro mini-grid project to explore the possibility of introducing additional productive loads and to examine the cost efficacy of generating energy from micro-hydro vis-à-vis solar.
- Discusses key aspects of and identifies key anomalies and distortions for off-grid electrification in Nepal.

The chapter is organised as follows. Section 2 spells out the study approach. Section 3 presents the macro energy scenario in Nepal. Section 4 highlights the renewable energy-based off-grid interventions in Nepal. Section 5 dwells on the policy, regulatory and institutional contours governing off-grid electrification in Nepal. Section 6 discusses the key elements of off-grid electrification in Nepal. Section 7 presents techno-economic analysis of a micro-hydro-based mini-grid project. Section 8 carries out a critical assessment of off-grid electrification by identifying the key anomalies and barriers. Section 9 offers some policy recommendations and the final section concludes the chapter.

2 Study Approach

The triangulation method of research design, built on both qualitative and quantitative research techniques, has been employed for the analysis. The method prioritises collecting, analysing and mixing both quantitative and qualitative data at different phases in the research cycle. While the emphasis of qualitative approach is to understand the critical nuances, actors and institutions associated with and processes involved with the off-grid energy development in Nepal, quantitative assessment supplements the qualitative analysis by carrying out critical evaluation of the primary and secondary information, gathered as part of the study.

A week-long visit was conducted in Nepal to carry out the survey and to gather information for the study. The survey was divided into two parts. First, key informant interviews were conducted with different stakeholders such as government officials, donor agencies, NGOs, various associations, researchers, private entrepreneurs and system manufacturers, bank officials to understand and assess the state of renewable energy-based off-grid electrification system in the country. Second, a field visit to a micro-hydro site was carried out to gather information about the project operation, management and aspects of project sustainability and to validate the findings from key informant interviews.

A host of data collection techniques namely research interviews, field research, stakeholder's analysis and focus group discussions (FGDs) were used to elicit the desired set of information. A semi-structured interview format with flexibility to accommodate changes was administered to conduct interviews with different stakeholders to obtain information. Interviews at the micro-hydro project site constituted transect walks, FGDs, and observational data gathering and semi-structured interviews with key local informants such as system technician, president of the management committee, village chief, school teachers, village shop keepers, productive end-users, and local health clinic staff. In order to identify the prospective stakeholders, a non-probabilistic purposive sampling method was used to select interviewees having knowledge and understanding of off-grid renewable energy in the country and having direct and indirect association with the sector. Key stakeholders interviewed are listed in the Annexure (Table A.1). The interview conducted was aimed to understand the multiple crucial dimensions of off-grid electricity sector development in Nepal such as growth and trend of renewable energy centric rural electrification, role of donor agencies, policy level supports and issues related to the subsidies and incentives, financial mechanisms, role of associations, NGOs and civil societies. The gathered information from multiple stakeholders that was recorded and coded for further analysis. Information gathered from field visits were used to carry out a comparative financial assessment of various electrification options for Nepal.

3 Macro Energy Setting in Nepal: An Overview

Nepal is one of the countries with lowest per capita electricity generation and consumption. Per capita energy consumption stood at 103 kWh in 2010 [31]. Total generation capacity of the country is about 714 MW [32], largely drawn from hydro sources. The energy sector in the country is characterised by slow growth in all the critical dimensions. International Energy Agency (IEA)'s Energy Development Index[1] (EDI) for Nepal ranks the country at 74 in 2012 with an EDI score of 0.08, echoing the poor state of energy development in the country [5]. State of energy access in Nepal exhibits the presence of heterogeneity in access to different forms of energy (Fig. 1). While overall electrification rate in Nepal is about 75 %, there exist wide disparities in electrification rates between urban and rural areas. 94 % of urban areas use electricity for lighting compared to 61 % in rural areas. About 22 % of rural households in Nepal rely on kerosene as their prime source of lighting. In addition, solar, largely in the form of solar home systems (SHSs) has also emerged as a potential alternative source of lighting energy and about 9 % of the rural households are using solar as a source of lighting [19]. This urban-rural disparity worsens when it comes to access to modern energy for cooking. Use of traditional biomass for cooking reveals that while 27 % of urban population rely on fire wood and cow dung for their cooking, the corresponding figure is 86 % for rural areas in Nepal [14].

Despite Nepal being the second hydro resource rich country after Brazil and having enormous potential for solar energy, three decades of research and development has not produced visible progress. In terms of availability of water resources, the country is endowed with around 6,000 rivers and rivulets with a theoretical potential of approximately 85,000 MW electricity generation capacities. Similarly, Nepal is also endowed with adequate solar resources with average radiation of 4.7 kWh m^{-2} day^{-1}. However, available hydro and solar resources have not been exploited optimally so far in the country. The electricity generation capacity of Nepal has remained almost stagnant for the last 20 years. This stagnation has resulted largely due to failure in completing the hydro project construction and commissioning in due time [27] and inadequate attention of donor agencies for the promotion of large-scale hydropower development of the country. In addition, the country has not been able to attract investors in this sector despite the liberalised environment created with the enactment of Hydro Power Development Policy 1992 [29]. On the other hand, rapid urbanisation and industrialisation has led the electricity demand to peak at 946 MW in 2011 and predicted to increase to 3,679 MW in 2027–2028 [13].

[1] Energy Development Index (EDI) of International Energy Agency (IEA) helps in understanding the role of energy in promotion of human development of a country [5]. It is constructed by a set of four indicators i.e. per capita commercial energy consumption, per capita electricity consumption in the residential sector, share of modern fuels in total residential sector energy use and share of population with access to electricity.

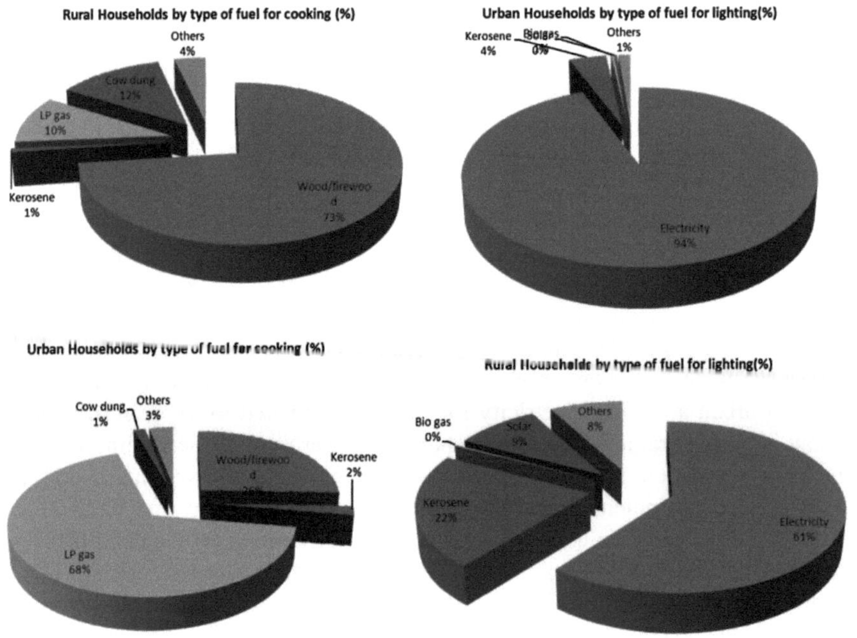

Fig. 1 Cooking and lighting energy scenario in urban and rural conurbations in Nepal. *Data source* Nepal Living Standard Survey [14]

In order to get a better picture of the state of energy scenario in Nepal, we present energy access index[2] for different districts in Nepal by combining access to electricity indicator and access to clean cooking fuel indicator as well as both the indicators separately for all the 75 districts in Nepal. It could be gleaned from Fig. 2 that clusters could be identified where high access to electricity is accompanied by high access to clean cooking energy and vice versa. However, there exist contrasting combinations suggesting specific policy thrusts on individual dimensions of access to energy.

The macro energy scenario in Nepal indicates the dismal state of energy availability and energy use. There exist enormous challenges of access to modern lighting and cooking energy. More importantly, given the limitations of the grid electrification system in the country, large sections of the rural population are still devoid of any form of modern energy. In view of this, renewable energy-based off-grid electrification options are now being prioritised as a supplementary route to the grid-based system. The following section dwells on the off-grid electrification systems in the country.

[2] We largely follow the methodology advocated by IEA for constructing such an index. However due to paucity of data, we have limited only to household level indicators of IEA. Variables are normalised applying the standard normalisation process. Equal weights to each individual indicator are assigned to construct the index.

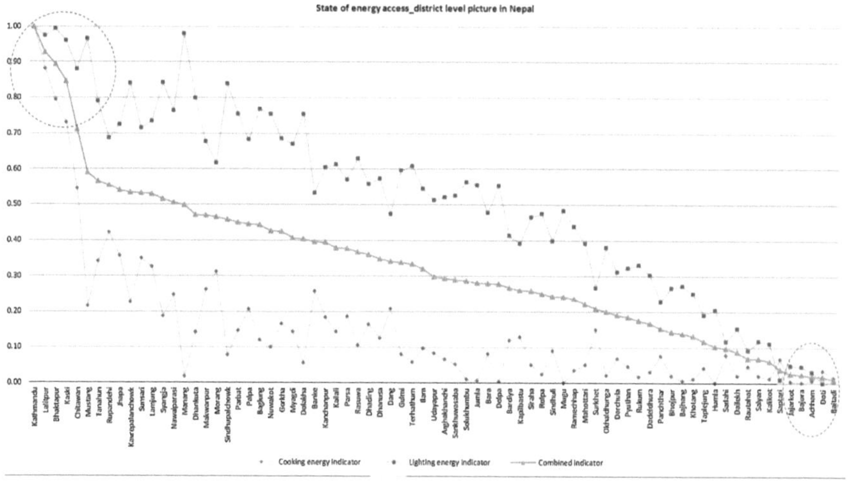

Fig. 2 State of energy access across districts in Nepal. *Data source* Nepal Living Standard Survey [14]

4 State of Off-Grid Renewable Electrification in Nepal

Development of alternative energy systems in Nepal could be traced back to the early 1970s. The off-grid renewable energy sector in Nepal has experienced a phased development process [23]. The initial focus was on adaptive research and technology transfer, followed by emphasis on pilot programmes and developing ad hoc policies for the promotion of renewable energy development during 1980s and 1990s. Next, emphasis was laid on setting targets, policy formulation, planning, resource allocation, capacity building and institution strengthening during 1990–2010, and the final phase was the period of scaling up, envisaging public–private-partnership (PPP) models, emphasising sectoral development, increasing the share of renewable energy in energy portfolio, linking energy with economy, upgrading technology, making coherent policies and linking with the environment [23].

Off-grid electrification is mainly targeted at electrifying remote rural areas of the country characterised by low population density, low load factor and geographical remoteness. In addition, on-going energy crisis also gave impetus for the greater use and exploitation of off-grid renewable energy resources of the country [4, 28]. Micro-hydro, pico-hydro, and SHSs are the preferred modes of off-grid electrification in Nepal, though a small amount of electricity is generated through small-scale wind energy systems. There exist about 3 million off-grid households in Nepal, out of which 97 % are in rural areas. About 12 % of the total population is electrified through renewable energy-based off-grid energy systems. While 22 MW generation capacities have been installed through micro-hydro and pico hydro schemes, 12 MWp capacities have been installed through solar PV schemes. Besides, micro-hydro, pico hydro and solar, about 20 kWe is generated through wind energy systems as well [1].

History of micro-hydro-based mini-grid systems in Nepal is the oldest and goes back to the early 1960s and being constantly developed since then. Initial focus of micro-hydro-based mini-grids was to create livelihood opportunities by utilising the electricity for agro processing and other allied activities. Gradually, micro-hydro-based mini-grids became a source of community electrification. These micro-hydro systems, during early days, were small in capacity ranging between 5 and 20 kW and largely supported by international donor agencies. These systems were providing lighting requirements with some productive activities like grinding, husking and oil-expelling. This was followed by installation of large-scale hydro-electric systems during mid-1980s. Development of micro-hydro systems was further accelerated through provision of loans and technical assistance and subsidies provided during early 1980s. It was further boosted by financial and fiscal incentives such as subsidies; supports extended through several donor agencies and a host of other factors. As per recent statistics, there are about 999 micro-hydro-based mini-grid systems constituting about 19 MW and about 1,480 pico-hydro projects totaling about 3.18 MW deployed by 2012. As far as the ownership of micro-hydro mini-grid projects are concerned, 95 % of the projects are community owned; whereas about 5 % are privately owned and a few are owned by the National Electricity Authority (NEA) [3].

Two major programmes supporting the development of micro-hydro mini-grid projects in the country are (1) Energy Sector Assistance Programme (ESAP) supported by DANIDA, KfW, DFID and Norwegian Government and (2) Rural Energy Development Programme (REDP) supported by UNDP and World Bank. While ESAP supported installation of about 290 micro-hydro mini-grid projects and about 402 pico-hydro projects, under REDP more than 300 micro-hydro mini-grid projects ranging from 10 to 100 kW have been deployed [1, 25]. Figure 3 portrays cumulative number of households connected and kW capacity of micro-hydro projects installed under the ESAP programme.

5 Policy, Regulatory and Institutional Contours

The need and significance of policies and institutional considerations has been highlighted by several scholars and experts. It is emphatically posited that deployment of off-grid renewable energy technologies requires concrete and plausible policies [17]. Potential to upscale renewable systems largely hinges on the country's institutional characteristics and policy landscape governing the sector [16, 33]. This section highlights the existing policy and institutional landscape governing the off-grid renewable energy sector in Nepal. Planned development process in Nepal also has recognised the importance of renewable energy-based off-grid systems in the country starting from the declaration of Seventh Development

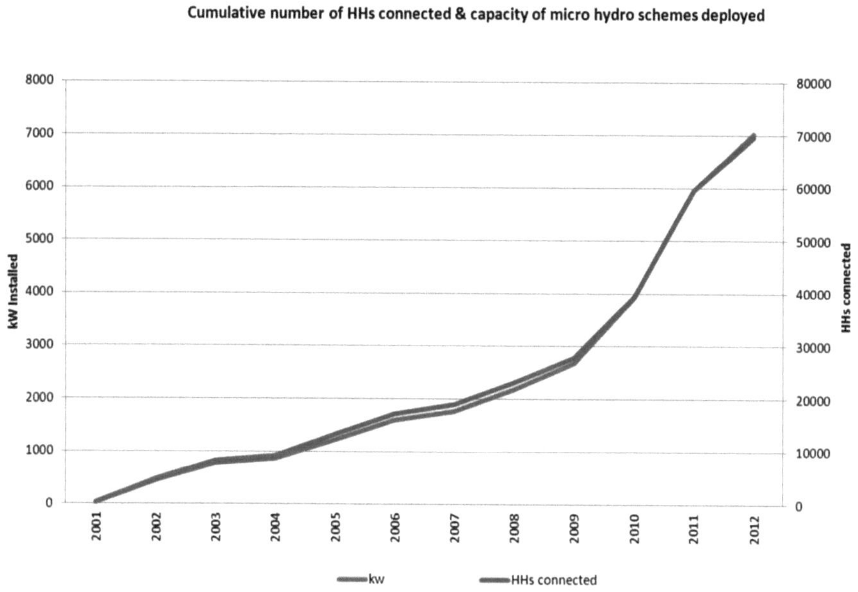

Fig. 3 Cumulative number of HHs connected and capacity (kW) deployed under ESAP programme, *Data source* AEPC [1]

Plan (1985–1990). A snapshot of the planned development process and its focus on renewable energy-based off-grid energy development in the country is presented in the Box 1.

Box 1: Planning process in Nepal and focus on renewable energy-based off-grid system development

The planned energy development in Nepal started with the declaration of Seventh Development Plan (1985–1990) with specific thrust on conservation of forest resources and upliftment of rural economy of the country. The Plan emphasised on the promotion of biogas, solar thermal, wind energy, improved cook stoves, small water turbines and improved water mills. Specific focus was laid on research and development aspects and tapping of private sector potential in the field. Priority is also assigned to incentivising the sector by giving grants and loans for large-scale dissemination of off-grid energy systems.

The Eighth Plan (1992–1997) continued its focus on renewable energy drawing from the experiences gained from the Seventh Plan. Specific thrust was laid on developing technical manpower and gathering basic data for development of biogas, solar energy and wind energy. Increase in use of solar-based systems like solar water heater, solar dryer, solar cooker, solar pump, solar generator, solar photovoltaic cells was prioritised. Emphasis was laid on attracting private investors. The plan proposed to develop a master plan for the diversification of use of solar energy. It declared a special provision for subsidies for PV household systems.

Ninth Plan (1997–2002) reiterated the need to develop renewable energy as an important element of national development agenda by putting emphasis on the creation of employment opportunities, improving rural livelihood conditions and prioritising environmental sustainability. Research and development was also kept high on the agenda to cut down the cost of generating power from alternative sources of energy.

Tenth Plan (2002–2007) laid emphasis on exploiting solar energy to meet the mounting energy demand. In addition, the need to electrify remote and rural areas through solar-based interventions was also assigned primacy. The Plan envisaged setting up of Rural Energy Fund (REF) to manage and channelise grants and loans for development of alternative energy systems in the country.

First 3 Year Interim Plan (2007–2010) focused on developing a long-term alternative energy agenda for the country with specific thrust on development of rural areas, creation of employment opportunities and sustainable development of the sector. Second 3 Year Interim Plan (2007–2010) set an ambitious target of procuring 10 % of energy from alternative sources.

Policy pronouncements for off-grid energy development in Nepal could be traced back to the declaration of Hydro Power Policy 1998, reformulated in 2001. Hydro Power Development Policy 2001 placed emphasis on development of rural economy through energisation and creation of favourable ground for private investors by devising appropriate policies and incentive schemes. One of the important policy initiatives in the off-grid sector was undertaken with declaration of Rural Energy Policy (REP) in 2006. This Policy prioritises access to clean, reliable, and appropriate energy in rural areas. The Policy sets the objective of reducing dependency on traditional sources of energy, conserving environment, generating employment and creating productive activities through development of rural energy resources. Priority is also laid on creating capacities, human resource development, strengthening local institutions and tapping private sector capabilities.

In addition to above policies, specific policies and mechanisms have been spelt out from time to time to disburse subsidies. First National Subsidy Policy 2000 envisaged providing subsidies to SHSs, and solar water pumps. Subsidy Policy 2009 broadened the scope by not only providing subsidies for SHSs and solar water pumps, but also extending subsidies to institutional solar PV systems. The Policy aimed at maximising service delivery and providing opportunities to the low income households in the rural areas, making use of grant assistance, and supporting and extending renewable energy markets. The new subsidy policy, i.e. Subsidy Policy 2013, inter alia, has set objectives like reducing cost of supply, encouraging productive use of energy, developing markets for renewable energy and contributing to the better health and education of people. In line with subsidy policies, subsidy delivery mechanisms are declared from time to time. For instance, while Subsidy Delivery Mechanism 2006 has spelt out the need for disbursing subsidies in a cost effective and easy access manner for the acceleration of renewable energy market, the Rural Energy Subsidy Delivery Mechanism 2010 has emphasised on setting subsidy criteria for various renewable energy resources and delivery mechanisms for disbursement of subsidies for different forms of renewable energy-based off-grid energy sources.

Thrust on promoting alternative energy in Nepal has also been recognised through several other policy pronouncements. For instance, National Adaptation Programme of Action (NAPA) of Nepal, 2010 has identified a list of priority adaptation options for the energy sector. Specific thrust was assigned on the promotion of alternative energy technologies and strengthening the institutional aspect of promoting alternative energy technologies [11]. Apart from policies, promotion of off-grid renewable energy sector is also done through several other fiscal and financial incentives such as exemption of taxes, reduction of tax amounts, etc.

While legal and policy systems of the country are designed to govern the sector, the state of institutional artefacts shape the system of governance of a country to a great extent. Organisational contours for off-grid energy sector in Nepal reveal a complex web of interrelationships between multitude of actors and entities.[3] At the ministerial level, the task is to formulate, implement, monitor and evaluate policies, plans, programmes. It is also the duty of the Ministry to carry out research and development activities, promote private sector participation in the sector and to deal with multilateral agreements. At the institutional level, Alternative Energy Promotion Centre (AEPC) being at the helm of the affairs of off-grid and renewable energy development in the country, is responsible for mainstreaming renewable energy-based interventions in the country. Created in 1996, AEPC is

[3] These actors and entities consist of a couple of ministries like the Ministry of Science, Technology and Environment (MoSTE) and the Ministry of Energy (MoE), several government originations and institutions like the National Planning Commission (NPC), the Water and Energy Commission Secretariat (WECS), Alternative Energy Promotion Centre (AEPC), Renewable Energy Test Station (RETS), several donor agencies e.g. Danish International Development Agency (DANIDA), European Union (EU), United Nations Development Programme (UNDP), Norwegian Government, Bank Aus Verantwortung (KfW), Department for International Development (DFID) of UK Government, Asian Development Bank (ADB), a couple of associations like the Nepal Micro-hydro Development Association (NMHDA), and the Solar Electric Manufacturers Association of Nepal (SEMAN), banking and credit lending institutions the Clean Energy Development Bank Limited (CEDBL), the Himalayan Bank Limited (HBL), and the Lakshmi Bank Limited (LBL) and a number of manufacturing and installation companies, NGOs, micro-finance groups, local NGOs, village co-operatives, research institutes and many more.

entrusted to carry out research and development for the promotion of renewable energy-based off-grid systems in the country, to manage and administer subsidy policies, and to act as an umbrella organisation for several other activities and initiatives such as UNDP led REDP, Danish and Norwegian Co-funded ESAP, and European Union funded Renewable Energy Project (REP) and Rural Energy Fund. The Centre has been responsible for the formulation of plans and policies, mobilising resources, monitoring and coordinating activities, keeping check on quality and executing all other necessary activities. AEPC also has local functionaries at the level of districts, namely District Energy and Environment Sections/Units and Regional Renewable Energy Service Centres (RRESC) to execute activities at the district level. There are also affiliate institutions like Renewable Energy Test Station (RETS), which conducts various quality tests of renewable energy products and qualifies companies to receive subsidies and get tax waivers.

Various donor agencies and organisations have been playing an instrumental role in off-grid energy development in Nepal supporting several off-grid renewable energy programmes. The focus and thrust of donor funded programmes differ and is largely driven by the guidelines and philosophies of donor countries. However, there seems to be lack of co-ordination and harmonisation among these donor funded programmes thereby often duplicating the efforts. Of late, efforts have been undertaken by AEPC to bring all the donor funded programmes under one umbrella through National Rural and Renewable Energy Programme (NRREP).

Of late, private sector banks have emerged as an effective channel to finance various off-grid renewable-based interventions in Nepal through innovative financing schemes. While public sector banks like Agricultural Development Bank Limited (ADBN) of Nepal was involved in micro-hydro project financing as early as 1980, several private sector banks such as Clean Energy Development Bank Limited (CEDBL), Himalayan Bank Limited (HBL), and Lakshmi Bank Limited (LBL) have recently ventured into the off-grid renewable energy space through various financing schemes. Some of these banks have set up separate dedicated cells to finance the energy projects. Banks are also entrusted responsibilities by donor agencies to manage special funds meant for promotion of off-grid electricity sector in Nepal. For instance, CEDBL and HBL are tasked to manage Micro-hydro Debt Fund which is supported by GIZ and anchored through AEPC. HBL has been able to finance six micro-hydro projects[4] through Micro-hydro Debt Fund scheme. Banks also act as financing agents for private renewable energy developers and energy service companies. For instance, CEDBL has financed four hydro power contractors, 8 micro-hydro installer companies, 28 solar companies, 6 biogas construction companies.

[4] These micro-hydro projects are Khani Khoka (20 kW) in Karve dirtict, Chari Tola (80 kW) in Ramechhap Dictrict, Thulo Khola (50 kW) in Okhaldhunga District, Swara Tap Khola (30 kW) in Khotang District, Lumju Khola (20 kW) in Khotang District, Midim Khola (100 kW) in Lamjung District [24].

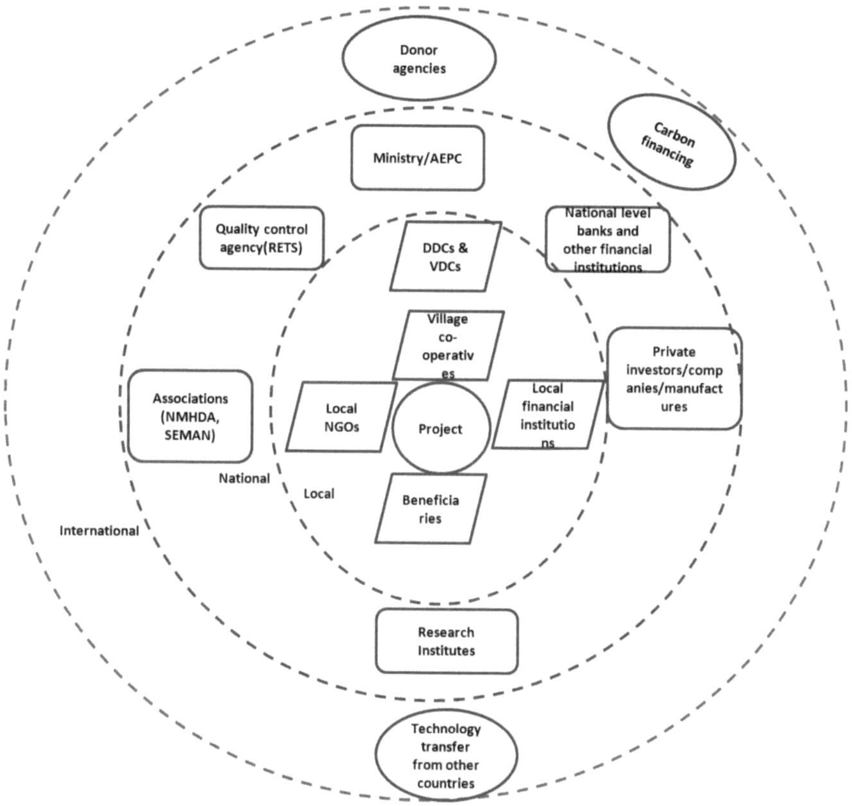

Fig. 4 Institutional contour governing renewable energy based off-grid energy system in Nepal, *Data source* Construed by authors

Associations formed by private companies like Nepal Micro-hydro Development Association (NMHDA), and Solar Electric Manufacturers Association of Nepal (SEMAN) also play a crucial role in driving the off-grid electricity sector in the country by creating necessary skills, expertise and by protecting the welfare of these private companies. General concerns of manufacturers and companies are addressed through these associations. These associations also conduct periodic training and capacity building programmes on various issues of importance. Apart from all of the above, there exist a number of research institutes and universities, private companies, NGOs, micro-financing institutions, village co-operatives contributing to the development of off-grid renewable energy systems in the country. A graphical exposition of various institutions and organisations placed at different scales governing the renewable energy-based off-grid energy sector in Nepal is presented in the figure below (Fig. 4).

6 Salient Features of Renewable Energy-Based Mini-Grid Systems in Nepal

This section analyses key features of renewable energy-based mini-grid electrification in Nepal.

6.1 Service Delivery Models

The dissemination of mini-grid projects in Nepal is done through a variety of delivery models. In majority of the cases, delivery models hinge upon the specific programme features and characteristics under which the project was developed. Deployment of micro-hydro-based mini-grid systems is largely done through community-managed delivery systems, albeit, in some cases by private entrepreneurs. In addition, private entrepreneurs also play an important role even in community-managed projects. Private manufacturing companies and private installer companies carry out the task of surveying, designing, installation of systems. These companies are pre-qualified by AEPC to channelise subsidies to the communities. Often deployment of projects largely follows programme protocols and guidelines. For instance, while Energy Sector Assistance Programme (ESAP) lays emphasis on improving the living conditions of the rural people through enhanced access to energy, Renewable Energy Development Programme (REDP) assigns primacy to community mobilisation aspects. Apart from community managed mini-grid schemes, other prevalent type is the private sector promoted projects, which constitutes about 5 % of the total mini-grids deployed [3].

6.2 Operational Modalities

Implementation of mini-grid projects in Nepal has been carried out through a public–private-partnership model. While the public sector performs several pertinent activities like capacity building, providing technical and financial assistance, and instituting mechanisms for quality control, private sector spearheads in manufacturing, supply and installation, after-sales service and internal quality checks. The first step is to identify the project site and then carry out demand assessment for such a project. Demand assessment is done through a scientific process of mapping by using the GIS tools and techniques. Next phase is the project approval phase, where the project gets approved at various stages through a structured approval system in place. The most immediate approval is required from District Development Committee (DDC)/Regional Renewable Energy Service Centres (RRESC), and then it gets approved by AEPC, where the project details are reviewed by a technical team known as Technical Review Committee

(TRC). Once it gets approved by the TRC, projects finally get in-principle approval for subsidy support. This is followed by installation and commissioning of the project. Once the project is installed and implemented, it goes through a quality control process. First level of quality control is in terms of monitoring and inspection of the project under construction. Next level of quality checking is done during testing and commissioning of the project. Final level of quality control is done through power output verification. As a process of quality check, 1 year guarantee is ensured by the project developer and consultant and 10 % of the project cost is kept as security money.

6.3 Financing of Projects

Project financing structures reveal some interesting patterns. In case of micro-hydro mini-grids, finance is mobilised through four major sources, e.g. government subsidy, community equity, contribution from local governments and contribution from other organisations.

In majority of the cases, subsidies from the Government of Nepal and donor grants constitute about 50 % of the project cost; of course variations exist depending on remoteness of the project area. The District Development Committees (DDCs) and the Village Development Committees (VDCs) contribute in terms of equity investment which is about 10 % of the total project costs; Rest of the amount is contributed by communities both in cash and kind terms. Contribution of communities in kind (in terms of labour and locally available construction material) constitutes 20 % of the total project cost and financial contribution by communities, either in the form of cash and/or bank loans forms rest 20 % of the project costs.

Subsidies have been instrumental in mainstreaming the renewable energy dissemination and wide scale deployment of micro-and pico-hydro projects and solar home systems (SHSs). Subsidy disbursement mechanisms are spelt out by the Government from time to time. Important aspect of this subsidy disbursement mechanism is the graded subsidy provisions based on the remoteness of the location. For disbursement of subsidies, VDCs have been grouped into different classes such as 'category A' VDCs, 'category B' VDCs and 'category C' VDCs on the basis of their remoteness, starting from very remote VDCs to accessible VDCs, respectively.

In order to enhance the access and instil a sense of commercialisation among the beneficiaries of mini-grid systems, innovative financing mechanisms have been devised by banks through support from multilateral and bilateral agencies and organisations. Micro-hydro Debt Fund—a dedicated fund to finance micro-hydro projects within the range of 10–100 kW has been operationalised by GIZ/AEPC. The funds have been routed through banks engaged in promotion of renewable energy in Nepal. GIZ/EnDev (German/Deutch collaboration) and NORAD have been supporting this initiative. For example, the Micro-hydro Debt Fund has a portfolio of €500,000 that is available for soft loan for micro-hydro projects. An additional €42,000 (in the form of technical assistance) is available for capacity building of local institutions, AEPC and rural communities. As mentioned in the previous section, loans from the fund are channelled through two banks, namely CEDBL and HBL of Nepal. Under this scheme, a maximum of 40 % debt financing is provided.

In majority of cases, decisions on tariff are made by the user committees. There are four major varieties of tariff structures prevailing in community managed mini-grid projects. These are largely monthly charges per household, monthly charges per bulb, charges based on per watt per month, and charges based on per unit (kWh) per month. Tariff rates are designed to meet the O&M cost of the project.

Of late, thrust has also been laid on carbon financing schemes, as an additional source of finance, to enhance the financial viability of projects. It is reported during the survey that about 650 micro-hydro-based mini-grid projects have been bundled by AEPC and registered in order to procure carbon financing. These projects are expected to generate about 40,535 tonnes of CO_2 emissions.

6.4 Quality Control Mechanisms

Another crucial dimension of renewable energy-based mini-grid development is the system of quality control. In order to ensure better quality of renewable energy products in the country, various quality control mechanisms have been put in place. For instance, disbursement of subsidies is linked to maintaining minimum standards of quality of renewable energy products and equipment [23]. Renewable Energy Test Station (RETS) has been created as a quality control arm of AEPC to ensure quality of renewable energy products and equipment. As of now, RETS has been primarily looking after the quality of photovoltaic-based systems as per the guidelines laid down in the Nepal Photovoltaic Quality Assurance (NEPQA) standard. Quality control of photovoltaic systems is done at three different stages, i.e. during pre-installation, during installation and post-installation stage. Quality control of micro-hydro-based mini-grid projects in Nepal is largely carried out directly by AEPC through periodic monitoring and evaluation of projects. For instance, Monitoring and Quality Assurance (MQA) Unit has been created within the technical support component of the recently launched National Rural Renewable Energy Programme (NRREP)[5] of AEPC. Lately, RETS has also been assigned the task of checking the quality of micro-hydro equipment and products.

[5] NRREP is a joint programme of Government Nepal and multiple donor agencies which brings all the individual programmes supported by multiple donor agencies under one umbrella and is executed by AEPC.

However, due to limited financial resources available at the disposal of the RETS, there has been limited progress achieved in terms of ensuring quality renewable energy products and equipment.

6.5 Capacity Building Efforts

Capacity building has always been considered a critical aspect for the sustainability of off-grid energy interventions. Capacity building is done through various ways and for different stakeholders. An important aspect of this capacity building effort is technical assistance schemes provided under various donor funded programmes. Major donor supported programmes such as ESAP and REDP have technical assistance components, seek to develop necessary knowledge and skill sets required for the operation and management of the systems. Technical support is provided through training, information, guidelines and quality assurances. In addition, enhancing the strength and ability of rural communities is also prioritised by several donor agencies. For instance, REDP programme gives emphasis on community mobilisation and communities are placed at the centre of the project operation, management and sustainability. Community mobilisation is considered pivotal and an important step in the project initiation process. Community mobilisation under REDP consists of organisation development, skill enhancement, capital formation, technology promotion, environment management, and empowerment of vulnerable groups.

Capacity building activities are also carried out by private associations like NMHDA and SEMAN. These associations conduct periodic training and capacity building programmes to build skilled manpower for the acceleration of off-grid renewable energy programmes in Nepal in association with AEPC under various donor funded programmes. For instance, Nepal Micro-Hydro Development Association (NMHDA) has been conducting surveyors training, managers training, quality and management aspects of MHP for installers, auto-cad training, output verification training, end use promotion training, operators training, advanced operators training and operators refresh training for last 10 years or so. NMHDA has conducted 42 training sessions till 2012. Importantly, these training and capacity building activities are supported by various donor-funded programmes like REDP, ESAP and RERL. The training and capacity building programmes of NMHDA have been able to generate a pool of skilled manpower to manage the micro-hydro-based mini-grids in Nepal. For instance, 44 operator trainings conducted by NMHDA since 1997 have trained around 883 numbers of skilled micro hydro operators in the country.

The above discussion brings out succinctly various aspects of renewable energy-based mini-grid systems in Nepal. It is evident that adequate measures have been taken in terms of creating necessary institutions, devising policy and incentive schemes, maintaining quality and building capacities to promote renewable energy-based mini-grid systems in the country. However, it would be more interesting to delve deep into the operational artefacts of mini-grid development in Nepal by conducting an in-depth analysis of a micro-hydro-based mini-grid project. The next section conducts such an analysis.

7 Case Study

In order to supplement the discussion above, we present here a case study drawing from the information gathered from our field visits and stakeholder interviews. While assessing the case study, we largely follow the framework suggested by Mishra and Sarangi [10] for mainstreaming renewable energy-based off-grid systems in developing countries.

The project, selected for the study, is located in Mahadevstan VDC in the Dhading district of Nepal. The project is serving about 265 households, out of which about 80 % are poor, and about 13 % are middle-income households and rest are rich households. Project cost decomposition reveals that about 75 % of cost of the project was covered through government subsidies, whereas the rest amount was mobilised in the form of private equity (both cash and kind), loans and contributions from local government, i.e. VDC and DDC. It must be noted here that since the project was supported by UNDP and World Bank funded Renewable Energy Development Programme (REDP), implementation modalities of this project largely follow the REDP programme guidelines. The details of the studied project are given in the Table (Table 1).

As suggested by Mishra and Sarangi [10], the very first task towards deployment of mini-grid system would be to carry out a needs assessment survey. It was revealed from our discussion with various stakeholders that a proper demand assessment was conducted as an important initial activity for the project installation. However, since REDP emphasises on mobilising communities as a pre-requisite for deployment of micro-hydro projects, the very first step undertaken was to form users committees and strengthen the community capability for the effective operation and management of the project. Once the user committees were formed, application was submitted through proper channel to DDC/RRESC and finally to AEPC. Second most important step carried out was the project identification, where a potential project site was identified through a scientific process by applying the GIS tools/techniques. The GIS mapping and detailed feasibility study (DFS) suggested the possible capacity of the project, monthly flow of water, catchment area with its land use type, geo-references position of major structures like intake, settling basin, fore bay, and power house, headrace canal, and penstock pipe length, and their head losses, the transmission lines and its segments and length. Next step was the project approval phase, where the project got approved through a structured approval system in place. The most immediate approval was received from DDC/RRESC, and then it got the approval from AEPC, where the project details were reviewed by a technical team known as Technical Review Committee (TRC). Finally, the project got in-principle approval for subsidy.

Table 1 Salient features of the project

Name of the project	Malekhu Khola
VDC	Mahadestan
District	Dhading
Capacity	26 kW
Number of HHs	265
Year of commissioning	2007

Source From the field

Since, the resource mapping at the community scale was only limited to micro-hydro schemes; it did not consider other potential resources. However, following Mishra and Sarangi [10], we have compared unit cost estimates of energy of the present technology with other technologies. Since, micro-hydro and solar are the key technologies used in Nepal for rural electrification [6, 20], we have limited our comparison to these two important technologies. In order to get a comparative picture of micro-hydro vis-à-vis solar energy, we estimated the levelised cost of electricity (LCOE)[6] for the micro-hydro project under consideration and a similar size solar PV project. Our analysis suggests that for same capacity plant, while LCOE for micro-hydro plant is coming out to be 0.07 USD, LCOE for solar plant is estimated to be 1.01 USD. Our findings are corroborated with the findings of the Mainali and Silveira [6], where authors have estimated that LCOE for solar projects ranges between 0.55 and 1.01 USD. The detailed parameters and estimated LCOE figures are presented in the table below (Table 2).

In addition, we also attempted to compare and contrast the existing case (base case) with possible scenarios with increased productive loads to examine the possible extent of unit cost reduction. Apart from the base case, we have envisaged three different scenarios based on increased productive end uses. This increase in productive loads is based on the assumptions taking into consideration the socio-economic profile of the villages being presently served by the project. This has been done primarily on the basis of recent concerns about low capacity utilisation of micro-hydro projects in Nepal due to poor productive end uses [3]. Load profiles for base case and different scenarios are presented in the Table 3 and Fig. 5.

It is evident from the table below (Table 3) that in the scenario 1 (S1) as we increase the productive load from 14 to 17 kW; LCOE comes down from 0.07 to 0.066 USD. In case of scenario 2 (S2), a further increase in the productive load by increasing the number of hours of productive end uses leads to further reduction in LCOE to 0.061 USD. Finally, in scenario 3 (S3), we have increased the productive load to 22 kW; this gives rise to further reduction in LCOE to 0.057 USD.

[6] For analysing the levelized cost of electricity (LCOE) one has to consider the total life time cost of the project which includes the capital costs, operation and maintenance costs, replacement cost, fuel cost and the environmental externalities costs and total electricity produced by the plant during its lifetime. The formula for estimating LCOE is, $\mathrm{LCOE} = \frac{\text{Total life time cost of the plant}}{\text{Total life time useful electricity produced}}$.

Table 2 Parameters for LCOE for 26 kW MHP and SPV plant

Parameters	MHP	SPV
Rated power capacity (kW)	26	26
Annual power generation (kWh)	51,496[b]	34,164
Life span of the plant (years)	15	20
Escalation factor (%)	5	5
Loan interest rate (%)	12	12
Inflation rate (%)	5	5
Emission factors (g/kWh)	0.01 (NO_x), 0.01 (SO_x), 5.92 (CO_2)	0.193 (NO_x), 0.322 (SO_x), 83.43 (CO_2)
Marginal external/damage cost ($/kg)[a]	1.5312 (NO_x), 5.8080 (SO_x), 0.0277 (CO_2)	1.5312 (NO_x), 5.8080 (SO_x), 0.0277 (CO_2)
Capital cost of the plant ($)	53937.00[b]	93600.00
Annual O&M cost ($)	1300.00[b]	2100.00
LCOE ($/kWh)	*0.07*	*1.01*

Source [a] Mainali and Silveira [6]; [b] From field surveys (Currency conversion rate: 1 USD = 80 NCR)

Financial efficacy of the project is also analysed by carrying out sensitivity analysis. Sensitivity analysis for the changing subsidies and its impact on LCOE is presented in the figure below (Fig. 6). It is evident from the figure below that with zero subsidy, the estimated LCOE is very high. However, micro-hydro is still financially attractive compared to the solar PV project.

It is evident from the analysis above that though micro-hydro is a cheaper option compared to solar PV; the challenge is to bring down the cost of supply by creating additional productive loads, which has been a concern in majority of off-grid micro hydro projects in the country [3, 28]. At present, productive loads such as two flour mills, one saw mill and one poultry firm are energised through this project. However, there still exists potential to create additional productive loads. Since the village economy is largely based on agriculture, productive loads of that nature could possibly be introduced. This needs designing policy thrusts to create productive loads which would not only optimise the plant capacity, but also enable income generation and employment at the local scale. Though specific subsidy schemes are put in place to enhance the end use applications, this has not been the case for every project.

In sum, it was found that the mini-grid project has been running successfully with a low annual average downtime of 9 %. The project has been able to generate positive impacts in terms of bettering the social infrastructure in the village by energising schools, health centres, improving the socio-economic conditions of local people by enhancing income and generating employment opportunities and empowering women through provision of modern lighting systems, thereby reducing their drudgery. The formation of a co-operative in the village to manage the project has also been able to create better social capital in the village. However, there exist a few challenges as far as sustainability of the project is

Table 3 Electricity demand constituents per HHs and scenarios

Items	Poor HH	Middle HH	Rich HH	Commercial load	Productive load
Base case	3 × 12 W lighting load for 5 h (5–10 pm)	3 × 12 W lighting load for 5 h (5–10 pm) and 1 × 80 W TV for 4 h	3 × 12 W lighting load for 5 h (5–10 pm), 1 × 80 W TV for 4 h and 1 × 500 W freezer for 6 h (10 am–4 pm)	420 W lighting and fan load for 5 h, 600 W computer for 1 h and 600 W refrigerator for 15 h	Two flour mills of 5 kW each and one saw mill of 3 kW for 3 h and 1 kW of poultry load for 6 h
S1	3 × 12 W lighting load for 5 h (5–10 pm)	3 × 12 W lighting load for 5 h (5–10 pm) and 1 × 80 W TV for 4 h	3 × 12 W lighting load for 5 h (5–10 pm), 1 × 80 W TV for 4 h and 1 × 500 W freezer for 6 h (10 am–4 pm)	420 W lighting and fan load for 5 h, 600 W computer for 1 h and 600 W refrigerator for 15 h	Two flour mills of 5 kW each and two saw mills of 3 kW each for 3 h and 1 kW poultry load for 6 h
S2	3 × 12 W lighting load for 5 h (5–10 pm)	3 × 12 W lighting load for h (5–10 pm) and 1 × 80 W TV for 4 h	3 × 12 W lighting load for 5 h (5–10 pm), 1 × 80 W TV for 4 h and 1 × 500 W freezer for 6 h (10 am–4 pm)	420 W lighting and fan load for 5 h, 600 W computer for 1 h and 600 W refrigerator for 15 h	One flour mill of 5 kW for 6 h, and one flour mill of 5 kW and two saw mills of 3 kW each for 3 h and 1 kW poultry load for 6 h
S3	3 × 12 W lighting load for 5 h (5–10 pm)	3 × 12 W lighting load for 5 h (5–10 pm) and 1 × 80 W TV for 4 h	3 × 12 W lighting load for 5 h (5–10 pm), 1 × 80 W TV for 4 h and 1 × 500 W freezer for 6 h (10 am–4 pm)	420 W lighting and fan load for 5 h, 600 W computer for 1 h and 600 W refrigerator for 15 h	One flour mill of 5 kW for 6 h, and one flour mill of 5 kW and two saw mills of 3 kW each for 3 h and 1 kW poultry load for 6 h and 5 kW furniture *industry* for 3 h

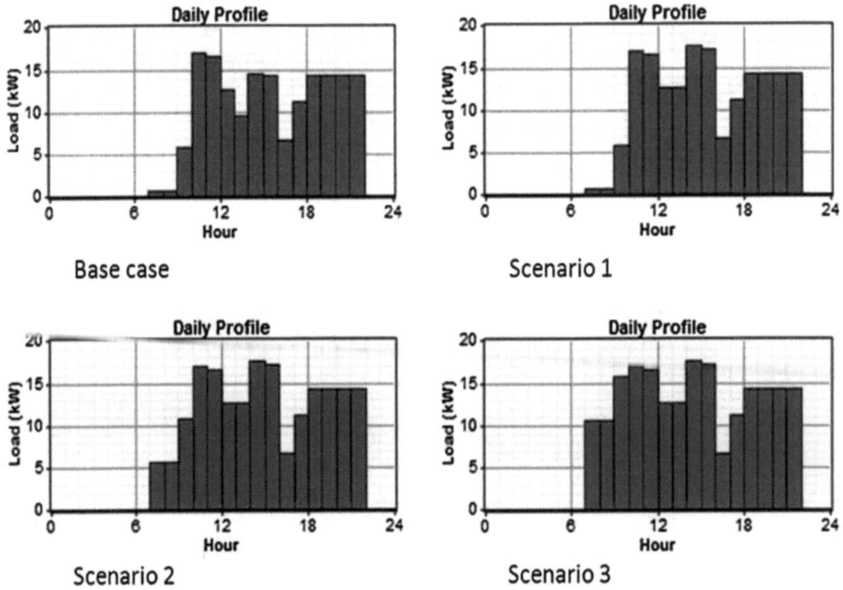

Fig. 5 Load profiles for different scenarios

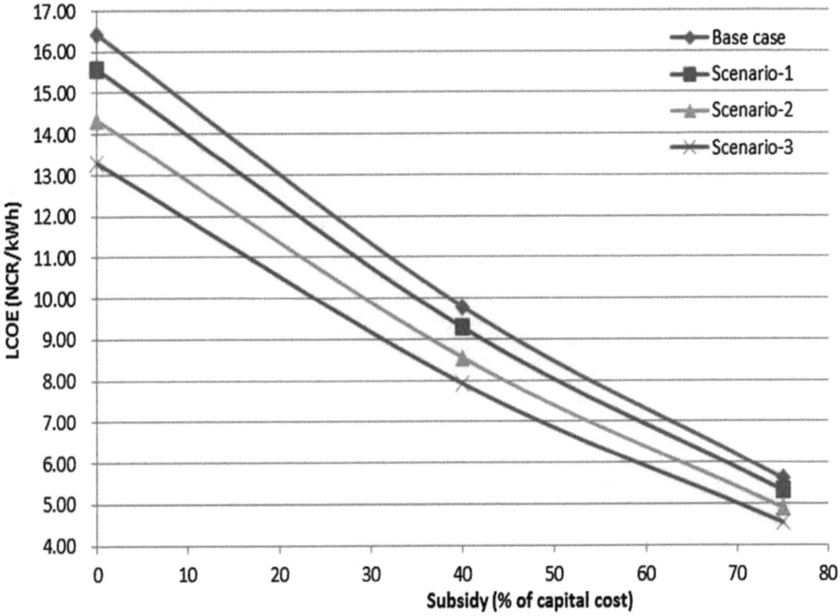

Fig. 6 Subsidy and its impact on LCOE—sensitivity analysis

concerned. Apart from the challenge of low load factor identified above, another related challenge is the lack of technical capacity at the local level to deal with unforeseen technical problems of the project. Additional challenge emerged during the period of project installation. It was difficult for communities to mobilise finance through loans from formal credit institutions due to lack of collateral. The next section discusses in a more detailed manner the generic set of challenges being confronted by the off-grid sector in the country.

8 Barriers to Scale Up

Despite policy thrusts, renewable energy-based mini-grid electrification in Nepal confronts multiple challenges and constraints. There exist multiple economic, social, political and institutional bottlenecks for up-scaling and wider dissemination of renewable energy-based mini-grid electrification projects in the country. The first 3 Year Interim Plan (2007–2010) of Government of Nepal recognised this slow progress of the alternative energy sector in the country and attributed it to the existing economic, social and institutional handicaps.

On the policy front, lack of strong legal framework in terms of an overarching act or policy, absence of clearly spelt out long-term realistic targets and lack of integrated rural development plans retard the growth of renewable energy-based mini-grid electrification process in the country. Most of these targets are ad hoc in nature as spelt out only in the annual budgets of the Government or are set by donor funded programmes. Subsidy policies declared from time to time are argued to have flawed design features thereby leaving scope for manipulation. For instance, subsidy schemes introduced in 1981–1982 resulted in undesirable cost escalation [16]. In overall, subsidy policies in Nepal have not succeeded in delivering the desired outcomes largely due to inherent complexities in the delivery mechanisms [7]. Subsidy mechanisms need to be simpler and should have provisions of gradual phasing out.

The weak institutional structure is characterised by lack of centralised energy planning, duplication of efforts resulting from lack of co-ordination, and disputes between local and national level institutions and cumbersome decision-making processes [16, 28]. Even renewable energy programmes implemented by donor agencies lack co-ordination and harmonisation. Multiple organisations continue to work on alternative energy space and are often having overlapping mandates. Incongruent legal provisions add further woes. For instance, while MHP and SHP systems are de-licensed on one hand, stipulations of Local Self-governance Act 1998, assign power to the local authorities to prioritise the use of water in their jurisdiction, generating potential conflict of interest among different stakeholders. In addition, local level institutions like village/community scale organisations lack managerial capacity to manage micro-hydro systems, thereby pose threat to the sustainability of these projects. This necessitates the need for continuous community mobilisation and capacity building.

Often societal constraints transpire into technical handicaps for the mini-grid projects. Low load factors emanating from low electricity demand have been a major concern for the sector in majority of the cases. It is posited that a majority of community-based micro-hydro projects have a maximum load of 20–25 % [3]. Because of this low load factor, private entrepreneurs do not consider this as a potential business avenue, and generally averse to invest in the sector. There is a need to identify and assess the potential of energy-operated cottage industries in areas where these mini-grids are deployed.

Challenges associated with the financing of mini grid projects in the country require targeted approach. Access to credit has been found to be the major hindrance for promotion of renewable energy-based rural electrification in the country [23]. Poor access to credit by rural entrepreneurs is largely due to lengthy bureaucratic procedure, lack of collateral and due to associated high transaction costs. In addition, financial institutions providing loans for the renewable energy sector in Nepal suffer from problems of bad debt primarily due to financially weak and vulnerable community structures and loose contractual links between equipment providers and their local agents [8]. Thrust should be laid on carbon financing as an additional source of finance to enhance the financial viability of mini-grids. Though some efforts have been undertaken in this direction, there still exist avenues to mobilise additional financial resources through carbon financing. Transaction costs of administering the financing of small projects are high, thereby deterring banks to venture into small-sized projects. In addition, absence of collateral makes it difficult for the banks to provide loans for these small-scale interventions.

There has also been a host of regulatory hurdles encountered by renewable energy sector in the country. Though, AEPC is placed at the helm of affairs as far as renewable energy sector is concerned, it also plays as a regulator for purposes like quality checking, disbursement of subsidies, and waiving of taxes and duties. Regulatory power of AEPC is limited primarily because of dominance of the nodal Ministry, i.e. Ministry of Science, Technology and Environment (MoSTE). Another regulatory challenge comes in the form of regulatory uncertainty about grid extension thereby stifling the private entrepreneur spirit to venture into the field. Studies point that almost 27 % MHP projects are within the vicinity of 5 km of grid electrification, thereby threatened by the possibility of grid extension in the near future [26]. In addition, monopolised structure of NEA has also been putting additional constraints. NEA is unwilling to connect MHPs with the centralised grid systems largely because of small capacity of the projects and due to technical and managerial constraints arising out of connecting these small projects with the grid. Another major regulatory challenge is associated with the lengthy and difficult project approval process. Approvals in the form of water source use license, company registration and tax registration derail the process of project approval [3]. It emanates from the discussion that though a liberalised environment has been created for the development of micro-hydro projects by delicensing projects up to 1,000 kW [21], it appears that the benefits of this liberalised environment has been translated into practice due to several associated constraints.

Inadequate capacities have also been posing as a threat for the successful growth of the sector. At the macro level, limited testing capacities of renewable energy test station (RETS) is creating sort of a 'technology lock in'. This has been primarily due to lack of adequate funds to equip the centre with advanced testing equipment combined with lack of adequate number of professionals to carry out the test.

9 Policy Recommendations

It emerges from the discussion above that renewable energy-based mini-grid systems have become the prime vehicle of electrification, especially in rural areas of the country. Though some laudable efforts have been undertaken by Government of Nepal through AEPC and by several associated actors and institutions, the sector still requires some specific policy thrusts and focus.

One of the challenges encountered by the sector is the lack of adequate investment, private investment in particular. Declaration of a long-term policy for the sector with necessary incentives and benefits could go a long way in attracting private investors into the field. In addition, sustainability of these projects requires mobilising small-scale financing through micro-financing and micro-credit routes. Another pertinent concern is about policy and regulatory certainty regarding grid extension. Demarcation of off-grid villages/localities by the Government could address policy uncertainties about grid extension, as has been done in Sri Lanka very recently.

Techno-economic comparative assessment of solar energy systems with micro-hydro systems reveals that solar-based systems are relatively expensive compared to hydro-based systems. However, a major concern with small-scale micro-hydro systems is the unutilised and underutilised capacity. Though, subsidy schemes exist for better end use applications, this has not been very successful in creating the required income and employment opportunities to sustain the productive end uses. Therefore, there is a need for better planning and designing from the beginning of the project initiation for the optimal use of energy keeping in consideration the present and future energy requirements of the local people.

Institutionally, the sector requires better co-ordination and harmonisation among various ministries, agencies, and donor agencies and other actors. AEPC by combining the entire donor funded programmes under one umbrella, i.e. National Rural and Renewable Energy Programme has been able to address this co-ordination problem to some extent. However, there still exist legal entanglements across policies and legislations, which require focused attention.

On the policy front, a long-term credit disbursement path should be declared with phased reduction of subsidies in order to develop sustainable off-grid energy sector in the country. Pockets should be identified, where private entrepreneurs could take lead roles in promoting off-grid energy systems. In addition, in the name of quality control, private entrepreneurs should not be demotivated to introduce better and advanced technologies. Given limited capacity of quality control authorities, quality ranges should be spelt out with flexibility to allow private entrepreneurs to innovate and introduce new technologies.

10 Conclusion

Renewable energy-based mini-grids in Nepal could serve as an effective alternative to the crisis ridden centralised electricity system. Substantial progress in this front has been achieved in the country primarily due to presence of a strong and focused engagement of several key institutions and agencies like AEPC, donor organisations, financial institutions like banks, private associations and, moreover, due to emergence of strong market elements nurtured by private entrepreneurs in the field. Financial evaluation of a micro-hydro mini-grid project reveals there exists room for cost reduction through enhancement of productive uses of energy and creation of livelihood opportunities. Comparative assessment of a micro-hydro project with similar capacity of solar project suggests that micro-hydro projects have cost advantage over solar projects. However, the sector still suffers from several anomalies and imperfections leading to slow progress of the sector. Political instability and uncertainty stand as a major roadblock. Adhocism and changing focus of donor funded programmes are distorting the very foundation of the sector. On the regulatory front, uncertainty about grid extension leads to underutilisation of private sector potential. In addition, poor access to credit and absence of formal financial institutions at the local level debar the ability of private entrepreneurs to venture into the sector. Importantly, lack of post installation evaluation of projects produces only dry statistics on the number of projects deployed without any information about their sustainability. It is pertinent to address all these concerns in a systematic and comprehensive manner to drive the sector on a sustainable trajectory.

Annexure

Table A.1 Key stakeholders interviewed

S. No.	Name	Organisation
1	Mr. Saroj Rai	SNV, Netherlands, Nepal
2	Prof. Govind Raj Pokharel	Alternative Energy Promotion Centre (AEPC), Nepal
3	Mr. Ram Prasad Dhital	Alternative Energy Promotion Centre (AEPC), Nepal
4	Mr. Madhusudhan Adhikari	Alternative Energy Promotion Centre (AEPC), Nepal
5	Mr. Jagadish Kumar Khoju	Alternative Energy Promotion Centre (AEPC), Nepal
6	Mr. Bhupendra Shakya	Renewable Energy for Rural Livelihood Programme (RERL)
7	Mr. Dilli Prasad Ghimire	National Association of Community Electricity Users-Nepal (NACEUN)
8	Prof. Tri Ratna Bajracharya	Centre for Energy Studies (CES), Institute of Engineering, Tribhuvan University, Nepal
9	Dr. Shree Raj Shakya	Centre for Energy Studies (CES), Institute of Engineering, Tribhuvan University, Nepal
10	Mr. Vishwa Bhushan Amatya	Practical Action, Nepal
11	Mr. Vijaya P Singh	UNDP, Nepal
12	Mr. Satish Gautam	Renewable Energy for Rural Livelihood Programme (RERL)
13	Ms. Anupa Rimal Lamichhane	UNDP, Nepal
14	Sanjay Kumar Gokhali	GIZ, Nepal
15	NEA Nepal	
16	Mr. Purna N Ranjitkar	Nepal Micro Hydropower Development Association (NMHDA)
17	Mr. Raj K Thapa	Solar Solutions Private Limited
18	Mr. Prem Bdr. Basnet	Renewable Energy Test Station (RETS), Nepal
19	Mr. Rudra Mani Pokharel	Renewable Energy Test Station (RETS), Nepal
20	Ms. Barsha Shrestha	Clean Energy Development Bank Ltd., Nepal
21	Mr. Nabin Bhujel	Suryodaya Urja Pvt., Ltd.,

References

1. AEPC (2012a) Mini-grid outlook 1999–2012. Alternative Energy Promotion Centre, Kathmandu
2. Bhandari R, Stadler I (2011) Electrification using solar photovoltaic systems in Nepal. Appl Energy 88:458–465
3. Ghale BB (2013) Private sector on micro-hydro power for energy services in rural society: experience of Nepal. Paper presented at energy for all mini-grid workshop, 6–7 Feb 2013, Kathmandu, Nepal
4. Gurung A, Ghimeray AK, Hasan SHA (2012) The prospects of renewable energy technologies for rural electrification: a review from Nepal. Energy Policy 40:374–380
5. IEA (2012) World energy outlook 2012. International Energy Agency, Paris
6. Mainali B, Silveira S (2013) Alternative pathways for providing access to electricity in developing countries. Renewable Energy 57:299–310

7. Mainali B, Silveira S (2012) Renewable energy markets in rural electrification: country case of Nepal. Energy Sustain Dev 16:168–178
8. Mainali B, Silveira S (2011) Financing off-grid rural electrification: country case Nepal. Energy 35:257–265
9. Ministry of Finance, Government of Nepal (2012) Economic survey of Nepal 2011–12
10. Mishra A, Sarangi GK (2011) Off-grid energy development in India: an approach towards sustainability. OASYS Working Paper 2011, No 12
11. NAPA (2010) National Adaptation Programme of Action (NAPA) to climate change. Ministry of Environment, Government of Nepal
12. NEA (2006) Fiscal Year 2006–07—a year in review. National Electricity Authority, Kathmandu
13. NEA (2011) Annual report 2011. National Electricity Authority, Kathmandu
14. NLSS (2011) Nepal living standard survey. Ministry of Statistics, Government of Nepal
15. Nepal R, Jamasb T (2012) Reforming small electricity systems under political instability: the case of Nepal. Energy Policy 40:242–251
16. Nepal R (2012) Roles and potentials of renewable energy in less developed economies: the case of Nepal. Renew Sustain Energy Rev 16:2200–2206
17. Nguyen KQ (2007) Alternatives to grid extension for rural electrification: decentralized renewable energy technologies in Vietnam. Energy Policy 35:2579–2589
18. NPC (2007) Three year interim plan (2007/08–2009/10). National Planning Commission, Government of Nepal
19. NPHC (2012) National population and housing census 2011. National Planning Commission Secretariat, Central Bureau of Statistics, Government of Nepal, Kathmandu
20. Palit D, Chaurey A (2013) Off-grid rural electrification experiences from South Asia. In: Bhattacharyya SC (ed) Rural electrification through decentralised off-grid systems in developing countries. Springer-Verlag, London
21. Palit D, Hazarika N (2004) Renewable energy programme in Northeast India: issues, constraints and prospects. In: Behera MC (ed) Development dilemma in the era of globalization—reflections from North East India. Mittal Publications, New Delhi
22. Parajuli R (2011) Access to energy in the Mid/Far west region—Nepal from the perspective of energy poverty. Renewable Energy 36:2299–2304
23. Pokharel G (2013) Mini-grid promotion and scaling up in Nepal. Paper presented at energy for all mini-grid workshop, 6–7 Feb 2013, Kathmandu, Nepal
24. Rana A (2012) Renewable energy financing: banks perspective. Paper presented at energy for all mini-grid workshop, 6–7 Feb 2013, Kathmandu, Nepal
25. REDP (2011) Achievements of REDP. Rural Energy Development Programme (REDP), Government of Nepal
26. Shakya B, Sharma SK (2013) Mini-grid a step ahead in micro-hydro sector: experience from Baglung mini-grid pilot project. Paper presented at the energy for all mini-grid workshop, 6–7 Feb 2013, Kathmandu, Nepal
27. Shrestha RS (2010) Electricity crisis (load shedding) in Nepal, Its manifestations and ramifications. Hydro Nepal 6:7–17
28. Sovacool B, Bambawale MJ, Gippner O, Dhakal S (2011) Electrification in the mountain kingdom: the implications of the Nepal Development Project (NPDP). Energy Sustain Dev 15:254–265
29. The Asia Foundation (2012) The political economy analysis of electricity tariff restructuring in Nepal. The Asia foundation. http://asiafoundation.org/resources/pdfs/PEAelectricitytariffrestructuringinNepal.pdf
30. UNDP (2013) Human development report 2013—the rise of South: human progress in the diverse world. United Nations Development Programme, New York
31. World Bank (2013a) Electric power consumption. The World Bank, Washington D.C. http://data.worldbank.org/indicator/EG.USE.ELEC.KH.PC?order=wbapi_data_value_2010+wbapi_data_value&sort=asc

32. World Bank (2013b) Project Appraisal Document 2013. The World Bank, Washington D.C.
33. Yadoo A, Alexandra G, Hetaher C (2011) Low carbon off-grid electrification for rural areas in the United Kingdom: lessons from the developing world. Energy Policy 39:6400–6407
34. Zahnd A, Kimber HM (2009) Benefits from a renewable energy village electrification system. Renewable Energy 34:362–368

Viability of Husk-Based Mini-Grids in South Asia

Subhes C. Bhattacharyya

Abstract As South Asia is a major producer of rice, this chapter analyses the financial viability of rice husk-based power generation in South Asia. The chapter first presents the business models of Husk Power Systems (HPS) and Decentralised Energy Systems India (DESI Power), two enterprises that have successfully provided electricity access by generating power using rice husks. It then presents financial analysis of two alternative supply options, namely a small 20 kW plant serving some 400 consumers under different demand scenarios, and a 20 kW plant serving a rice mill and a rural community of 400 consumers. It then explores the viability of a larger 200 kW plant serving a rice mill and a cluster of rural communities. We show that serving low electricity consuming customers alone leads to part capacity utilisation of the electricity generation plant and results in high cost of supply. But the tariff plan based on contracted capacity (Watts) rather than their electric energy use has so far ensured sufficient revenue generation for HPS to sustain its operations. The financial viability improves as some consumers use more electricity but the declining block tariff used to promote such consumption behaviour benefits high consuming customers at the cost of poor consumers. The integration of rice mill demand, particularly during the off-peak period, with a predominant residential peak demand system improves the viability and brings the levelised cost of supply down. Finally, using larger plant sizes to take advantage of economies of scale brings down the cost significantly and can be quite competitive with alternative sources of supply. But the higher investment need and the risks related to monopoly supply of husk from the rice mill, organisational challenges of managing a larger distribution area and plant operation challenges (or risk of failure) can adversely affect the investor interest. Moreover, the regulatory uncertainties in respect of off-grid electrification in general and the coverage of larger geographical areas in particular can hinder business activities in this area.

S. C. Bhattacharyya (✉)
Institute of Energy and Sustainable Development, De Montfort University, Leicester LE1 9BH, UK
e-mail: subhesb@dmu.ac.uk; subhes_bhattacharyya@yahoo.com

S. C. Bhattacharyya and D. Palit (eds.), *Mini-Grids for Rural Electrification of Developing Countries*, Green Energy and Technology,
DOI: 10.1007/978-3-319-04816-1_14,

1 Introduction

Rice is the staple food in many South Asian countries and this region is a major rice producer, contributing about 30 % of global rice production [3]. Accordingly, rice cultivation and the subsequent processing of rice (i.e. milling) are major economic activities in the rural areas of the region. Rice production process generates considerable amounts of wastes in the form of rice straw and rice husks. Although a part of the waste is used as fodder (or in animal food preparation), as roofing material and in brick kilns, most of it is used as a source of energy, mainly for cooking purposes and for parboiling of rice. However, the conversion efficiency of such applications tends to be very low (below 10 % indicated by Kapur et al. [5] as they rely on primitive systems that burn the residue in simple stoves. Although researchers have considered the potential and commercial viability of improved utilisation of the resource for energy purposes, particularly for electricity generation in various contexts (e.g. [5, 10] among others), the successful delivery of off-grid electricity by an Indian firm, Husk Power Systems (HPS), has renewed the commercial interest in this waste to electricity conversion. Since its inception in 2007, HPS has installed 80 mini power plants of 20–100 kW size in various Indian villages of Bihar and provided access to electricity to more than 200,000 people. As rice is produced in many parts of South and South-East Asia, the lessons from such successful commercial ventures can allow wider application of this waste to electricity technology.

The purpose of this chapter is to analyse the business model and techno-economic feasibility of rice husk-based electricity generation to understand the basic conditions required for developing a viable husk power business. Moreover, the issue of rice husk-based power in South Asia has not been widely analysed so far. This chapter tries to bridge the above gaps. The chapter is organised as follows: Sect. 2 presents a review of the HPS business model and DESI Power model; Sect. 3 considers rice-husk based electricity generation for rural applications with and without rice mill demand and checks for the viability of such a system. It then expands the system size to consider the viability of operation at the village cluster (or block) level. Finally, some concluding remarks are presented in the last section.

2 Husk-Based Power Systems

Husk has been used and promoted as an energy resource for electricity generation by two private businesses, namely the Husk Power Systems and DESI Power. We review their business models in this section.

2.1 The HPS Model

The Husk Power Systems calls itself a ‘rural empowerment enterprise’ that enables rural development by providing access to electricity while ensuring environment protection, well-being of local population and empowerment of local communities. Headquartered in Patna, Bihar, one of the least electrified states in India, the HPS was developed in 2007 by a group of locals who wanted to combine low cost electricity supply with high quality service to challenge the conventional wisdom that categorises rural electricity supply as a non-viable business. Their search for a viable, small-scale electricity generation option led them to an abundant local resource in the rice-growing region of the country, which was hitherto treated as a waste, the husk that can be procured at very low cost for conversion to electricity. However, given the high silica content and silica-cellulose structure of husk, it does not burn easily and causes wear and tear of components coming in contact with it [5]. This required customised, proprietary design of gasification technology that can be built and maintained locally without high level technical expertise.

The HPS organised its business synergizing a number of factors that taken together produced the desired outcome (see Fig. 1). The power plants are located in the paddy growing area and more importantly, close to rice mills. In some cases, the company has integrated the rice mill business to ensure business viability [6] and to internalise the symbiotic relationship with the power plant. The mill provides a steady supply of husk, while the power plant supplies electricity that is reliable and perhaps at a cheaper rate than the alternative sources like diesel generators. The rice mill also provides a base load demand for the power plant and thus helps achieve a better plant utilisation rate, which in turn reduces the average cost of supply.

Moreover, the HPS has exploited other potential income generation opportunities and created a community support system to ensure better integration with the community. The char obtained from burning the husk is used for incense stick making where local women are employed. It is reported by Sevea [9] that a 32 kW plant produces 6 tonnes of incense sticks per year. Silica precipitation is sold for mixing with cement. The HPS has registered a Programme of Activities (PoA) under the Clean Development Mechanism[1] for its electrification activity that aims at providing electricity to non-electrified areas through renewable energy sources using biomass gasification, solar PV systems (both AC and DC) and hybrid systems combining solar PV and biomass gasification. It has estimated that 215 Mt of CO_2 emissions will be reduced per year. The PoA will remain active for 28 years (until 2040) and will generate an additional stream of revenue support for the plants. The innovative approach towards revenue generation from various products surely helps in improving its financial position.

[1] See http://cdm.unfccc.int/ProgrammeOfActivities/poa_db/3Z2JFO1WYTASQLUB0GE54XM6IDHKN9/view for details.

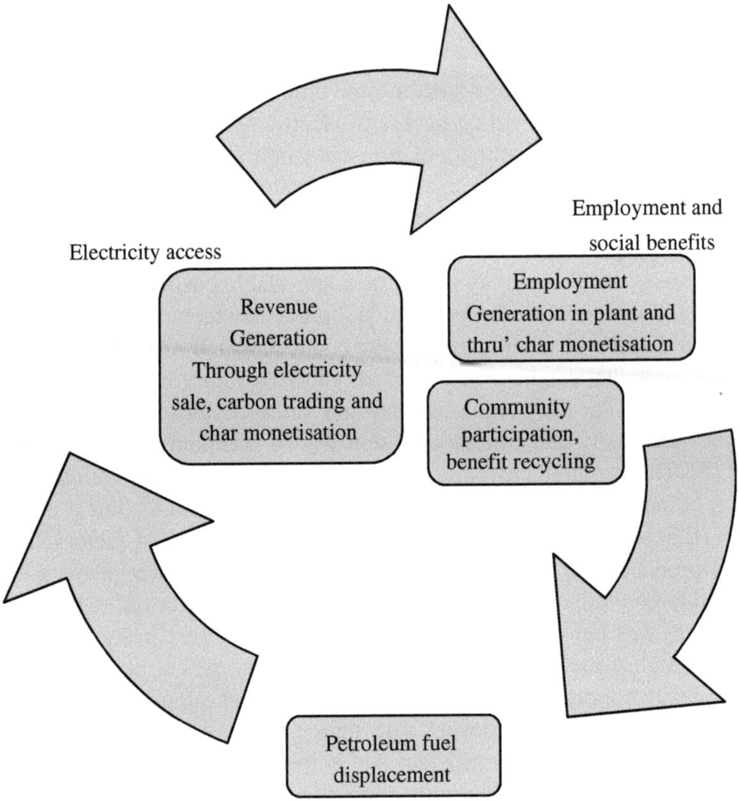

Fig. 1 The HPS business model. *Source* Based on Pandey [8]

In addition, the company has been relying on smart technologies to reduce operating costs and potential revenue losses. The distribution mini grid uses smart features for remote monitoring of the system. The company uses smart metres for billing purposes and the bill collectors use hand-held data recorders to keep record of the collection made from door-to-door bill collection rounds. The use of insulated cables hoisted on bamboo poles reduces the potential for electricity theft while reducing the capital investment required for the distribution network.

The company has also developed a system of direct mutual benefits within the local community where its plants are located. Each plant engages 3 to 4 staff—a plant operator, an electrician, a husk loader and a bill collector, who are taken from local youths and trained by the company. Through this employment scheme, about $400/month is recycled in the local economy. The rice mills supplying husk to the power plant receive about $25/tonne of husk (or about $2500/year for a 32 kW plant). This is an extra source of income for the mills and some of them share this with their customers by offering a reduced fee for milling. Their waste disposal problem is also taken care of in the process. The incense stick making activity mentioned previously also provides earning opportunities to local women. In addition, in some cases, the bill collector also acts a "travelling salesman" who takes orders from the households, procures them in bulk from the nearby town and delivers to the households for a small commission.[2] This ensures an extra income for the bill collector and the households get their goods at wholesale rates. The HPS claims that in the process it returns more to the local community than it collects through its electricity bills.

As a consequence, the company has been successful in extending its business to create more than 80 plants in 300 villages to provide electricity to more than 200,000 people. The process for installing a new plant is normally initiated on receipt of a request for such a service from a village or the local authority. The HPS requires an initial deposit from the interested villagers to cover up to three months cost of electricity. This is taken in advance and once a suitable number of consumers have expressed interest, the feasibility of a biomass-based plant is undertaken, which essentially identifies a secure source of fuel supply for the plant. The site selection, which plays an important role, is thus driven by economic viability of the business. Critics point out that HPS only operates in niche areas where villages had been receiving diesel-based electricity from local entrepreneurs and the relatively rich consumers in those areas were already paying high charges for their electricity. HPS has thus displaced diesel-based generation by offering electricity at a cheaper rate.

The installation process takes about three months and a local team is set up to operate the system on a daily basis. The supply is given for a fixed period of time, normally for 6–7 h in the evening using a 3 phase 220 V system. Consumers willing to gain access have to pay a connection charge and a flat monthly fee (varying between \$2 and \$2.5) for the basic level of service (2 CFLs and a mobile charging point, called the 30 W package). However, customised packages are also available and consumers with a higher level of consumption benefit from a lower unit rate. A typical plant can serve, depending on the size of the village and willing consumers, up to four villages with about 400 consumers within a radius of 1.5 km of the plant. Small commercial enterprises are also supplied with electricity, but they generally pay a higher flat rate of \$4 to \$4.5/month due to higher demand.

The HPS aims to provide electricity to 10 million people in 10,000 villages by installing 3,000 plants by 2017.[3] It has successfully managed to secure funds from a variety of sources in the past, including charitable sources and financial institutions. Although the plants initially followed the Build, Own, Operate and Manage (BOOM) model, the HPS is also employing other modes of operation,

[2] However, the response to this practise has not been very encouraging and the company appears to have stopped this arrangement.

[3] This is indicated in the company website and in the Programme of Activities Design Document available from the UNFCCC website.

namely the Build, Own and Maintain (BOM) and Build and Maintain (BM) lately to grow faster. In the BOOM model, the company looks after the entire chain of the business, which in turn requires a dedicated set of staff that needs growing with new plants. The overhead can be high and the company faces the investment challenge. In the BOM model, the business is partly shared with an entrepreneur who makes a small contribution to capital (about 10 %). The HPS maintains the plant and gets a rental fee but the operational aspects are taken care of by the entrepreneur. This reduces some of the management tasks for the HPS, and builds a local network of entrepreneurs but the HPS still faces the investment challenge. Moreover, verifying the quality of the local entrepreneur is a challenging task and the speed of replication using this approach remains unclear. The company transfers the ownership after a specified period of time, upon recovering the cost of investment. The Build and Maintain model essentially transforms the HPS into a technology supplier where its role is limited to supply of the equipment for a fee and maintaining the plant through a maintenance contract. The supply business is undertaken by a local entrepreneur and the HPS does not get involved in this activity, although the entrepreneur uses the HPS brand for the supply. Thus, the business uses the franchisee model in this case and as long as the franchisee is able to finance the investment and is capable of running it effectively, the business can grow. Although this is a proven approach in many other businesses, in the context of rural electricity supply this has not been widely used. This model requires a strong quality control and standardisation of the business operation but it is not clear whether or to what extent this has been developed in HPS. It appears that HPS is planning to focus more on technology supply and maintenance as part of its expansion strategy but in rural electricity market, the challenge of operation and revenue collection remains. HPS has been a successful model in the entire chain but more attention to the upstream activities can leave a significant vacuum in the business, thereby affecting the overall performance and expansion of the activity.

Thus, a rapid replication of activities, which is necessary for achieving the company target of electrifying 10 million people by 2017, depends to a large extent on how the above business models work. This expansion demands significant energy resources, financial resources, management capabilities, skilled local staff and commensurate manufacturing capabilities. It is not clear whether the company can ensure all the success factors to ensure a rapid growth. It is reported that the husk price has significantly increased since its plants have started operation as its availability has become an issue due to pressures from alternative uses. Moreover, the niche areas for its operation where consumers can afford high tariffs may be difficult to find in the future, which can limit the growth prospect. In addition, the plant size will range between 5 and 250 kW,[4] which fails to exploit the economies of scale and scope and affects the business prospects. Clearly, the replication issue requires further investigation.

[4] Mentioned in the PoA DD indicated above.

2.2 DESI Power

The Decentralised Energy Systems India Private Limited (DESI Power)[5] is a not-for-profit company registered in Delhi and was promoted by Development Alternatives (DA) and DASAG Seuzach, a Swiss renewable energy company[6] in 1996. Its aim is to provide affordable and clean decentralised energy to rural communities for rural development by offering an integrated solution. To achieve this objective, DESI Power builds and operates decentralised power plants, creates rural service infrastructure through mini/micro grids, engages with the local community for establishing partnership models and organisation structures for community-based management of the services, and provides training for capacity building in rural areas for micro-enterprise and business development. It integrates two core activities, namely rural electricity supply through renewable energy sources and rural development through rural enterprising using the following arrangements:

(1) DESI Power (DP) for generation and supply of electricity to village consumers;
(2) DESI Power Gramudyog (DPG) for village level businesses and enterprises;
(3) DESI Mantra for training and capacity building; and
(4) Joint Ventures and partnerships for energy service and village enterprises.

Like HPS, DESI Power has also relied on biomass gasification systems for rural electricity supply and the first plant was set up in 1996 at Orcha, Madhya Pradesh in TARA gram campus of DA village in Madhya Pradesh. It has set up 16 power plants in total by 2012. These power plants have installed capacities ranging between 11 and 120 kW, and use biogas and dual fuel systems. Unlike HPS, most of these plants are used for captive or own energy use purposes by small industries that would otherwise rely on diesel generators for their electricity supply to complement unreliable grid supply. Typically, in such cases it acts as a rural power generator and enters into a power purchase agreement with the buyers' organisation (which could be an individual entity, a co-operative society or an association of buyers). The company also enters into biomass purchase agreements with local suppliers (who can be villager groups or commercial suppliers). It also assists in the development of micro-enterprises, often linked to agriculture.

However, beyond this niche area of operation, DESI Power has also installed four mini-grid systems to supply electricity to households, micro-enterprises and mobile phone towers. In order to achieve cost effective operation, an anchor demand (like the mobile phone towers) is generally included in the system that offers the base load and increases the financial security for the operation. Until

[5] This section is based on the following sources: company website at http://www.desipower.com/; company profile at http://www.desipower.com/downloads/DESI-Power-Company-Profile.pdf; and IFMR [4].

[6] They acquired the licence for the gasifier technology developed by the Indian Institute of Science, Bangalore.

2012, 10 mobile phone towers have been connected to its existing power plants and it plans to expand this to another 20 towers in 2 years. Moreover, the emphasis is on generating as much electricity as possible through the inclusion of micro-enterprises. This reduces the average cost of supply that in turn enhances viability of the micro-enterprises. This interdependence is exploited to ensure affordable power as well as rural economic development.

In contrast to HPS, which relies on temporary networks, DESI Power has installed underground cables to connect consumers. Although underground cabling is less prone to theft and is more secure, it is a costlier option for the company. DESI Power also offers a range of bill collection options—daily for small households and micro-enterprises and monthly for bigger industrial/institutional consumers. Although this appears to be working for them at the moment, the daily collection of revenue is a labour intensive, costly option. Moreover, it follows the Build-Operate-Transfer model of operation wherein it hands over the plant to the local community or village groups after a period of operation.

DESI Power follows a pricing policy that mimics the charges imposed by diesel-based electricity suppliers. Normally, for a light point of 60 W a fixed rate of Rs 5/day is charged, while micro-enterprises pay a fixed fee for the service. For example, 1 h of irrigation water supply from a 5 HP pump is charged at Rs 60 [4]. Actually they sell the irrigation service and not energy, and so the tariff is based on water supply and not how much energy is consumed. This was so designed as farmers can understand it better and link with the outcome.

To take advantage of carbon credits, DESI Power has registered a small-scale project with the CDM Board for establishing 100 biomass gasifier-based decentralised, power plants in the District of Araria in Bihar state (India). The plants will be of 50 kW capacity with the exception of a few 100 kW plants. In total, 5.15 MW of capacity was expected to be installed, which will reduce about 360 kilo tonnes of CO_2 emission over the first ten years of the project.[7] However, it appears that only a few plants have been set up so far, thereby significantly underachieving in terms of emissions reduction and capacity addition targets. Although the company expansion plan maintains that it aims to achieve its 100 village target in 3–5 years, the outlook remains uncertain.

Apparently, the investment challenge is the most important barrier faced by DESI Power. It estimates that a 100 kW gasifier plant (running on pure producer gas) costs $80,000, while the micro-enterprise development requires another $73,000. The cost for a 50 kW plant is indicated as $45,000. The investment requirement for a 20 × 5 HP diesel generator is just $10,000. This capital cost difference affects the electricity generation cost particularly at low plant utilisation factors.[8] In addition, the village co-operatives or associations have limited

[7] For details, see the CDM PDD https://cdm.unfccc.int/filestorage/0/B/G/0BGM0Q8ZJQQXFMD2YOWWDGIYOD7CS0/PDD%20revised.pdf?t=dU98bW4zbTYzfDCbexG61m04_xsFtcH5OpIy.

[8] CDM PDD indicated in footnote 7.

borrowing capacity and do not have the required deposit or bank guarantees for availing any debt finance. Similarly, in the absence of a bankable agreement with the co-operatives or the buyers, the company cannot finance its projects. This constraint appears to be having a significant effect on the business expansion of the company. In addition, the technical capacity to deliver plants and human capacity to operate and maintain them are also constrained.

3 Business Case of Power Generation from Husk

The proprietary nature of information does not allow us to analyse the economic and financial dimensions of the rice husk-based power business accurately. Instead, we present a set of cases based on available information and realistic assumptions about electricity generation from husk. The analysis is presented for a plant size of 20 kW for different levels of electricity demand and alternative capital structures. It is then extended to include electricity supply to the rice mill and to explore the possibility of supplying to a wider geographical area using bigger plants.

3.1 Providing Access to Electricity with a 20 kW Plant

We assume that a plant generally serves about 400 households, most of whom may be consuming a minimum amount of electricity for lighting and mobile phone charging. We consider alternative scenarios of demand as indicated in Table 1.

It can be noticed from Table 1 that in Scenario 1, when all 400 households use the basic level of electricity and if there are 20 small commercial units serviced at 75 W/unit, the demand can be met by a 20 kW system but the plant runs at part load (66 % of its full capacity). In the second scenario where 90 % of the households use the basic level of demand, the rest 10 % use a moderate level of electricity at 75 W/household, and 30 commercial units are considered instead of 20 units, the demand increases marginally but a 20 kW plant still can service the load at 78 % loading. The third scenario modifies the residential load slightly to ensure a 100 % loading of the plant. However, as the plant runs for a fixed period of 6 h, its overall capacity utilisation in the above scenarios does not exceed 25 %. This is relatively low for a power generating plant.

For the financial analysis of the 20 kW generator plant, the following assumptions are made:

(a) The cost per kW of capacity is taken as $1,300 following Sevea [9]. Pandey [8] suggests a slightly lower value, but the above cost appears to be in line with similar plant costs. DESI Power also suggests a lower capital cost of $800/kW, but this may be underestimating the capital requirement. The cost includes civil construction costs.

Table 1 Alternative demand scenarios

Description	Scenario 1	Scenario 2	Scenario 3
Total number of households serviced	400	400	400
% of HH using basic 30 W service	100 %	90 %	85 %
% of HH using a medium level of 75 W	0 %	10 %	10 %
% of HH using high demand of 250 W	0 %	0 %	5 %
Number of commercial units	20	30	30
Demand by commercial units	75 W	75 W	75 W
Hours of service	6	6	6
Days of operation	365	365	365
Electricity demand (kWh/year)	28,908	34,164	43,800
Required Plant utilisation (for a 20 kW plant)	16.5 %	19.5 %	25 %
Plant loading (for 6 h of operation)	66 %	78 %	100 %

(b) The cost of distribution network per kilometre is taken as $2,000. IFMR [4] indicates that the distribution network accounted for 8–9 % of the total cost of DESI Power installations. This cost can vary depending on the quality of the network, materials used, terrain and cost of labour. For underground cables, the cost will be significantly higher, while for distribution systems using bamboo poles, it may be lower. For this analysis, it is considered that the consumers are within a radius of 2 km of the plant.

(c) The monthly operating cost is considered to be about $100 [9]—or about 4 % of the capital cost.

(d) Each plant employs four employees with a salary of $100/month on average, which is close to the monthly salary cost of $380 indicated in Sevea [9].

(e) The plant life is taken as 15 years and the plant operates 6 h/day, every day of the year.

(f) A price of $25/t is used for husk and it is assumed that the fuel price remains unchanged over the project lifetime.

(g) The calorific value of husk is taken as 12.6 MJ/kg (Kapur et al. 1996) and the conversion efficiency of gasifier is taken as 20 %.

(h) The cost of debt is taken at 5.5 %, while the rate of return on equity is taken as 10 %.[9] The weighted average cost of capital is used to determine the discount rate.

(i) A straight line depreciation is used after allowing a 10 % salvage value for the asset at the end of its life. Where grant capital is used, it is assumed that the grant capital reduces the capital required for investment and the depreciation charge is reduced accordingly. Although the grant capital can be treated differently in accounting terms, the above provides a simple treatment of the grant.

[9] The analysis uses US dollar as the currency and the cost of debt and equity appropriate for dollar dominated funding is used here.

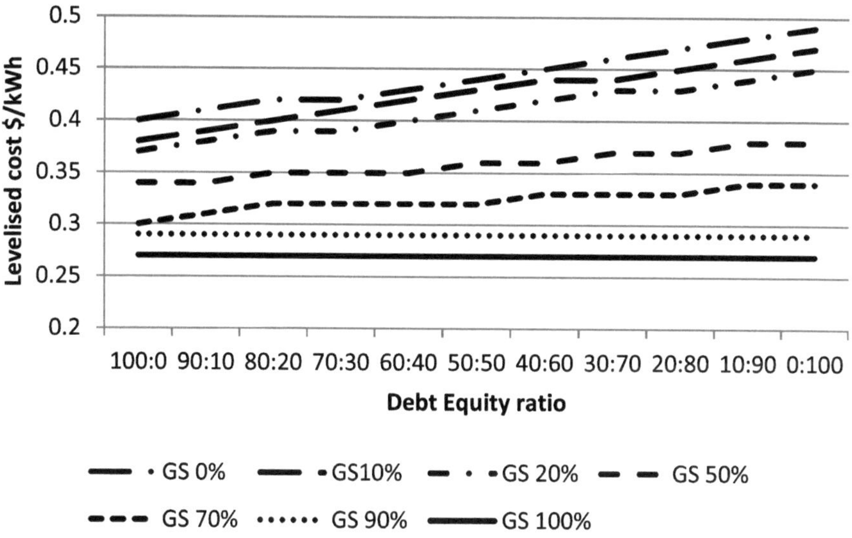

Fig. 2 Levelised cost of electricity supply for Scenario 1. *Note GS* Grant share

(j) It is assumed that the company is not paying any tax, and hence the tax benefit arising from debt capital does not apply here.

Considering alternative debt equity combinations and grant capital share, the levelised cost of electricity supply is estimated. For Scenario 1, the result of the levelised cost analysis is shown in Fig. 2. As expected, the lowest levelised cost is obtained when the entire capital requirement comes from grants and the cost for this scenario comes to $0.27/kWh. But if no grant is received, the cost of supply that has to be borne by the consumers varies between $0.4 to $0.49/kWh depending on the share of debt and equity. This clearly shows that part load operation of the system is a costly option despite the low capital cost per kW compared to other technologies (such as solar PV or wind). Clearly, both HPS and DESI Power have realised this and used adequate households and/or micro-enterprises to ensure high plant capacity utilisation.

However, the important issue is whether or not a flat rate charge of $2 or $2.5/month/household can recover the expenses in Scenario 1. As the consumers use only 5.5 kWh/month, their effective tariff varies between $0.36 and $0.46/kWh depending on $2 and $2.5 monthly charges, which is considerably higher than the prevailing rate for grid-based electricity. Therefore, as long as the levelised cost of electricity supply is below the above tariff, the business becomes viable in this scenario. If the company charges $2.5/month, even without subsidy it can operate the business profitably as long as the debt–equity ratio is not worse than 50:50. If it charges $2/month, the company needs at least 50 % capital grant subsidy to run the business, unless other sources of income can make up for the loss. As other income tends to be limited in nature, it becomes clear that providing access to poor households with limited demand remains a vulnerable business.

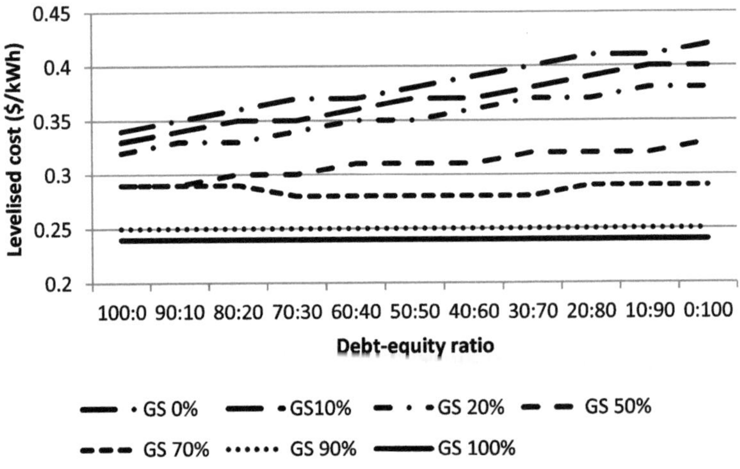

Fig. 3 Levelised cost of electricity supply for scenario 2. *Note GS* Grant share

In Scenario 2 as the plant utilisation rate improves, the cost of supply reduces to \$0.24/kWh for a capital subsidy of 100 %, while the cost varies between \$0.34 and \$0.42/kWh for no capital subsidy (see Fig. 3). However, although the cost per kWh of electricity reduces, the income would not change if all residential consumers are charged at a flat rate. Consequently, when different consumer categories use different levels of electricity, a single flat rate will not be sufficient to recover the cost. But a higher flat rate for the high-end consumers could generate adequate revenue in this case and those consuming higher quantities will end up paying a lower average rate, due to higher consumption. The tariff per Watt instead of watt-hours is thus a simple but effective way of passing higher charges to poorer consumers in disguise.

In the third scenario, the levelised cost reduces even further due to higher demand. In fact, this scenario, as expected, produces the lowest cost of supply (see Fig. 4). It needs to be mentioned here that the levelised cost of supply still remains quite high compared to the grid-based supply particularly when there is no capital subsidy, which arises due to low efficiency of the system and a comparatively higher capital investment required per kW of capacity. But compared to other renewable energy sources (such as solar PV or wind), the cost of supply is lower. This supports the claim made by HPS that they are in an advantageous position compared to other renewable technologies. However, the tariff for grid-based supply may not be a true comparator given the unreliable and poor quality of supply. Consumers tend to spend considerably higher amounts for alternative sources of supply (e.g. from generator sets). Accordingly, the willingness to pay for a reliable supply is likely to be higher than the tariff for grid supply, particularly for commercial and industrial consumers.

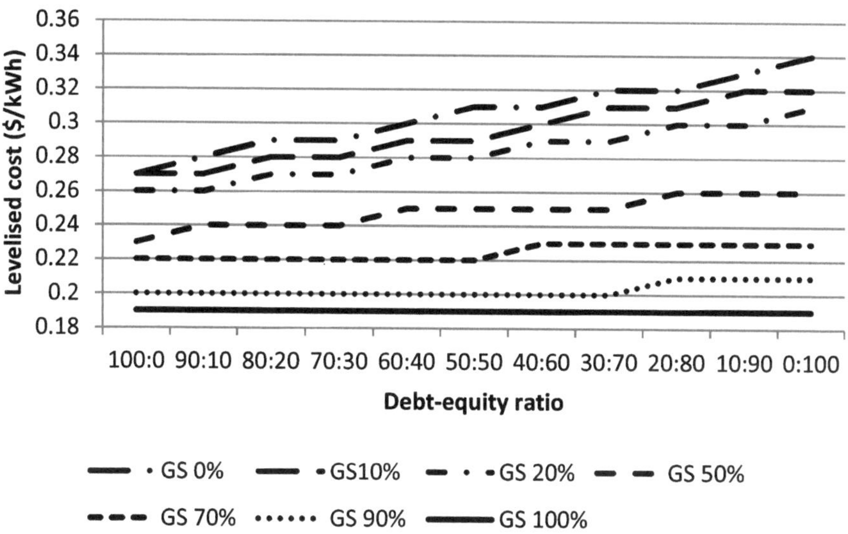

Fig. 4 Levelised cost of electricity supply for Scenario 3. *Note GS* Grant share

Once again, a differential tariff will be required to ensure adequate revenue generation. From the company's perspective, running the plant near its full load will ensure higher profitability, and clearly this will ensure that the operation can be sustained with limited or no financial support. But grant capital surely contributes towards risk mitigation and acts as an incentive for the supplier.

It can be concluded that limited electricity supply for a fixed duration can be effectively provided using the rice husk-based system. The cost-effective operation of the system however requires that the plant is operated near full load by enlisting adequate number of consumers, preferably with some demanding more than just the basic level of supply (30 W/household). Although the cost of supply remains higher than the prevailing grid-based supply, the business can be run viably with a suitably designed tariff system. The difference in the approach between HPS and DESI Power can be understood from this analysis. HPS has ensured viability by enlisting adequate number of residential customers, whereas DESI Power enlisted the support of micro-enterprises. This avoids reliance on a large number of very small consumers as the business or commercial load tends to be much higher than the basic level of residential demand. However, the cost per kWh incident on the poor tends to be higher than those consuming more in the absence of any cross-subsidy or direct subsidy.[10] Hence any support for additional income generation will surely be beneficial.

[10] This tends to be true in any electricity system—more so in a privately owned and operated system, but mitigating measures are often used through direct social safety nets and/or subsidised supply schemes.

3.2 Providing Electricity Access to Households and Supplying Electricity to the Rice Mill

In the previous case, the power plant just procures the husk from one or more rice mills, but does not supply electricity to the mill. However, given the poor quality of power supply in many places, the possibility of supplying power to the rice mill can also be considered. Clearly, the energy demand for a rice mill will depend on its size, processing activities involved, level of automation, operation time and such factors.

For the purpose of this analysis, the following assumptions are made:

(a) The rice mill capacity is chosen in such a way that adequate husk can be sourced from the mill to meet the demand for electricity generation;
(b) Small mills in a village or small town location tend to be indigenously made and tend to consume more energy. It is assumed that the electricity consumption requirement per tonne of raw rice processed is 43 kWh/tonne [10].[11]
(c) Rice mills in India can be categorised into two broad groups: small sized ones with less than 1 tonne/h processing capacity and bigger mills. Small mills generally operate a single shift of 6–7 h for about 200 days (i.e. 1,200 h of annual operation), while larger mills run two shifts (between 2,400 and 3,000 h of annual operation). In this case, we assume a single shift operation for 1,200 h/year.
(d) Husk availability is estimated considering a husk to paddy (or raw rice) ratio of 0.2. This is a conservative estimate. Baruah and Jain [1] indicate a husk-to paddy ratio of 0.267.
(e) It is assumed that the rice mill operates during day time when the residential demand is not serviced. This in effect extends the hours of operation of the power plant. Electricity demand is unlikely to be constant for the entire period of operation. It is likely that the evening load may be higher than the day load. For the sake of simplicity of financial analysis, an equivalent plant loading is used that generates the total amount of electricity required to meet the total demand.
(f) Scenario 3 from the previous section is used for electricity demand for non-mill purposes.
(g) The power plant operates two shifts of 6 h and instead of four employees uses six employees, each receiving a monthly wage of $100. This is logical given that the work for bill collector and the plant technician does not increase proportionately with hours of plant operation.

The rice mill has to be such that it produces enough rice husk in a year to meet the electricity needs of the mill and the village community. Given that the electricity demand corresponding to Scenario 3 is 43,800 kWh, and considering

[11] Depending on the processes used in a rice mill, the energy requirement changes.

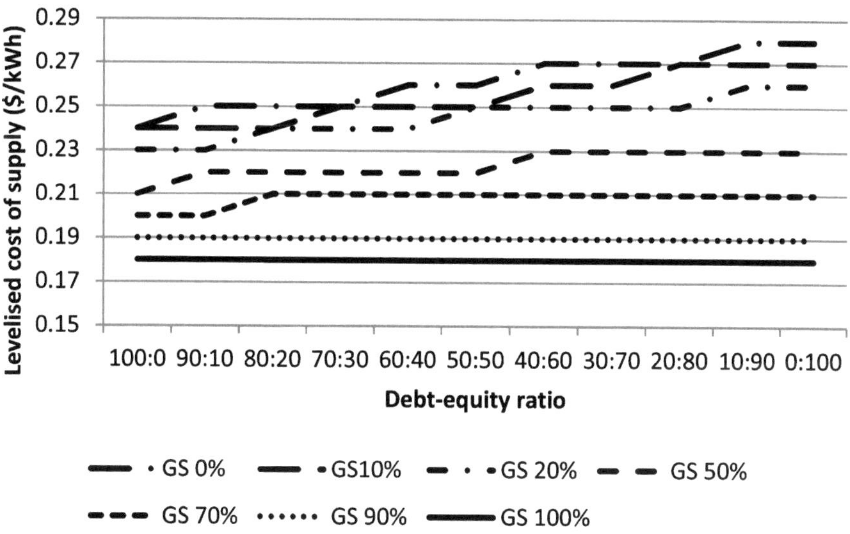

Fig. 5 Levelised cost of electricity supply for integrated operation. *Note GS* Grant share

43 kWh electricity required for processing 1 tonne of rice, we find that a rice mill of 0.4 t/h capacity operating in a single shift of 6 h for 200 days in a year will produce sufficient rice husk. The rice mill will require 20,640 kWh of electricity and the power plant needs to produce at least 64,440 kWh/year. The rice mill will process 480 tonnes of raw rice per year and will produce 96 tonnes of husks per year. The power plant will require approximately 93 tonnes of husk for its operation, which can be procured from the rice mill directly.

Figure 5 presents the levelised cost of supply for the integrated power supply operation to the rice mill and the village community. As can be seen, the cost of supply reduces considerably in this case due to higher plant utilisation rate. The lower end prices with capital subsidy will be quite attractive to most consumers. Even otherwise, the cost of supply reduces significantly. Hence, it makes economic sense to extend the supply to the rice mills, particularly when the operation does not coincide with the peak demand. This will benefit the rice mill by reducing its dependence on grid electricity, and providing a reliable supply at a reasonable price. Other consumers also benefit from this integration as the overall cost of supply reduces.

Although rice mills can install power generating stations for own use, such installations do not qualify for government support schemes for rural electricity supply. Moreover, the skill requirement is very different for operating a power plant and electricity distribution business compared to running a rice mill. In organisational terms, it makes better sense to have separate entities dealing with two separate businesses but linked to each other through contracts for fuel supply and electricity supply. Such contractual arrangements are important to ensure risk sharing, bankability of investments and reliability of business operations. The captive power supply model used by DESI Power is a good example.

4 From kW-Scale to Larger Power Generating Plants

India produced about 144 Mt of raw rice (or paddy) in 2010 [3] and has a total rice milling capacity of about 200 Mt/year [7]. In addition, Bangladesh produced 50 million tonnes of paddy in 2010, while Sri Lanka, Nepal and Pakistan produced about 15 Mt of paddy in 2010 [3]. Thus, South Asia produces more than 200 Mt of paddy per year. Rice milling takes place both at the household level (using hand pounding or pedal operated systems) and in rice mills. Generally, a small amount of raw rice is processed at the household level, mostly for own consumption. The processing of raw rice takes two forms: dry hulling which tends to account for a small share of total paddy processing and processing of parboiled rice. Rice milling in the region was a licenced activity for a long time that reserved the activity to small and medium-scale industries. This resulted in the proliferation of small mills throughout the region. However, these mills tend to be inefficient and produce poor quality output (higher percentage of broken rice). Moreover, because many of them fall under the unorganised sector, there is no systematic information about the number, distribution and size of rice mills. However, it is generally believed that the mini mills can process 250–300 kg of paddy per hour, small mills have a capacity of 1 t/h whereas larger, modern mills have capacities ranging from 2 to 10 t/h. Smaller mills operate a single shift of 6 h, while modern mills operate 2 shifts or even 3 shifts, but tend to have a seasonal operation.

Assuming a 2-shift operation of modern rice mills for 200 days/year, and considering that about 30 % of the electricity that can be produced from the husk can be used to meet the energy needs of the mill, a simple estimation is made of potential excess electricity and the potential number of consumers that can be served to meet the basic demand of 30 W/consumer for 6 h a day for every day of the year (see Table 2). It can be seen that thousands of consumers can be served by such power plants and a large cluster of villages (or blocks) can be considered as the basic unit of electrification. Alternatively, excess electricity from the mills can also be sold to the grid, if mills are grid connected or can be sold to a small number of local productive users (e.g. irrigation pumps, flour mills, food storage, etc.). Such larger plants thus open up the possibility of including productive applications of electricity beyond rice mill use, which in turn can catalyse economic activities at the village level. Although agriculture is the main rural activity in South Asia, food processing and other agro-based industrial activities (such as storing and warehousing), play a limited role yet due to lack of infrastructure and reliable electricity supply. While small-scale generating plants can only provide limited supply to households and small commercial consumers, larger plants can act as an agent for rural development.

In terms of cost of supply, two opposing forces are expected to operate. On one hand, the unit cost of generating plant ($/kW) is likely to reduce as the size increases. On the other, the fuel cost, distribution cost and wages would increase. The fuel cost increases proportionately with power generation. The area to be served may increase disproportionately and the extension of low voltage lines over

Table 2 Potential for serving large consumer bases

Mill capacity (t/h)	Husk production (t/year)	Potential electricity output (kWh)	Mill consumption (kWh)	Excess electricity (kWh/year)	Number of basic demand consumers that can be served
2	960	672,000	206,400	465,600	7,087
3	1440	1,008,000	309,600	698,400	10,630
4	1920	1,344,000	412,800	931,200	14,174
5	2400	1,680,000	516,000	1,164,000	17,717
6	2880	2,106,000	619,200	1,486,800	22,630
8	3840	2,688,000	825,600	1,862,400	28,347
10	4800	3,360,000	1,032,000	2,328,000	35,434

long distances will increase distribution losses and affect power quality. This will require a distribution system at 11 kV or higher voltage level and accordingly, the cost will increase. Finally, the staff requirement will increase in proportion with the area being serviced. Billing and collection cost can increase rapidly. Accordingly, the accurate cost estimation is rather difficult in this case.

However, to obtain a rough idea about the economic viability of a larger plant, we consider husk obtained from a 2 tonne/h rice mill. This can feed an electricity plant of 200 kW. For the cost calculations, we assume the following:

(a) The capital requirement per kW to be $1,000 for a 200 kW plant;
(b) Twenty five staff will be employed for generation, distribution and supply management;
(c) The distribution system is extended over a distance of 20 km;
(d) Other assumptions remain unchanged.[12]

The levelised cost of electricity for no subsidy case comes to $0.19/kWh. The cost reduces further with different levels of subsidy (see Fig. 6). The levelised cost in this case is the lowest of all options considered in this study. Clearly, this shows that as long as sufficient number of willing consumers can be enlisted, and the power supply company can manage to run its village cluster level operations, a bigger business can be profitably run. Alternatively, the excess power can be sold to captive users or to the grid at a break-even price of $0.19/kWh to make the venture viable. However, the tariff offered by the utility for buy-back is not as remunerative as this, which hinders financial viability of such power plants.

In addition, such plants will also qualify for carbon credits under the CDM programme. Either, a Programme of Activities can be considered following the example of Husk Power or a bundle of small projects as registered by DESI Power can also be appropriate. Although the carbon credit may not contribute

[12] It is possible to consider 24 h operation of the power plant but in this case, the available rice husk can support a smaller power plant capacity. Moreover, a husk-based plant is unlikely to operate continuously for 24 h. In this case, a back-up will be required. This can be looked into separately.

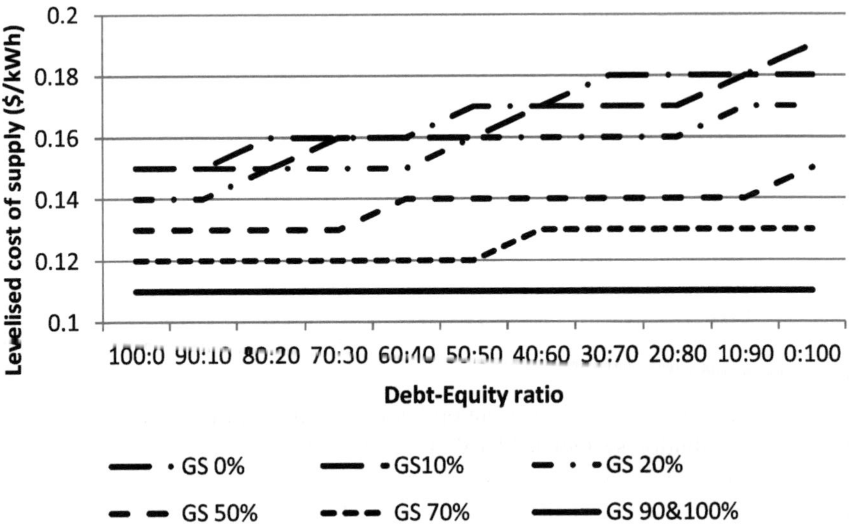

Fig. 6 Levelised cost of electricity supply for a 200 kW plant

significantly to the revenue stream of the project, it will nonetheless improve the project viability.

Clearly, there are risks involved in any business and a full-scale electricity supply business envisaged above cannot escape them. The dependence on the rice mill for husk is a major risk. As the generating plant capacity increases, the fuel requirement increases considerably. A 200 kW plant would require about 1,000 tonnes of husks per year for its operation and if the rice mill breaks down or goes bankrupt, the power supply business will also break down. For smaller plants, it may be possible to procure husk from other sources, but larger plants will struggle to procure the fuel from other sources. In addition, transportation and storage of husk can be problematic for large plants. The transportation cost of feedstock can easily increase the fuel supply cost and render the electricity supply less cost-effective. In addition, due to seasonal availability of husk, power plants may need to store significant volumes of husk to avoid shortage in fuel supply.

Similarly, as the plant serves a larger area, any fault with the generating plant will result in a power supply disruption in the entire area. It may be possible to address this by installing two 100 kW plants, but this may increase the cost of supply to some extent. In addition, the plant would need regular maintenance on a daily basis to ensure proper cleaning of the gas filters, and this makes it difficult to run the system continuously. It is likely that a back-up system will be required to meet the essential demand for a limited period of time. Depending on the fuel or technology used, the back-up system can increase the overall cost of electricity supply.

As the number of small consumers increases, the transaction costs related to retail business increase. A number of customers may not pay on time, thereby creating bad debt problems. To mitigate this, it may be appropriate to find anchor loads that can provide some business stability. DESI Power has been experimenting with mobile phone towers such as anchor loads. Similarly, local agro-based activities can also act as anchor loads. However, the anchor load may also increase business risk depending on its credibility as a reliable customer.

The investment requirement for a plant of 200 kW can easily reach $250,000. This is a substantial investment in a rural location, and companies willing to enter into such businesses will need to muster adequate financial resources and relevant experience. As there is no bankable sales agreement with most of the consumers, project financing of such mini-grids looks unlikely. Securing long-term debt funds from the financial institutions can be a major challenge as many of them require more than 100 % guarantee for such loans. In addition, the loan term (period and interest rate) may not be favourable to this type of businesses. Any support from the government and international agencies in facilitating finances through credit facilities, grants and guarantees can be helpful.

A further issue arises due to regulatory uncertainties in the area of off-grid electrification. As indicated in Bhattacharyya and Dow [2], the supply of electricity through a local off-grid network can be considered to qualify for the conventional utility regulatory supervision due to the possibility of monopolistic exploitation of the consumers, supply quality concerns and potential disputes between the supplier and the consumers. However, in India, the regulatory arrangement is not quite clear—it appears that the Electricity Act of 2003 has given some sort of exemption to certain organisations from the application of licence requirements for rural electricity supply. However, the conditions for such waiver and the roles and responsibilities of parties involved in the activities have neither been clearly specified in the Act nor in the regulations. More importantly, the rural areas covered by off-grid supply still come under the jurisdiction of the utility providing the central grid-based supply. Any decision to extend the grid subsequent to the installation of the off-grid plant can make the off-grid business unviable and stranded. Given the volume of investment involved and non-remunerative rate of grid buy-back of power, the investment can easily loose its alternative use. Thus, the regulatory uncertainty needs urgent consideration.

5 Conclusions

This chapter has considered off-grid electrification through electricity generated from rice husk in South Asia. The Husk Power Systems has successfully used rice husks to provide decentralised electricity in rural areas of India, and has so far installed 80 plants to electrify 300 villages. The success of the HPS can be traced to their choice of technology that is less capital intensive compared to other renewable energy options, their innovative approaches towards system cost reduction (e.g. using temporary structures made of bamboo poles for distribution network and local manufacturing of gasifiers) and additional income generation (e.g. use of carbon offsets and monetisation of byproducts), careful tariff design linked to Watts of demand instead of Watt-hours of energy used and careful siting of plants where about 400 customers are willing to pay for the service. DESI Power on the other hand has placed emphasis on productive use of power and used husk-based systems to displace diesel-based electricity supply to micro-enterprises. It has also used anchor loads (such as supply to mobile telephone towers) to improve the financial viability of the business.

The financial analysis of rice husk-based power generation shows that the levelised cost remains high compared to the supply from the centralised grid when just the basic demand (of 30 W) of households is met. This is due to low plant utilisation factor but the tariff based on Watts helps generate the required revenue to run the system. As the system utilisation improves either due to higher electricity consumption by some or by integration of the supply system to the rice mill, the levelised cost of supply reduces. However, the benefits of such cost reduction are enjoyed by those who consume more when an inverted block tariff system is used. The integration of rice mill's electricity demand brings the costs down considerably due to extended use of the facility during off-peak hours. Such integration can ensure an anchor load and can be beneficial for the electricity supplier. The rice mill on the other hand benefits from a reliable supply at a comparable price and reduces its cost arising out of electricity disruption. While the rice mill can develop a power plant for its own consumption, it is better to allow a specialised, separate entity to deal with the power generation business and develop contractual arrangements for fuel and power supply.

The extension of the analysis to include larger power plants for electricity distribution to a cluster of villages results in the cheapest cost of supply due to realisation of economies of scale. The cost of supply in such a case can be very competitive even without any capital grants. This suggests that it makes economic and financial sense for a supply company to extend the business to cover larger areas as long as there are sufficient willing customers and adequate supply of rice husks from rice mills. This also can promote economic activities in rural areas and promote economic development urgently needed to reduce rural poverty. Yet, the regulatory uncertainty, limited access to financial resources and markets, increased complexity of the distribution network (i.e. it may require higher voltage permanent network systems to reduce losses), and higher dependence on a single or limited fuel supply source would have to be carefully considered. Such bigger systems would require careful system design to ensure adequate system reliability, appropriate maintenance and limited line loss in distribution.

Being a major rice producing region, South Asia surely has a significant potential of utilising a major agro-waste to produce electricity for rural supply and rural development. However, to realise the potential the barriers mentioned above need to be addressed. In addition, the potential for using rice straw alongside rice husk can also be considered for power generation. Similarly, the potential for replication of this business in South Asia needs further analysis.

References

1. Baruah DC, Jain AK (1998) Distribution of agricultural crop residues in India. J Agric Eng 35(1):7–12
2. Bhattacharyya SC, Dow SR (2013) Regulatory issues related to off-grid electricity access, Chap. 11. In: Bhattacharyya S (ed) Rural electrification through decentralised off-grid systems in developing countries. Springer, London
3. FAO (2013) Rice Market Monitor, XVI (1), January 2013, Food and Agricultural Organisation of the United Nations, Rome, Italy
4. IFMR (2010) Empowering villages: a comparative analysis of DESI power and husk power systems: small-scale biomass power generators in India, Rural Market Insight Brief. Institute for Financial Management and Research, Chennai, India
5. Kapur T, Kandpal TC, Garg HP (1996) Electricity generation from rice husk in Indian rice mills: potential and financial viability. Biomass Bioenergy 10(5–6):393–403
6. Krithika PR, Palit D (2013) Participatory business models for off-grid electrification, Chap. 8. In: Bhattacharyya S (ed) Rural electrification through decentralised off-grid systems in developing countries green energy and technology. Springer, London
7. NSDC (undated) Human Resource and Skill Requirements in the Food Processing Sector, National Skill Development Corporation, New Delhi
8. Pandey G (2011) Rural electrification using biomass—husk power systems. In: Presentation at the OASYS South Asia workshop on decentralised off-grid electricity generation in developing countries: business models for off-grid electricity supply, TERI University, New Delhi, 6 January 2011
9. Sevea (2013) Husk power systems—power to empower, Synergie pour l'Echange et la Valorisation des Entrepreneurs d'Avenir (SEVEA), France
10. Sookkumnerd C, Ito N, Kito K (2007) Feasibility of husk-fuelled steam engines as prime mover of grid-connected generators under the Thai very small renewable energy power producer (VSPP) program. J Clean Prod 15:266–274

Conclusions

Subhes C. Bhattacharyya and Debajit Palit

Abstract This chapter summarises the main findings of the book and presents the concluding remarks. Various chapters of the book show that with proper technical design, appropriate choice of delivery options and a supporting environment, mini-grids can be an effective solution for enhancing electricity access in rural areas of developing countries. The experience from several successful examples of mini-grids in South Asia suggests that success in each case was achieved through a unique combination of critical factors. All successful cases have managed the technology well, ensured effective delivery and followed a business-like approach.

1 Summary of Findings

As part of the research project on Off-grid Access Systems in South Asia (OASYS-South Asia), a detailed investigation on the mini-grids as a viable option for electrification in South Asia was undertaken. In 13 chapters (excluding Introduction and the Conclusion) this book reports various dimensions of the study and provides the theoretical technological underpinning, practical design aspects, case studies and business-related considerations. In two parts, the book has strived to combine both the background knowledge required to appreciate the mini-grid based electrification in rural areas and in-depth research studies that demonstrate the integration of a multidimensional analytical framework into research. One of the main gaps in the literature relates to inadequacy of information about

S. C. Bhattacharyya (✉)
Institute of Energy and Sustainable Development, De Montfort University, Leicester, UK
e-mail: subhesb@dmu.ac.uk; subhes_bhattacharyya@yahoo.com

D. Palit
The Energy and Resources Institute, New Delhi, India
e-mail: debajitp@teri.res.in

S. C. Bhattacharyya and D. Palit (eds.), *Mini-Grids for Rural Electrification of Developing Countries*, Green Energy and Technology,
DOI: 10.1007/978-3-319-04816-1_15, © Springer International Publishing Switzerland 2014

successful mini-grid examples. This book bridges this gap through detailed case studies of examples from South Asia. Moreover, given the emphasis on productive use of electricity for rural livelihood generation, this book reports the step-by-step process that was followed to develop the demonstration project set up in a cluster of five villages in Dhenkanal district of Odisha (India), where lighting provision for residential use has been combined with livelihood generation activities and other support initiatives for social capital development. It thus contributes to the energy-plus agenda of energy access theme and demonstrates the delivery of an integrated activity that links rural electricity with rural development.

Establishing the link with the previous book published from this research project, "Suite of Off-Grid Options in South Asia" provided an overview of a suite of decentralised options available in South Asia to address the electricity access challenge. Although South Asia has gained some experience in decentralised options, the stand-alone solar home systems (SHS) received greater attention. But the limited service provided by the SHS makes it a temporary solution and the relatively high initial investment requirement limits the choice to the richer section of the population. There is some experience with mini-grids as well and alternative business models have been explored but the regulatory environment is not conducive yet to ensure long-term viability of the options.

The technical options for mini-grids in rural areas, covered in "Technical Aspects of Mini-Grids for Rural Electrification", demonstrate the near cost-effectiveness of a number of technologies for which the potential of renewable energy resources exists in delivering electricity in rural areas through mini-grids. The chapter explained how such potential can be harnessed through appropriate technologies. Referring to the debate on AC versus DC supply, the chapter suggests that, with technological improvements, the DC options should also be considered because appliances using DC supply are now not limited to basic applications such as lighting and mobile charging but are being extended to those requiring electric motors, which offer precise speed and torque control thereby resulting in energy saving. In rural areas, a DC system eliminates the conversion loss and reduces the capital investment requirement but the distance limitation with a 24 V supply system restricts its use to closely distributed households with limited demand. But in a hybrid system where loads and generating sources can be segregated in AC and DC, a dual bus system with a bi-directional converter can be used. As the applications increase, the costs will decline, making mini-grids even more cost-effective and a reliable option in the future.

"Smart Design of Stand-Alone Solar PV System for Off Grid Electrification Projects: Technical Aspects of Mini-Grids for Rural Electrification" supplements the technical discussion of "Technical Aspects of Mini-Grids for Rural Electrification" through a detailed presentation of solar PV mini-grid design in an intelligent way. The entire range of activities of a PV system design is covered with the objective of providing a reliable but affordable supply at a high efficiency. The smart use of load profiling, resource profiling and load categorisation suggested here can lead to cost saving through better capacity choices and management. Similarly, the balance of the systems such as batteries and inverters can also be better managed for cost saving through a smart configuration of the system. In addition, by providing for the grid connectivity option, the system can be protected from becoming a stranded asset in the future when the central grid is extended eventually. The step-by-step guideline provides the practical advice for potential users while the smart features make the contribution state-of-the-art.

The integrated framework presented in "Analytical Frameworks and an Integrated Approach for Mini-Grid-Based Electrification" goes beyond the conventional techno-economic analysis of mini-grids normally found in the academic literature. The detailed review of the literature shows that a plethora of analytical approaches has been applied to study rural electrification in general and mini--grid-based decentralised electrification in particular. Yet, the practice-based literature lacks academic rigour and the academically oriented studies are not always grounded in practical reality. Moreover, the frameworks tend to neglect the iterative process of project development and incorporation of feedback from the ground. The framework presented in this chapter remedies the above limitations and provides a sound yet practically grounded approach to analyse mini-grid-based electrification that has been applied in various case studies reported in the book.

The field-level analysis presented here confirms that the success of an off-grid project depends on factors beyond techno-economic considerations. The policy environment, social acceptance, linkages with income-generating activities and technological appropriateness play a defining role. Accordingly, the macro-economic policy instruments need to be aligned with the local context; the technology choice and scale of operation are determined by the nature of the local energy service, which in turn depends on the economic condition and social structure while influencing the financing options of the initiative. This emphasis on local context and the feedback from it plays an important role in creating a viable operation.

Although demand-side management is often discussed in relation to central grid systems, it has a vital role in small-scale decentralised systems where the resources are limited, demand diversity is less pronounced and supply has to be balanced with demand anyway. This assumes greater importance in systems where the energy source is variable in nature. This important issue is captured in "Demand Management for Off-Grid Electricity Networks", providing a review of alternative options for demand management in off-grid systems. It suggests that dynamic current limits such as those provided by an electronic trip are appropriate for generator-limited systems while an energy budgeting scheme is appropriate for a battery-limited system.

Electricity supply through mini-grids in rural areas represents a paradigm shift and the commercial aspects of such a business require a careful consideration. Creating a viable business is a challenging task in a difficult micro-environment exemplified by limited customer base, locational disadvantage, poor paying capacity and lack of skilled manpower. The macro-environment may not be supportive either, thereby creating impediments to any commercial viability. This may arise from lack of regulatory clarity about such business prospects, hurdles in accessing finance, short-term nature of funding and non-conducive policy environment. Accordingly, "Business Issues for Mini-Grid-Based Electrification in

Developing Countries" suggests that choosing an appropriate scale of the business in such a situation is clearly vital. If the rural area does not have a ready market, a limited supply for lighting only may be attempted. Where more productive uses exist or dormant demand exists, a more ambitious system (light-plus) can be considered. Where an anchor load is readily available, the supply viability can improve due to the contractual arrangement with the dominant buyer. However, in each case, there are specific challenges particularly in terms of funding, ensuring reliable supply and managing the business locally. The choice of technology, supply chain management and careful operation and maintenance of the system in rural areas is therefore important to economise on costs and investment needs. Moreover, appropriate charges for the supply and efficient collection arrangement are also essential to run a healthy business.

The range of case studies included in the second part shows the practical dimension of mini-grid-based electricity development in South Asia. As part of the OASYS South Asia project, a number of demonstration efforts are being undertaken. The implementation of such a project in a cluster of five villages in Dhenkanal district of Odisha is reported in "Approach for Designing Solar Photovoltaic-Based Mini-Grid Projects: A Case Study from India". This demonstration project showcases a small-scale lighting-plus system using solar PV as the source of electricity generation. The supply will be locally managed by the community with the support of an NGO. Three larger villages are connected with AC grids whereas two smaller ones are connected with DC grids, thereby demonstrating two distribution options in a single site. While the households in each case were provided with a connection of two LED lamps of 3 W each and a mobile charging connection, the project is also catering to commercial loads in the main village to support livelihood activities, community loads (e.g. street lights, water purifier and facilities in the community hall) and water pumping for agriculture. The implementation activity reinforces that in a community-based system, it is important to engage regularly with the community from the beginning of the work and take them into confidence. The remoteness of the site and limited infrastructure availability need to be considered in the planning process. The difficulties in transporting the equipment and materials, as well as finding suitable semi--skilled workers for various construction activities can lead to cost and time over-runs unless properly taken care of. The technology selection needs to consider ease of operation and maintenance while the system design should keep the reliability of supply and system operation in mind. Moreover, once electricity reaches through the mini-grid, users may like to demand more and some provision for additional demand as demand management is important as well.

Application of the integrated framework of analysis in two chapters ("Renewable Energy-Based Mini-Grid for Rural Electrification: Case Study of an Indian Village" and "From SHS to Mini-Grid-Based Off-Grid Electrification: A Case Study of Bangladesh") marks a departure from the conventional analysis. The analysis has paid greater attention to demand estimation and has used alternative demand scenarios, whereby plausible cases were developed to capture various possibilities. The analysis has relied on the simulation, optimisation and sensitivity analyses using HOMER to identify the appropriate technology combinations to meet the demand. The work then extends into the business-related analysis to see if a viable business can be identified. The iterative process of analysis is also captured here. Similarly, an application of multicriteria decision making for off-grid technology choice is presented in "Application of Multi-criteria Decision Aids for Selection of Off-Grid Renewable Energy Technology Solutions for Decentralised Electrification" taking the perspective of some stakeholders involved in off-grid electricity supply.

The case study on Solar PV mini-grids in the state of Chhattisgarh in India is another contribution to knowledge. Although the system has been working effectively for some years, it has not received due attention in the academic circle. Our detailed case study shows that a top-down approach can also deliver and replicate if the critical factors are well managed. Chhattisgarh has developed a successful delivery model through a dedicated agency involvement, appropriate technology choice, technology standardisation, appropriate financial support from the government, and a clear delivery model with focused duties and responsibilities. While the financial viability has not been the main aim here, nonetheless successful operation of a large number of mini-grids in difficult conditions is not a mean achievement.

The mini-hydro-based local grid system in Nepal on the other hand provides a contrasting picture. Availability of a cost-effective renewable resource alone does not ensure success of a mini-grid system. Although Nepal has made visible progress towards electricity access, it still faces weak institutions, unclear policies, limited access to funding and limited demand from productive activities. Although mini-hydro plants tend to produce cheaper electricity, it seems to be losing out to solar PV.

Finally, the analysis of the business potential of electrifying rural South Asia using rice husks as a source of energy, following the example of Husk Power Systems, reveals that a viable supply can be developed for various scales of operation. The cost of electricity using rice husk is likely to be lower than that achievable with solar PV but grid price parity may not be easily achievable without capital subsidy and/or debt capital at reduced interest rates. Yet, the business can operate successfully through a suitably designed tariff structure. Integration of productive demand and higher capacity utilisation improves the financial viability of the business. Economies of scale through a bigger plant size bring down the cost of supply and the supply can become very competitive but greater attention is required for distribution system design, plant operation and supply chain management.

The main message that emanates from the studies reported here is that mini-grids can effectively complement the grid extension in rural areas and through proper technology choice, system design, enabling business environment and support mechanism, rural electricity supply can be offered through various delivery options. There is scope for a range of entities to be involved in the process: the private sector can achieve financial viability by economising on costs, developing an innovative supply and designing a careful tariff structure; public sector operators or community organisations that may not be looking for profitability as such can take advantage of public funds and aim for operational success by recovering running costs and managing the system effectively. Different scales of operation are likely to be appropriate in different locations and possibilities of clustering systems to strengthen the local supply will also arise. However, in all cases, local context requires careful understanding, which then needs to be translated into a suitable technical design, supported by an appropriate delivery mechanism and a conducive business environment. Absence or hindrances to these critical elements often endanger the process, causing failures and loss of interest in servicing this bottom of the pyramid.

2 Policy Recommendations and Way Forward

The challenge of achieving universal electrification by 2030 as envisaged by the UN Secretary General's initiative on Sustainable Energy for All requires commensurate solution strategies. While electrification through mini-grids offers such a strategy, achieving significant electrification through this mode demands leapfrogging to the rapid deployment stage from the pilot/demonstration project stage in which mini-grids are found at the moment. Such a transformation requires major efforts in various areas, some of which are highlighted below:

(a) *Create a supportive environment*: In order to expand rural supply through mini-grids, a supportive environment is required. In the policy sphere, a clearly spelt out rural electrification policy is required with well-defined roles and responsibilities for various organisations, clearly demarcated role for decentralised options, integration of off-grid electrification with local development policies and a transparent monitoring and implementation mechanism.
(b) *Ensure regulatory certainty*: As noted in "Analytical Frameworks and an Integrated Approach for Mini-Grid-Based Electrification", regulatory uncertainty and vacuum can stifle growth of mini-grid projects and the current paradigm of ad hoc and piecemeal dispensation to regulation produces sub-optimal outcomes. The mini-grid deployment as a miniature version of the utility model may not be legally viable unless its regulatory arrangement and scope is properly defined. In addition, a clear understanding with the incumbent utility is required in respect of grid extension, as this threatens and deters new investment in mini-grids. Similarly, clear regulatory arrangements are required in respect of licence or permissions for doing business, tariff systems, connection to the grid, safety requirements and quality of supply.
(c) *Guarantee technology neutrality*: The local context of rural areas in terms of resource potential, demand profile and paying capacity has a deciding influence on the technical choice. It is therefore important to avoid any a priori technology bias in the decision-making process. This in turn requires that the policy or regulatory environment should ensure technology neutrality and allow the most appropriate option to be developed without any hindrance.

(d) *Develop appropriate funding arrangements*: Access to funding is one of the main bottlenecks in expanding the business. As discussed in "Approach for Designing Solar Photovoltaic-Based Mini-Grid Projects: A Case Study from India", limited access to capital, short debt tenure, small size of projects (which excludes large financial institutions' direct involvement in such activities), limited credit record of rural entities, limited guarantee for debts and high interest rates are some of the financial barriers faced by rural communities and investors. These challenges have to be overcome to promote mini-grid-based electrification. Governments need to develop appropriate credit lines, project appraisal capabilities and support systems through rural electrification agencies, rural banks and similar financial institutions. The arrangements have to ensure neutrality in terms of borrowing entity to allow rural community organisations, cooperatives and private entities to gain access to such funds as long as the business looks credible.

(e) *Support rural infrastructure development*: As mini-grid development entails developing local electricity generation and distribution, it forms part of the rural infrastructure development. The state has supported investments in such infrastructure projects everywhere in the world and the developing countries are no exception. Our earlier work has suggested that countries which have made significant progress towards universal electricity access have all supported this effort. Accordingly, financial support for rural generating capacity and network development is justified and required. Although 100 % grant funding may not be desirable, a significant support can reduce the cost of supply and make the service affordable to rural communities.

(f) *Build required capacity*: The mini utility model of electrification is a demanding business that requires a set of skills to manage the system and run the activities effectively. As these skills are unlikely to be readily available in rural areas and perhaps lacking in urban areas, it is important to build capacity in anticipation. These skills will include, among others, technical expertise in running and maintaining the systems, basic accounting and record-keeping skills, project design and implementation, regulatory skills and financial skills to deal with financial institutions. South–South cooperation in this respect can be considered and the best practices from successful examples can be used to replicate such models.

This book has made contribution in most of the above areas and we hope that the readers will benefit from the research presented here.

Any study of this nature faces certain constraints and this work is no exception. Accordingly, we suggest a few areas for further investigation. First, although we have provided some case studies from South Asia, we are aware of rapid development in mini-grid-based electrification elsewhere in the world. An extension of the work to cover such experience from the rest of the world will surely be a valuable work. Second, we have mainly focused on village level mini-grids and have not considered the issue of replication and scaling-up in detail. This is an area worth investigating. Third, related to the previous issue, is the possibility of integration of local grids in the future, with an eventual formation of a central grid that can be connected to the main grid. This shift from an isolated operation to a centralised operation deserves careful attention and advance planning to avoid costly refurbishment in the future. Fourth, we have not considered MW scale generation projects located in rural areas which may serve local population and export the balance power to the central grid. Some such projects exist or are being developed in some Asian and African countries. As they are generally grid connected, they are not off-grid as such and did not come under our scope. But such projects can offer important lessons, particularly in terms of business models. A future study could include them. In addition, although we have considered a few business models, there may be other mini-grid models successfully running in India, Sri Lanka, Bangladesh, Nepal or elsewhere. With the limited resources and time, it was not possible to undertake an exhaustive study of all operating mini-girds. There is surely some scope for additional work here.

Also, while we have covered smart design of solar PV plants and have provided case studies of solar and gasifier projects, we have not covered the hybrid systems to a great extent. Such projects are limited in practice and have not become very popular yet. However, the hybrid systems may play a greater role in the future and further work in this area is required.

Finally, the regulatory arrangements and models for mini-grids have not been well covered in this book, as this was covered in the previous book published by the project (see [1]). Yet, a specific study on business models and regulatory arrangements has some potential. We shall endeavour to cover some of the above aspects in our upcoming works.

Reference

1. Bhattacharyya SC (ed) (2013) Rural electrification through decentralized off-grid systems in developing countries. Springer, London

About the Editors

Prof. Subhes Bhattacharyya of Institute of Energy and Sustainable Development of De Montfort University is an internationally renowned energy specialist with more than 25 years of experience. He leads a consortium of research organisations from the UK and India working on off-grid electricity access in South Asia. He has written widely on energy issues in developing countries and particularly on energy access. He is the author of a best-selling book on energy economics and has edited a book on rural electrification through decentralised off-grid systems.

Mr. Debajit Palit of TERI has around 15 years of experience working in the field of clean energy access, rural electrification policy and regulation, distributed generation, solar photovoltaics and biomass gasification. He has also written widely on energy access and rural electrification issues, particularly related to South Asia, and authored a book on rural electricity distribution franchisees. He has a wide national and international experience, working in projects spread across Asia and Africa. He holds a Master degree in Physics and is currently pursuing Ph.D. in rural electrification policy from TERI University.

S. C. Bhattacharyya and D. Palit (eds.), *Mini-Grids for Rural Electrification of Developing Countries*, Green Energy and Technology,
DOI: 10.1007/978-3-319-04816-1,

Index

S. C. Bhattacharyya and D. Palit (eds.), *Mini-Grids for Rural Electrification of Developing Countries*, Green Energy and Technology,
DOI: 10.1007/978-3-319-04816-1,

MIX
Papier aus verantwortungsvollen Quellen
Paper from responsible sources
FSC® C105338

If you have any concerns about our products,
you can contact us on
ProductSafety@springernature.com

In case Publisher is established outside the EU,
the EU authorized representative is:
Springer Nature Customer Service Center GmbH
Europaplatz 3, 69115 Heidelberg, Germany

Printed by Libri Plureos GmbH
in Hamburg, Germany